Load Testing of Bridges

Structures and Infrastructures Series

ISSN 1747–7735

Book Series Editor:

Dan M. Frangopol

Professor of Civil Engineering and
The Fazlur R. Khan Endowed Chair of Structural Engineering and Architecture
Department of Civil and Environmental Engineering
Center for Advanced Technology for Large Structural Systems (ATLSS Center)
Lehigh University
Bethlehem, PA, USA

Volume 13

Load Testing of Bridges

Proof Load Testing and the Future of Load Testing

Editor

Eva O.L. Lantsoght

Politécnico, Universidad San Francisco de Quito, Quito, Ecuador
Concrete Structures, Department of Engineering Structures, Civil Engineering
and Geosciences, Delft University of Technology, Delft, The Netherlands

CRC Press
Taylor & Francis Group
Boca Raton London New York

CRC Press is an imprint of the
Taylor & Francis Group, an **informa** business

A BALKEMA BOOK

Colophon

Published by:
CRC Press/Balkema
Schipholweg 107C, 2316 XC Leiden, The Netherlands

First issued in paperback 2023

© 2019 by Taylor & Francis Group, LLC
CRC Press/Balkema is an imprint of Taylor & Francis Group, an informa business

No claim to original U.S. Government works

ISBN-13: 978-0-367-21082-3 (Volume 12) (hbk)
ISBN-13: 978-0-367-21083-0 (Volume 13) (hbk)
ISBN-13: 978-1-03-257079-2 (Volume 13) (pbk)
ISBN-13: 978-1-138-09198-6 (set)

DOI: https://doi.org/10.1201/9780429265969

Typeset by Apex CoVantage, LLC

Library of Congress Cataloging-in-Publication Data

Names: Lantsoght, Eva O. L., editor.
Title: Load testing of bridges : proof load testing and the future of load testing /
    editor: Eva O.L. Lantsoght.
Description: Leiden : CRC Press/Balkema, [2019] | Series: Structures and infrastructures series,
    ISSN 1747-7735 ; volumes 12-13 | Includes bibliographical references and index.
Identifiers: LCCN 2019016183 (print) | LCCN 2019018556 (ebook) | ISBN 9780429265426 (ebook) |
    ISBN 9780367210823 (volume 1 : hbk) | ISBN 9780429265426 (volume 1 : e-Book) |
    ISBN 9780367210830 (volume 2 : hbk) | ISBN 9780429265969 (volume 2 : e-Book) |
    ISBN 9781138091986 (set of 2 volumes : hbk)
Subjects: LCSH: Bridges—Testing. | Bridges—Live loads. | Dynamic testing.
Classification: LCC TG305 (ebook) | LCC TG305 .L63 2019 (print) | DDC 624.2/52—dc23
LC record available at https://lccn.loc.gov/2019016183
Structures and Infrastructures Series: ISSN 1747–7735
Volume 13

Visit the Taylor & Francis Web site at
http://www.taylorandfrancis.com

and the CRC Press Web site at
http://www.crcpress.com

Table of Contents

Part V   Conclusions and Outlook   359

Editorial

Welcome to the book series *Structures and Infrastructures*.

Our knowledge to model, analyze, design, maintain, manage, and predict the life-cycle performance of structures and infrastructures is continually growing. However, the complexity of these systems continues to increase and an integrated approach is necessary to understand the effect of technological, environmental, economic, social, and political interactions on the life-cycle performance of engineering structures and infrastructures. In order to accomplish this, methods have to be developed to systematically analyze structure and infrastructure systems, and models have to be formulated for evaluating and comparing the risks and benefits associated with various alternatives. We must maximize the life-cycle benefits of these systems to serve the needs of our society by selecting the best balance of the safety, economy, and sustainability requirements despite imperfect information and knowledge.

In recognition of the need for such methods and models, the aim of this book series is to present research, developments, and applications written by experts on the most advanced technologies for analyzing, predicting, and optimizing the performance of structures and infrastructures such as buildings, bridges, dams, underground construction, offshore platforms, pipelines, naval vessels, ocean structures, nuclear power plants, and also airplanes, aerospace, and automotive structures.

The scope of this book series covers the entire spectrum of structures and infrastructures. Thus it includes, but is not restricted to, mathematical modeling, computer and experimental methods, practical applications in the areas of assessment and evaluation, construction and design for durability, decision-making, deterioration modeling and aging, failure analysis, field testing, structural health monitoring, financial planning, inspection and diagnostics, life-cycle analysis and prediction, loads, maintenance strategies, management systems, nondestructive testing, optimization of maintenance and management, specifications and codes, structural safety and reliability, system analysis, time-dependent performance, rehabilitation, repair, replacement, reliability and risk management, service life prediction, strengthening, and whole life costing.

This book series is intended for an audience of researchers, practitioners, and students worldwide with a background in civil, aerospace, mechanical, marine, and automotive engineering, as well as people working in infrastructure maintenance, monitoring, management, and cost analysis of structures and infrastructures. Some volumes are monographs defining the current state of the art and/or practice in the field, and some are textbooks to be used in undergraduate (mostly seniors), graduate and postgraduate courses. This book

DOI: https://doi.org/10.1201/9780429265969

series is affiliated to *Structure and Infrastructure Engineering* (www.tandfonline.com/toc/nsie20/current), an international peer-reviewed journal which is included in the Science Citation Index.

It is now up to you, authors, editors, and readers, to make *Structures and Infrastructures* a success.

Dan M. Frangopol
Book Series Editor

About the Book Series Editor

Dan M. Frangopol is the first holder of the Fazlur R. Khan Endowed Chair of Structural Engineering and Architecture at Lehigh University, Bethlehem, Pennsylvania, USA, and a Professor in the Department of Civil and Environmental Engineering at Lehigh University. He is also an Emeritus Professor of Civil Engineering at the University of Colorado at Boulder, USA, where he taught for more than two decades (1983–2006). Before joining the University of Colorado, he worked for four years (1979–1983) in structural design with A. Lipski Consulting Engineers in Brussels, Belgium. In 1976, he received his doctorate in Applied Sciences from the University of Liège, Belgium, and holds four honorary doctorates (Doctor Honoris Causa) from the Technical University of Civil Engineering in Bucharest, Romania (2001); the University of Liège, Belgium (2008); and the Gheorghe Asachi Technical University of Iaşi, Romania (2014); and the Polytechnic University of Milan (Politecnico di Milano), Milan, Italy (2016).

Dr. Frangopol is an Honorary Professor at 13 universities (Hong Kong Polytechnic, Tongji, Southeast, Tianjin, Dalian, Hunan, Chang'an, Beijing Jiaotong, Chongqing Jiaotong, Shenyang Jianzhu, Royal Melbourne Institute of Technology (RMIT), Changsha University of Science and Technology, and Harbin Institute of Technology) and a Visiting Chair Professor at the National Taiwan University of Science and Technology. He is a Distinguished Member of the American Society of Civil Engineers (ASCE), Foreign Member of the Academia Europaea (Academy of Europe, London), Foreign Member of the Royal Academy of Belgium for Science and the Arts, Honorary Member of the Romanian Academy, Honorary Member of the Romanian Academy of Technical Sciences, Inaugural Fellow of both the Structural Engineering Institute and the Engineering Mechanics Institute of ASCE, Fellow of the American Concrete Institute (ACI), Fellow of the International Association for Bridge and Structural Engineering (IABSE), Fellow of the International Society for Health Monitoring of Intelligent Infrastructures (ISHMII), and Fellow of the Japan Society for the Promotion of Science (JSPS). He is the President of the International Association for Bridge Maintenance and Safety (IABMAS), Honorary Member of the Portuguese Association for Bridge Maintenance and Safety (IABMAS-Portugal Group), Honorary Member of the IABMAS-China Group, Honorary Member of the IABMAS-Australia Group, Honorary Member of the IABMAS-Japan Group, Honorary President of the IABMAS-Italy Group, Honorary President of and the IABMAS-Brazil Group, Honorary

DOI: https://doi.org/10.1201/9780429265969

President of the IABMAS-Chile Group, Honorary President of the IABMAS-Turkey Group, and Honorary President of the IABMAS-Korea Group.

He is the initiator and organizer of the Fazlur R. Khan Distinguished Lecture Series (www.lehigh.edu/frkseries) at Lehigh University. He is an experienced researcher and consultant to industry and government agencies, both nationally and abroad. His main research interests are in the development and application of probabilistic concepts and methods to civil and marine engineering, including structural reliability; life-cycle cost analysis; probability-based assessment, design, and multi-criteria life-cycle optimization of structures and infrastructure systems; structural health monitoring; life-cycle performance maintenance and management of structures and distributed infrastructure under extreme events (earthquakes, tsunamis, hurricanes, and floods); risk-based assessment and decision making; multi-hazard risk mitigation; infrastructure sustainability and resilience to disasters; climate change adaptation; and probabilistic mechanics.

According to ASCE (2010), "Dan M. Frangopol is a preeminent authority in bridge safety and maintenance management, structural systems reliability, and life-cycle civil engineering. His contributions have defined much of the practice around design specifications, management methods, and optimization approaches. From the maintenance of deteriorated structures and the development of system redundancy factors to assessing the performance of long-span structures, Dr. Frangopol's research has not only saved time and money, but very likely also saved lives."

Dr. Frangopol's work has been funded by NSF, FHWA, NASA, ONR, WES, AFOSR, ARDEC, and numerous other agencies. He is the Founding President of the International Association for Bridge Maintenance and Safety (IABMAS, www.iabmas.org) and the International Association for Life-Cycle Civil Engineering (IALCCE, www.ialcce.org), and is Past Director of the Consortium on Advanced Life-Cycle Engineering for Sustainable Civil Environments (COALESCE). He is the Past Vice-President of the International Association for Structural Safety and Reliability (IASSAR), Past Vice-President of the Engineering Mechanics Institute (EMI) of ASCE, and Past Member of its Board of Governors. He is also the Founding Chair of the ASCE-SEI Technical Council on Life-Cycle Performance Safety, Reliability and Risk of Structural Systems and of the IASSAR Technical Committee on Life-Cycle Performance, Cost and Optimization. He has held numerous leadership positions in national and international professional societies including Chair of the Technical Activities Division of the 20,000+ members of the Structural Engineering Institute (SEI) of ASCE, Chair of Executive Board of IASSAR, Vice-President of the International Society for Health Monitoring of Intelligent Infrastructures (ISHMII), and Chair of IABSE Working Commission 1 on Structural Performance, Safety, and Analysis.

Dr. Frangopol is the recipient of several prestigious awards including the George W. Housner Medal, the 2016 ASCE OPAL Award for Lifetime Accomplishments in Education, the 2016 ASCE Alfredo Ang Award, the 2016 ASCE-Lehigh Valley Section Civil Engineer of the Year Award, the 2015 ASCE Noble Prize, the 2014 ASCE James R. Croes Medal, the 2012 IALCCE Fazlur R. Khan Life-Cycle Civil Engineering Medal, the 2012 ASCE Arthur M. Wellington Prize, the 2012 IABMAS Senior Research Prize, the 2008 IALCCE Senior Award, the 2007 ASCE Ernest Howard Award, the 2006 IABSE OPAC Award, the 2006 Elsevier Munro Prize, the 2006 T. Y. Lin Medal, the 2005 ASCE Nathan M. Newmark Medal, the 2004 Kajima Research Award, the 2003 ASCE Moisseiff Award, the 2002 and 2016 JSPS Fellowship Award for Research in

Japan, the 2001 ASCE J. James R. Croes Medal, the 2001 IASSAR Research Prize, the 1998, 2004 and 2019 ASCE State-of-the-Art of Civil Engineering Award, and the 1996 Distinguished Probabilistic Methods Educator Award of the Society of Automotive Engineers (SAE). Among several awards he has received at the University of Colorado, he is the recipient of the 2004 Boulder Faculty Assembly Excellence in Research Scholarly and Creative Work Award, the 1999 College of Engineering and Applied Science's Research Award, the 2003 Clarence L. Eckel Faculty Prize for Excellence, and the 1987 Teaching Award. He is also the recipient of the Lehigh University's 2013 Eleanor and Joseph F. Libsch Research Award and of the Lehigh University's 2016 Hillman Award for Excellence in Graduate Education. He has given plenary keynote lectures at numerous major conferences held in Asia, Australia, Europe, North America, South America, and Africa.

Dr. Frangopol is the Founding Editor in Chief of *Structure and Infrastructure Engineering* (Taylor & Francis, www.tandfonline.com/toc/nsie20/current), an international peer-reviewed journal. This journal is dedicated to recent advances in maintenance, management, and life-cycle performance of a wide range of structures and infrastructures. He is the author or coauthor of 3 books, 50 book chapters, more than 380 articles in refereed journals, and over 600 papers in conference proceedings. He is also the editor or co-editor of 48 books published by ASCE, Balkema, CIMNE, CRC Press, Elsevier, McGraw-Hill, Taylor & Francis, and Thomas Telford, and is an editorial board member of several international journals. Additionally, he has chaired and organized several national and international structural engineering conferences and workshops.

Dr. Frangopol has supervised 45 Ph.D. and 55 M.Sc. students. Many of his former students are professors at major universities in the United States, Asia, Europe, and South America, and several are prominent in professional practice and research laboratories.

For additional information on Dr. Frangopol's activities, please visit www.lehigh.edu/~dmf206/.

Preface

Load testing of bridges is a practice as old as bridge engineering. From the early days, when a load test prior to opening had to convince the traveling public of the safety of the bridge, engineers and the general public have seen the value of load testing. The purpose of this work, divided across two volumes, is to study load testing from different perspectives. As such, this work deals with the practical aspects related to load testing and the current scientific developments related to load testing of bridges, and it offers an international perspective on the topic of load testing of bridges. You can find general recommendations, advice, and best practices in these books along with detailed case studies, so that this work can be a guide for the engineer preparing a load test. You can also find open research questions, topics for future research, and the latest research findings related to load testing in these books, so that this work can be a guide for researchers who want to identify interesting topics to study.

The work is divided in eight parts across two volumes (Volumes 12 and 13). Part I of Volume 12 gives a background to bridge load testing, including the historical perspectives and currently governing codes and guidelines. The background is discussed from an international perspective, outlining the history of load testing in North America and Europe, and summarizing the current codes and guidelines for load testing from Germany, the United States, the United Kingdom, Ireland, Poland, Hungary, Spain, the Czech Republic and Slovakia, Italy, Switzerland, and France.

Part II of Volume 12 deals with general practical aspects of load testing, which are valid for the different types of load tests (diagnostic and proof load tests). These practical aspects cover the entire stage of a load testing project: from preparation, to execution, to post-processing and reporting about the load test. This first topic in this part contains general considerations for each load testing project: which type of load test is suitable for the project, and whether load testing is the best option for the considered bridge. If the decision is made to load test a bridge, the next step is to prepare this load test. Therefore, the next topic discusses the elements of the preparation of a load test: inspection, preliminary calculations, and planning of the project. Then, general aspects of the execution of the load test are discussed, with regard to loading equipment, measurement equipment, and practical aspects of communication and safety. The last topic of this part deals with post-processing of load testing data, and reporting of load tests.

Part III of Volume 12 focuses on diagnostic load testing of bridges. It discusses the specific aspects of diagnostic load testing during the preparation, execution, and post-processing of a diagnostic load test and the general methodology for diagnostic load tests. Chapters describing detailed examples of diagnostic load tests from North America

DOI: https://doi.org/10.1201/9780429265969

and Europe are included. One chapter that deals with the particularities of testing pedestrian bridges is included as well.

Part I of Volume 13 focuses on proof load testing of bridges. It discusses the specific aspects of proof load testing during the preparation, execution, and post-processing of a proof load test. Important topics in this part are the interpretation of the measurements in real time during the experiment and the determination of the target proof load. Since high loads are applied in proof load tests, the risk of collapse or permanent damage to the structure exists. For this reason, careful instrumentation and monitoring of the bridge during the proof load test is of the utmost importance. Criteria based on the structural response that can be used to evaluate when further loading may not be permitted are discussed. Chapters describing examples of proof load test from North America and Europe are included.

Part II of Volume 13 describes how the practice of load testing can also be applied to buildings. Whereas the main focus of this book is on bridges, the same principles can be used for load testing of buildings. Buildings can require a load test prior to opening when there are doubts about the adequate performance of the building, or during the service life of the building when there are doubts regarding the capacity as a result of material degradation or deterioration, or when the use and associated live load of the building is to be changed.

Part III of Volume 13 discusses novel ideas regarding measurement techniques used for load testing. Methods using non-contact sensors, such as photography- and video-based measurement techniques are discussed. With acoustic emission measurements, signals of microcracking and distress can be observed before cracks are visible. Fiber optics are explored and applied as a new measurements technique. The topic of measurements through radar interferometry is also discussed. These chapters contain the background of these novel measurement techniques and recommendations for practice, and identify topics for further research and improvement of these measurement techniques.

Part IV of Volume 13 discusses load testing in the framework of reliability-based decision-making and in the framework of a bridge management program. The first topic of this part deals with the topic of updating the reliability index of a bridge after load testing and discusses the required proof load magnitude for this purpose, as well as systems reliability considerations. The second topic deals with the effect of load test results on the estimation of the remaining service life of a tested bridge, taking into account the effects of degradation. Finally, two chapters from the perspective of the bridge owner discuss how load testing fits in the framework of a bridge management program.

Part V of Volume 13 brings all information from the previous parts in Volumes 12 and 13 together and discusses the current state of the art on load testing. An overview of open questions for research is given, along with generally practical recommendations for load testing.

Besides the general structure in eight parts (three parts in Volume 12 and five parts in Volume 13), this work contains 25 chapters (12 chapters in Volume 12 and 13 chapters in Volume 13) written by a collective of international experts on the topic of load testing. Chapter 1 of Volume 12, by Lantsoght, introduces the general topic of load testing of structures. In this chapter, the reader can find a short background to the topic of load testing of structures, the scope of thiswork, the aim of this work, and a short discussion of the structure and outline of the two volumes making up this work.

Chapter 2 of Volume 12, by ElBatanouny, Schacht, and Bolle, provides an overview of the historical development and current practices of bridge load tests in Europe and the United States. The use of load tests to prove the capacity of structures is as old as mankind and plays an important role in the historical development of reinforced concrete design and construction. Historically, load tests provided a proof that a structure can carry a certain load and, therefore, they were likely used to convince the people of the bearing capability of bridges. With the development of static calculations and acceptable design rules, load tests became unnecessary for new structures. However, currently, strength evaluation of existing structures is becoming more and more important and the advantages of experimental assessments are used. Chapter 2 of Volume 12 describes the development of the technology of load testing and shows the European and American way of practice.

In Chapter 3 of Volume 12, Lantsoght reviews the existing codes and guidelines for load testing of structures. A summary of the main requirements of each existing code is provided, with a focus on the determination of the required load and measurements. The requirements for load testing of bridges and buildings are revised, for new and existing structures. An international perspective is given, revising the practice from Germany, the United Kingdom, Ireland, the United States, France, the Czech Republic, Slovakia, Spain, Italy, Switzerland, Poland, and Hungary. The chapter concludes with a short overview of the current developments and a discussion of the differences between the available codes and guidelines.

In Chapter 4 of Volume 12, Lantsoght and Schmidt discuss aspects that should be considered prior to every load test. The first questions that need to be answered are "Is this bridge suitable for load testing?" and "If so, what are the goals of the load test?" To answer these questions, information must be gathered and preliminary calculations should be carried out. In order to evaluate if a bridge is suitable for load testing, different types of testing are shown, the topic of when to load test a bridge is discussed, and structure type considerations are debated. Finally, some safety precautions that should be fulfilled during a load test are discussed.

Subsequently, Lantsoght and Schmidt discuss in Chapter 5 of Volume 12 the aspects related to the preparation of load tests, regardless of the chosen type of load test. After determination of the test objectives, the first step should be a technical inspection of the bridge and bridge site. With this information, the preparatory calculations (assessment for existing bridges and expected behavior during the test) can be carried out. Once the analytical results are available, the practical aspects of testing can be prepared: planning, required personnel, method for applying the load, considerations regarding traffic control and safety, and the development of the sensor and data acquisition plan. It is good practice to summarize all preparatory aspects in a preparation report and to provide this information to the client/owner as well as to all parties involved with the load test.

Then in Chapter 6 of Volume 12, Lantsoght and Schmidt discuss the aspects related to the execution of load tests, regardless of the chosen type of load test. The main elements required for the execution of the load test are the equipment for applying the load and the equipment for measuring and displaying the structural responses (if required). This chapter reviews the commonly used equipment for applying the loading and discusses all aspects related to the measurements. The next topic is the practical aspects related to the execution. This topic deals with communication on site during the load test and important safety aspects.

Finally, in Chapter 7 of Volume 12, Lantsoght and Schmidt discuss the aspects related to processing the results of a load test after the test. The way in which the data are processed depends on the goals of the test. As such, the report that summarizes the preparation, execution, and post-processing of the load test should clearly state the goal of the load test, how the test addressed this goal, and what can be concluded. Typical elements of the post-processing stage include discussing the applied load, the measured structural responses, and then evaluating the bridge based on the results of the load test.

Chapter 8 of Volume 12, by Lantsoght, Bonifaz, Sanchez, and Harris, deals with the methodology for diagnostic load testing. All aspects of diagnostic load testing that are shared with other load testing methods have been discussed in Part II of Volume 12. In this chapter, the particularities of diagnostic load testing of new and existing bridges are discussed. These elements include loading procedures, monitoring behavior during the test, reviewing test data, calibrating analytical models, and evaluating the test results.

In Chapter 9 of Volume 12, Hernandez and Myers illustrate an example diagnostic load test. This chapter introduces an example diagnostic load test conducted on the superstructure of Bridge A7957, built in Missouri, USA, to illustrate how experimental in-situ parameters can be included in the estimation of a bridge load rating. The experimental load rating was less conservative than the analytical load rating.

In Chapter 10 of Volume 12, Diaz Arancibia and Okumus illustrate an example diagnostic load test. They argue that bridge load testing is widely used for assessing bridge structural behavior and may be preferred over other means, since it is capable of capturing the actual response of structures. However, load testing is complex and requires careful consideration of several activities that precede its execution. They describe the planning, coordination, scheduling, execution, and data analysis of a load test, using an example of a highly skewed, prestressed concrete girder/reinforced concrete deck bridge under service loads. The load test allowed the evaluation of the effects of high skew angles and mixed pier support fixity arrangements on girder load distribution behavior and deck performance.

In Chapter 11 of Volume 12, Olaszek and Casas present principles and justification of diagnostic load tests of bridges as performed from the European point of view. Normally, diagnostic tests serve to verify and adjust the predictions of an analytical model. However, as presented in this chapter, the results of a diagnostic load test in a bridge can also serve other objectives. Several examples of application are presented with the main objective not only to show how the tests are carried out, but also to introduce which are the main issues to take into account to obtain accurate and reliable results that can be used in the assessment of the actual capacity of the bridge. In the case of static tests, conclusions regarding the measurement stabilization time are presented. For dynamic tests in railway bridges, the extrapolation to lower and higher speeds is also discussed.

In Chapter 12 of Volume 12, Bačinskas, Jakubovskis, and Kilikevičius show how diagnostic load tests can be applied to pedestrian bridges. This chapter contains basic information related to static and dynamic testing of pedestrian bridges. A brief overview of test objectives and test classification is presented. The chapter also covers the stages of preparation for testing, aspects of test program creating, and methods of static and dynamic loading of footbridges. A sequence of test organization and execution is also presented. The second part of the chapter discusses the aspects of processing and evaluation of static and dynamic test results as well as aspects of comparing them with obtained by the theoretical modeling of the bridge. The chapter presents requirements for pedestrian bridges specified in the design codes and recommendations of various countries. These need to be taken into account in assessing the results of the tested bridges. The chapter ends with a discussion on the presentation of the test results and aspects of assessment of the bridge condition according to the test data. The presented material may be useful to researchers and experts involved in the design, construction, and maintenance of pedestrian bridges.

Chapter 1 of Volume 13, by Lantsoght, deals with the methodology for proof load testing. All aspects of proof load testing that are shared with other load testing methods have been discussed in Part II of Volume 12. In this chapter, the particularities of proof load testing are discussed based on the current state of the art. These elements include the determination of the target proof load (which is still a topic of research), the procedures followed during a proof load test (loading method, instrumentation, and stop criteria), and the post-processing of proof load test data, including the assessment of a bridge after a proof load test.

In Chapter 2 of Volume 13, Jauregui, Weldon, and Aguilar show that load rating of concrete bridges with no design plans is currently an issue in the United States and other countries. Missing or incomplete design documentation creates uncertainties in establishing the safe load limits of bridges to carry legal vehicles. Guidance for evaluating planless concrete bridges, in particular prestressed structures, is limited in the AASHTO Manual for Bridge Evaluation and very few state departments of transportation have rating procedures for such bridges. The authors developed a multi-step load rating procedure for planless prestressed bridges that includes (1) estimating the material properties from past specifications and amount of prestressing steel using Magnel diagrams; (2) verifying the steel estimate by rebar scanning; (3) field testing at diagnostic and/or proof load levels based on strain measurements; and (4) rating the bridges using the proof test results. Three prestressed concrete bridges (including a double T-beam, box beam, and I-girder bridge) are evaluated using nondestructive load testing and material evaluation techniques to illustrate the procedures. Rating factors are determined for AASHTO and New Mexico legal loads using the proof test results for the serviceability limit state (SLS) (i.e. concrete cracking). Using load rating software, rating factors are also computed for the strength limit state (i.e. shear or flexural capacity) based on the measured bridge dimensions and estimated material properties. Load ratings for serviceability and strength are finally compared to establish the bridge capacities.

Lantsoght, Hordijk, Koekkoek, and Van der Veen describe a case study of a proof load test in Chapter 3 of Volume 13. The viaduct Zijlweg was proof load tested at a position that is critical for bending moment and at a position that is critical for shear. The viaduct Zijlweg has cracking caused by an alkali-silica reaction, and the effect of material degradation on the capacity is uncertain. Therefore, the assessment of this viaduct was carried out with a proof load test. This chapter details the preparation, execution, and evaluation of viaduct Zijlweg.

In Chapter 4 of Volume 13, Schacht, Bolle, and Marx show how the principles of load testing of bridges also can be used for building constructions, by sharing the background and main recommendations from the guideline for load testing of existing concrete structures of the Deutscher Ausschuss für Stahlbeton (DAfStb). Thanks to this guideline, published in 2000, the experimental proof of the load-bearing capacity became a widely accepted method in Germany. Since then, over 2000 proof load tests have been carried out and a lot of experience exists in the usage of the guideline. Nevertheless, there are still reports about load tests carried out with mass loads or using mechanical measurement techniques. To prevent misuse of the guideline in the future, to take into account the great experiences existing in the evaluation of the measuring results and the bearing condition, the DAfStb decided to revise the existing guideline. In recent years there has also been increased research activity focusing on the development of evaluation criteria for possible brittle failures in shear and to extend the safety concept for the evaluation of elements that have not been directly tested. These results will also be considered in the new guideline. The authors give an overview of the rules of the existing guideline and discuss these in the background of the gained experience over the past two decades. The new criteria for evaluating shear capacity are explained, and thoughts about the new safety concept for load testing are introduced.

In Chapter 5 of Volume 13, Alipour, Shariati, Schumacher, Harris, and Riley show that in order to capture the deformation response of a bridge member during load testing, the structure has to be properly instrumented. Image- and video-based measurement techniques have significant advantages over traditional physical sensors in that they are applied remotely without physical contact with the structure, they require no cabling, and they allow for measuring displacement where no ground reference is available. The authors introduce two techniques used to analyze digital images: digital image correlation (DIC) and Eulerian virtual visual sensors (VVS). The former is used to measure the static displacement response and the latter to capture the natural vibration frequencies of structural members. Each technique is introduced separately, providing a description of the fundamental theory, presenting the equipment needed to perform the measurements, and discussing the strengths and limitations. Case studies provide examples of real-world applications during load testing to give the reader a sense of the advances and opportunities of these measurement techniques.

Chapter 6 of Volume 13, by ElBatanouny, Anay, Abdelrahman, and Ziehl, discusses the use of acoustic emission (AE) measurements for load testing. Several methods for analyzing AE data to classify damage in reinforced and prestressed concrete structures during load tests were developed over the past two decades. The majority of methods offer relative assessment of damage – for example, classifying cracked versus uncracked conditions in prestressed members during or following a load test. In addition, significant developments were made in various AE source location techniques, including one-, two-, and three-dimensional source location as well as moment tensor analysis, which allow for accurate location of damage through advanced data filtering techniques. This chapter provides an overview of the acoustic emission technique for detecting and classifying damage during load tests. Recent efforts to apply the method in the field are also presented along with recommended field applications based on the current state of practice.

Chapter 7 of Volume 13, by Casas, Barrias, Rodriguez Gutiérrez, and Villalba, presents the application of fiber optic sensor technology in the monitoring of a load test. First, fiber optic technology is described with main emphasis in the case of distributed optical fiber sensors (DOFS), which have the potential of measuring strain and temperature along the fiber with different length and accuracy ranges. After that, two laboratory tests in reinforced and prestressed concrete specimens show the feasibility of using this technique for the detection, localization, and quantification of bending and shear cracking. Finally, the technique is applied to two real prestressed concrete bridges: in the first case, during the execution of a diagnostic load test; and in the second case, for the continuous monitoring in time and space of a bridge subjected to rehabilitation work. These experiences show the potential of this advanced monitoring technique when deployed in a load test.

In Chapter 8 of Volume 13, Gentile shows that recent advances in radar techniques and systems have favored the development of microwave interferometers suitable to the non-contact measurement of deflections on large structures. The main characteristic of the new radar systems, entirely designed and developed by Italian researchers, is the possibility of simultaneously measuring the deflection of several points on a large structure with high accuracy in either static or dynamic conditions. In this chapter, the main radar techniques adopted in microwave remote sensing are described, and advantages and potential issues of these techniques are addressed and discussed. Subsequently, the application of microwave remote sensing in live-load static and ambient vibration tests performed on full-scale bridges is presented in order to demonstrate the reliability and accuracy of the measurement technique. Furthermore, the simplicity of use of the radar technology is exemplified in practical cases, where the access with conventional techniques is difficult or even hazardous, such as the stay cables on cable-stayed bridges.

In Chapter 9 of Volume 13, Frangopol, Yang, Lantsoght, and Steenbergen review concepts related to the uncertainties associated with structures and discuss how the results of load tests can be used to reduce these uncertainties. When an existing bridge is subjected to a load test, it is known that the capacity of the cross section is at least equal to the largest load effect that was successfully resisted. As such, the probability density function of the capacity can be truncated after the load test, and the reliability index can be recalculated. These concepts can be applied to determine the required target load for a proof load test to demonstrate that a structure achieves a certain reliability index. Whereas the available methods focus on member strength and the evaluation of isolated members, a more appropriate approach for structures would be to consider the complete structure in this reliability-based approach. For this purpose, concepts of systems reliability are introduced. It is also interesting to place load testing decisions within the entire life cycle of a structure. A cost-optimization analysis can be used to determine the optimum time in the life cycle of the structure to carry out a load test.

In Chapter 10 of Volume 13, Val and Stewart demonstrate the determination of the remaining service life of reinforced concrete bridge structures in a corrosive environment after load testing. Reinforced concrete (RC) bridge structures deteriorate with time, and the corrosion of reinforcing steel is one of the main causes of bridge deterioration. Load testing alone is unable to provide information about the extent of deterioration and the remaining service life of a structure. In this chapter, a framework for the reliability-based assessment of the remaining service life of RC bridge structures in corrosive environments will be described. Existing models are considered for corrosion initiation, corrosion-induced cracking, and effects of corrosion on stiffness and strength of RC members. Special attention is paid to the potential effects of a changing climate on corrosion initiation and propagation in these structures. Examples illustrating the framework application are provided.

Chapter 11 of Volume 13, by Elfgren, Täljsten, and Blanksvärd, provides the perspective of bridge owners on load testing by discussing the experience in Sweden. Load testing of new and existing bridges was performed regularly in Sweden up to the 1960s. It was then abandoned due to high costs as opposed to the small amount of extra information obtained. Most bridges behaved well in the serviceability limit range, and no knowledge of the ultimate limit stage could be obtained without destroying the bridge. At the same time, methods for calculating the capacity were developed and new numerical methods were introduced. Detailed rules were given on how these methods should be used. Some decommissioned bridges were tested to their maximum capacity to study their failure mechanisms and to calibrate the numerical methods. In this chapter, some examples are given on how allowable loads have increased over the years and how tests are being performed. Nowadays, load testing may be on its way back, especially to test existing rural prestressed concrete bridges, where no design calculations have been retained

In Chapter 12 of Volume 13, De Boer gives an overview of load testing within the framework of bridge management in the Netherlands. In the Netherlands, field tests on existing bridges have been carried out over the past 15 years. The tests consists of SLS load level tests (before the onset of nonlinear behavior, also known as proof load tests) on existing bridges, ultimate limit state tests (collapse tests) on existing bridges, and laboratory tests on beams sawn from the existing bridges. The goal of these experiments is to have a better assessment of the existing bridges. The assessment is combined with material sampling (concrete cores, reinforcement steel samples), for a better quantification of the material parameters, and with nonlinear finite element models. In the future, a framework will be developed in which the concrete compressive strength, determined from a large number of similar bridges from which core samples are taken, is first used as input for an assessment with a nonlinear finite element model according to the Dutch non-linear finite element analysis guidelines. Then it can be determined if sampling of the actual viaduct is needed, following recommendations on how to handle concrete cores and how to scan reinforcement, and if a load test should be used to evaluate the bridge under consideration. For this purpose, guidelines for proof load testing of concrete bridges are under development, but further research on the topic of load testing for shear is necessary for the development of such guidelines.

Finally, in Chapter 13 of Volume 13, Lantsoght brings all information from the previous chapters in Volumes 12 and 13 together and briefly restates the topics that have been covered. The topics to which more research energy has been devoted are discussed, as are the remaining open questions. Based on the current knowledge and state of the art, as well as the practical experiences presented by the authors in this book, general practical recommendations are presented.

These 25 chapters across two volumes provide a well-rounded overview of the state of the art internationally and the open research questions related to load testing of bridges, and how these insights can be applied to other structures such as buildings. The target audience of this book is researchers, students, practicing engineers, and bridge owners. With the inclusion of practical examples as well as open research questions, this book could be useful for academics as well as practitioners.

As book editor, I would like to thank all authors for their contributions to this work. Developing one (or more) book chapters is a large commitment in terms of time and effort, and I am grateful for all the authors who collaborated on this work. I would also like to thank the reviewers of the book proposal for their ideas on how to improve the contents of these books. Finally, I would like to acknowledge the work of the team at Taylor & Francis who made the publication of this work possible. In particular, I would like to thank Mr. Alistair Bright for his support throughout the development of this work, and his kind guidance along the way.

Eva O. L. Lantsoght
Quito, Ecuador, December 2018

About the Editor

Eva O. L. Lantsoght is a full professor at Universidad San Francisco de Quito in Quito, Ecuador; a part-time researcher at Delft University of Technology, Delft, the Netherlands; and a structural engineer at ADSTREN, Quito, Ecuador. She obtained a Master's Degree in Civil Engineering from the Vrije Universiteit Brussels, Brussels, Belgium in 2008; a Master's Degree in Structural Engineering from the Georgia Institute of Technology, Atlanta, USA, with scholarships from the Belgian American Educational Foundation (BAEF) and Fulbright in 2009; and a Ph.D. in Structural Engineering from Delft University of Technology, Delft, the Netherlands in 2013.

The field of research of Dr. Lantsoght is the design and analysis of concrete structures and the analysis of existing bridges. Her work has focused on the following topics: second-order effects in reinforced concrete columns, shear in one-way slabs under concentrated loads close to supports, fatigue in high-strength concrete, assessment of reinforced concrete slab bridges, plastic design models for concrete structures, proof load testing of concrete bridges, shear and torsion in structural concrete, and measurement methods and monitoring techniques. She has published more than 50 indexed journal and conference papers on the aforementioned topics.

Dr. Lantsoght is an active member of the technical committees of the Transportation Research Board in Concrete Bridges (AFF-30) and Testing and Evaluation of Transportation Structures (AFF-40). She serves on the technical committees of the American Concrete Institute and Deutscher Ausschuß für Stahlbeton Shear Databases (ACI-DAfStb-445-D), the joint ACI-ASCE (American Society of Civil Engineers) committee on Design of Reinforced Concrete Slabs (ACI-ASCE 421), and the ACI Committee on Evaluation of Concrete Bridges and Concrete Bridge Elements (ACI 342), and she is an associate member of the committees on Shear and Torsion (ACI-ASCE 445), and on Strength Evaluation of Existing Concrete Structures (ACI 437). She is a guest associate editor for *Frontiers in Built Environment – Bridge Engineering* for the research topic "Diagnostic and Proof Loading Tests on Bridges" and was the editor of the ACI Special Publication "Evaluation of Concrete Bridge Behaviour through Load Testing – International Perspective." She has served on the scientific committee of numerous conferences, and has organized a number of special sessions and mini-symposia on topics related to load testing and monitoring of bridges. She is a reviewer for many scientific journals, an academic editor for *PLOS One*, and the editor in chief of ACI *Avances en Ciencias e Ingenierías*.

DOI: https://doi.org/10.1201/9780429265969

As a professor, Dr. Lantsoght teaches undergraduate and professional courses in the field of structural engineering and enjoys working with undergraduate and graduate students on their thesis and/or research projects. She is dedicated to sharing knowledge about existing structures to professionals in Ecuador and beyond in order to improve the safety and sustainability of the built environment.

Dr. Lantsoght's interests include also science communication, science outreach, higher education policy, and the improvement of doctoral education. She is the editor and main author of "PhD Talk," a blog for PhD students, and the author of the free e-book *Top PhD Advice from Start to Defense and Beyond* and the textbook *The A–Z of the PhD Trajectory – A Practical Guide for a Successful Journey* in the Springer Texts in Education series.

Author Data

Abdelrahman, Marwa A.
Wiss, Janney, Elstner Associates, Inc.
330 Pfingsten Road,
Northbrook, IL 60062
Tel: (1) 847.753.6382
Email: mabdelrahman@wje.com

Aguilar, Carlos V.
HDR, Inc.
2155 Louisiana Blvd. NE
Suite 9500
Albuquerque, NM 87110
Tel: (1) 505-830-5411
Email: Carlos.Aguilar@hdrinc.com

Alipour, Mohamad
Department of Civil and Environmental Engineering
University of Virginia
351 McCormick Road
Charlottesville, VA, 22904-4742, USA
Tel: (1) 434-924-3072
Email: ma4cp@virginia.edu

Anay, Rafal N.
University of South Carolina
300 main street
Columbia SC, 29208, USA
Tel: +1(803)7416699
Email: ranay@email.sc.edu

Barrias, António
Department of Civil and Environmental Engineering
Technical University of Catalonia – BarcelonaTech
c/ Jordi Girona 1-3
08034 Barcelona, Spain
Tel: +34 934016513
Email: antonio.jose.de.sousa@upc.edu

DOI: https://doi.org/10.1201/9780429265969

Blanksvärd, Thomas
Structural Engineering
Department of Civil, Mining and Natural Resources Engineering,
Luleå University of Technology
SE-97187 Luleå, Sweden
Tel: +46 920-491642
Email: Thomas.Blanksvard@ltu.se

Bolle, Guido
Hochschule Wismar
Civil Engineering Department
23952 Wismar, Germany
Tel: +49 3841 753 72 90
Email: guido.bolle@hs-wismar.de

Casas, Joan Ramon
Technical University of Catalunya, UPC-BarcelonaTECH
Jordi Girona 1-3. Campus Nord. Modul C1
08034 Barcelona, Spain
Tel: +34-934016513
Email: joan.ramon.casas@upc.edu

de Boer, Ane
de Boer Consultancy
6823 NM Arnhem
the Netherlands
Tel: +31 6 53811617
Email: ane1deboer@gmail.com

ElBatanouny, Mohamed K.
Wiss, Janney, Elstner Associates, Inc.
330 Pfingsten Road
Northbrook, IL 60062
USA
Tel: (1) 847.753.6395
Email: melbatanouny@wje.com

Elfgren, Lennart
Structural Engineering
Department of Civil, Mining and Natural Resources Engineering,
Luleå University of Technology
SE-97187 L uleå, Sweden
Tel: +46 920 493 660
Email: Lennart.Elfgren@ltu.se

Frangopol, Dan M.
The Fazlur R. Khan Endowed Chair of Structural Engineering and Architecture
Department of Civil and Environmental Engineering
Center for Advanced Technology for Large Structural Systems (ATLSS Center)
Lehigh University
117 ATLSS Drive, Imbt Labs
Bethlehem, PA 18015-4729, USA
Tel: +1-610-758-6103
Email: dan.frangopol@lehigh.edu

Gentile, Carmelo
Politecnico di Milano, Department of Architecture, Built environment
and Construction engineering (DABC)
Piazza Leonardo da Vinci
32 - 20133 Milan
Italy
Tel: +39 02 2399 4242
Email: carmelo.gentile@polimi.it

Harris, Devin K.
Associate Professor - Department of Civil and Environmental Engineering
Director - Center for Transportation Studies
Faculty Director of Clark Scholars Program
University of Virginia
351 McCormick Rd Charlottesville
VA 22904
USA
Tel: (434)924-6373
Email: dharris@virginia.edu

Hordijk, Dick A.
Concrete Structures
Civil Engineering and Geosciences
Delft University of Technology
2628 CN Delft
the Netherlands
Tel: (+31) (0)15 27 84434
Email: D.A.Hordijk@tudelft.nl

Jauregui, David V.
Department of Civil Engineering
New Mexico State University
3035 South Espina Street
Hernandez Hall, MSC 3CE
Las Cruces, NM 88003, USA
Tel: (1) 575-646-3801
Email: jauregui@nmsu.edu

Koekkoek, Rutger Tycho
BAM Infraconsult
H.J. Nederhorststraat 1
2801 SC Gouda
the Netherlands
Tel: 0031 (0)6 31 036 069
Email: rutger.koekkoek@bam.com

Lantsoght, Eva O.L.
Politécnico
Universidad San Francisco de Quito
Diego de Robles y Pampite
Sector Cumbaya
EC 170157 Quito, Ecuador
Tel: (+593) 2 297-1700 ext. 1186
Email: elantsoght@usfq.edu.ec

Concrete Structures
Civil Engineering and Geosciences
Delft University of Technology
Stevinweg 1
2628 CN Delft, the Netherlands
Tel: (+31)152787449
Email: E.O.L.Lantsoght@tudelft.nl

ADSTREN Cia Ltda.
Plaza del Rancho
Av. Eugenio Espejo 2410
Bloque 1, Oficina 203
Quito, Ecuador
Tel: +593 23 957 606
Email: elantsoght@adstren.com

Marx, Steffen
Leibniz Universität Hannover
Institute of Concrete Construction
Appelstraße 9A
30167 Hannover
Germany
Tel: + 49 511 762 3352
Email: marx@ifma.uni-hannover.de

Riley, Charles E.
Civil Engineering Department
Oregon Institute of Technology
3201 Campus Drive,
Klamath Falls, OR 97601, USA
Tel: 541-885-1922
Email: charles.riley@oit.edu

Rodríguez Gutiérrez, Gerardo
Institute of Engineering
National Autonomous University of Mexico UNAM
Circuito Escolar S/N Instituto de, Ingeniería, Ciudad Universitaria
04510 Ciudad de México, CDMX
Mexico
Tel: +52 (55) 56233600 Ext. 8417
Email: grog@pumas.ii.unam.mx

Schacht, Gregor
Marx Krontal GmbH
Uhlemeyerstraße 9+11
30175 Hannover
Germany
Tel: +49 511 51515423
Email: gregor.schacht@marxkrontal.com

Schumacher, Thomas
Department of Civil and Environmental Engineering
Portland State University
1930 SW 4th Avenue, Suite 200
Portland
OR 97201, USA
Tel: +1-503-725-4199
Email: thomas.schumacher@pdx.edu

Shariati, Ali
Applied Structural Technologies Inc
719 Cambridge St
Cambridge, MA 02141, USA
Tel: 484-714-9223
Email: shariati.ali@gmail.com

Steenbergen, Raphaël
TNO
Stieltjesweg 1
2628 CN Delft
the Netherlands
Tel: +31 88 866 34 23
Email: raphael.steenbergen@tno.nl

Ghent University
Faculteit Ingenieurswetenschappen en Architectuur
Jozef Plateaustraat 22
9000 Gent, Belgium
Tel: +32 9 264 55 35
Email: raphael.steenbergen@ugent.be

Stewart, Mark G.
Director, Centre for Infrastructure Performance and Reliability
School of Engineering
The University of Newcastle
Newcastle,   NSW, 2308, Australia
Tel: +61 2 49216027
Email: mark.stewart@newcastle.edu.au

Täljsten, Björn
Structural Engineering
Department of Civil, Mining and Natural Resources Engineering,
Luleå University of Technology
SE-97187 Luleå, Sweden
Tel: +46(0)70 537 43 70
Email: Bjorn.Taljsten@ltu.se

Val, Dimitri V.
Institute for Infrastructure & Environment, EGIS
Heriot-Watt University
Edinburgh EH14 4AS, UK
Tel: +44 (0)131 451 4622
Email: D.Val@hw.ac.uk

van der Veen, Cor
Concrete Structures
Civil Engineering and Geosciences
Delft University of Technology
2628 CN Delft
the Netherlands
Tel: (+31) (0)15 2784577
Email: C.vanderveen@tudelft.nl

Villalba, Sergi
Cotca, S.A.
C/ Balmes 200, 5^0 2^a
08006 Barcelona, Spain
Tel: +34 93.218.71.46
Email: sergi.villalba@cotca.com

Weldon, Brad D.
Department of Civil Engineering
New Mexico State University
3035 South Espina Street
Hernandez Hall, MSC 3CE
Las Cruces, NM   88003
Tel: (1) 575-646-1167
Email: bweldon@nmsu.edu

Yang, David Y.
Department of Civil and Environmental Engineering
Engineering Research Center for Advanced Technology for Large
Structural Systems (ATLSS Center)
Lehigh University
117 ATLSS Drive, Imbt Labs
Bethlehem, PA 18015-4729, USA
Tel: +1 610-758-5639
Email: yiy414@lehigh.edu

Ziehl, Paul
Assistant Dean for Research, College of Engineering and Computing
Professor, Dept. of Civil and Environ. Engineering
University of South Carolina
300 Main Street, C206
Columbia, SC 29208, USA
Phone: 803 467 4030
Email: ziehl@cec.sc.edu

Contributors List

Abdelrahman, Marwa A., *Wiss, Janney, Elstner Associates, Inc., Northbrook, IL, USA*

Aguilar, Carlos V., *HDR Inc., Albuquerque, NM, USA*

Alipour, Mohamad, *Department of Civil and Environmental Engineering, University of Virginia, Charlottesville, VA, USA*

Anay, Rafal, *University of South Carolina, Columbia, SC, USA*

Barrias, António, *Department of Civil and Environmental Engineering, Technical University of Catalonia, UPC, Barcelona, Spain*

Blanksvärd, Thomas, *Department of Civil, Mining and Natural Resources Engineering, Luleå University of Technology, Luleå, Sweden*

Bolle, Guido, *Hochschule Wismar, Wismar, Germany*

Casas, Joan R., *Department of Civil and Environmental Engineering, Technical University of Catalunya, UPC-BarcelonaTECH, Barcelona, Spain*

de Boer, Ane, *de Boer Consultancy, Arnhem, the Netherlands*

ElBatanouny, Mohamed K., *Wiss, Janney, Elstner Associates, Inc, Northbrook, IL, USA*

Elfgren, Lennart, *Department of Civil, Mining and Natural Resources Engineering, Luleå University of Technology, Luleå, Sweden*

Frangopol, Dan M., *Department of Civil and Environmental Engineering, Lehigh University, Bethlehem, PA, USA*

Gentile, Carmelo, *Politecnico di Milano, Department of Architecture, Built Environment and Construction Engineering (DABC), Milan, Italy*

Harris, Devin K., *Department of Civil and Environmental Engineering, University of Virginia, Charlottesville, VA, USA*

Hordijk, Dick A., *Concrete Structures, Delft University of Technology, Delft, the Netherlands*

DOI: https://doi.org/10.1201/9780429265969

Jauregui, David V., *Department of Civil Engineering, New Mexico State University, Las Cruces, NM, USA*

Koekkoek, Rutger T., *BAM Infraconsult, Gouda, the Netherlands*

Lantsoght, Eva O. L., *Politecnico, Universidad San Francisco de Quito, Quito, Ecuador; Concrete Structures, Delft University of Technology, Delft, the Netherlands; ADSTREN Cia Ltda., Quito, Ecuador*

Marx, Steffen, *Leibniz Universität Hannover, Hannover, Germany*

Riley, Charles J., *Civil Engineering, Oregon Institute of Technology, Klamath Falls, OR, USA*

Rodriguez Gutiérrez, Gerardo, *National Autonomous University of Mexico UNAM, Mexico City, Mexico*

Schacht, Gregor, *Marx Krontal, Hannover, Germany*

Schumacher, Thomas, *Civil and Environmental Engineering, Portland State University, Portland, OR, USA*

Shariati, Ali, *Applied Structural Technologies Inc, Cambridge, MA, USA*

Steenbergen, Raphaël D. J. M., *TNO, Delft, the Netherlands; Ghent University, Ghent, Belgium*

Stewart, Mark G., *Centre for Infrastructure Performance and Reliability, University of Newcastle, NSW, Australia*

Täljsten, Bjorn, *Department of Civil, Mining and Natural Resources Engineering, Luleå University of Technology, Luleå, Sweden*

Val, Dimitri V., *Institute for Infrastructure and Environment, EGIS, Heriot-Watt University, Edinburgh, UK*

Van der Veen, Cor, *Concrete Structures, Delft University of Technology, Delft, the Netherlands*

Villalba, Sergi, *Cotca, S.A., Barcelona, Spain*

Weldon, Brad D., *Department of Civil Engineering, New Mexico State University, Las Cruces, NM, USA*

Yang, David Y., *Department of Civil and Environmental Engineering, Lehigh University, Bethlehem, PA, USA*

Ziehl, Paul, *College of Engineering and Computing, and Department of Civil and Environmental Engineering, University of South Carolina, Columbia, SC, USA*

List of Tables

DOI: https://doi.org/10.1201/9780429265969

List of Figures

DOI: https://doi.org/10.1201/9780429265969

Part I

Proof Load Testing of Bridges

Chapter 1

Methodology for Proof Load Testing

Eva O. L. Lantsoght

Abstract

This chapter deals with the methodology for proof load testing. All aspects of proof load testing that are shared with other load testing methods have been discussed in Part II of Volume 12. In this chapter, the particularities of proof load testing are discussed. These elements include the determination of the target proof load, the procedures followed during a proof load test (loading method, instrumentation, and stop criteria), and the post-processing of proof load test data, including the assessment of a bridge after a proof load test.

1.1 Introduction

Proof load testing is a method of load testing in which a load representative of the factored live load is applied to the structure. If the structure can carry this load without signs of distress, the test is considered successful and the structure has been shown experimentally to fulfill the code requirements. Since for this type of load testing large loads are involved, the risk and cost involved with proof load testing is higher than for diagnostic load testing. The risk encompasses possible damage to the structure or its collapse (structural safety) as well as the risk for the executing personnel and the traveling public. To minimize the risk for the structure, the tested structure needs to be instrumented and the measurements need to be followed in real time. If a preset threshold for these measurements (a so-called stop criterion) is exceeded, further loading is not permitted even though the target proof load may not yet be achieved. To minimize the risk for the executing personnel and the traveling public, safety regulations need to be implemented and a safety plan needs to be developed prior to the load test.

To guarantee the structural safety, the proof load test needs to be carefully prepared. On the other hand, however, there is a need to develop methods to quickly carry out proof load tests so that with a minimum number of sensors and simplified loading protocol, an advice about the bridge can be given. This "quick and easy" type of proof load testing can only be applied to bridges of which the structural behavior is well understood and for ductile failure modes. The development of recommendations for this type of proof load testing is the subject of current research but is a promising application of proof load testing for the future.

The preparations for a regular proof load test include extensive calculations and predictions of the behavior of the bridge as well as the preparation of the sensor plan and the

DOI: https://doi.org/10.1201/9780429265969

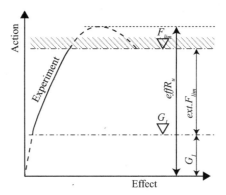

Figure 1.1 Use of stop criteria during proof load test

development of the threshold values for the stop criteria. At least two qualified testing engineers should be following and evaluating the measurements during the test.

The followed safety philosophy using stop criteria (Deutscher Ausschuss für Stahlbeton, 2000) is shown in Figure 1.1 (see also Chapter 3 of Vol. 12). If a stop criterion is reached, the corresponding load is F_{lim}. Loading beyond this point is not allowed. A proof load test is successful if the sum of the permanent loads G_1 and the target proof load F_{target} is smaller than the limiting load F_{lim}. If the opposite is found in a proof load test, the bridge does not fulfill the requirements of the code for the considered safety level that was the target for the proof load test, but it may still fulfill the requirements at a lower safety level.

The safety plan for the proof load test needs to be developed after a technical inspection of the bridge. This inspection should signal possible safety risks and should consider the access to the test site. All possible dangerous situations and all possible problems (technical, electrical, electronical, planning-related, etc.) that may arise during the on-site activities need to be reported, and possible solutions and backup plans for these events need to be presented (Koekkoek et al., 2015). During all activities on site, the safety of the personnel and traveling public needs to be evaluated by a safety engineer. In the Netherlands, it is required that the information of all emergency services is visible on site in case of a calamity, and that a first-aid kit and at least one person with first-aid training is always present on site. Access to the tested structure should be restricted during testing. Only the executing personnel are allowed access to the bridge during the test. During the test, nobody is allowed to go under the bridge unless this event has been communicated with all parties and the load is removed. If the bridge cannot be closed to traffic during the proof load test, special safety considerations should be taken for the traveling public. It is strongly advised to temporarily close the bridge when the target load is applied.

For those who are not directly involved with the proof load test but may be affected by it, such as the traveling public and inhabitants of houses near the bridge, information should be provided. Similarly, information should be available upon request for the press and other interested parties. It is recommended that the communications expert of the local road authority develop this information.

1.2   Determination of target proof load

1.2.1   *Dutch practice*

In Dutch practice, the target proof load should be equivalent to the factored loads from the required load combination with load factors corresponding to the governing code. This equivalence is described in terms of an equivalent sectional force or moment. For this purpose, a linear finite element model is used. In this finite element model, the applied loads correspond to the required load combination. For the assessment of bridges in the Netherlands, this load combination typically consists of the self-weight of the structure, the superimposed dead load, and the live loads according to the live load model from the code under consideration.

In NEN-EN 1991–2:2003 (CEN, 2003), the live load model consists of the distributed lane load and a design tandem for each lane (see Fig. 1.2). Outside of the notional lanes, a distributed load of 2.5 kN/m² (0.05 kip/ft²) is applied, and on the sidewalk a load of 5 kN/m² (0.10 kip/ft²) corresponding to pedestrian loading needs to be applied. The distributed lane load can be applied in a checkerboard pattern over the different spans in a continuous bridge to find the most unfavorable loading situation. For the design tandem, four wheel prints with an unfactored load of 150 kN (34 kip) each on a contact surface of 400 mm × 400 mm (1.31 ft × 1.31 ft) are used. The axle distance is 1.2 m (3.9 ft) and the transverse distance is 2 m (6.6 ft). The live load model is shown in Figure 1.2. The tandem is centered in the notional lane of 3 m (9.8 ft). Over the height direction, the wheel print can be distributed under 45° to the slab mid-depth when shell elements are used for the finite element model.

To make the load combination, the respective load factors should be used. For existing bridges in the Netherlands, besides the Eurocodes, the codes for existing structures NEN 8700:2011 (Code Committee 351001, 2011a) and the guidelines for the assessment of bridges from the Dutch Ministry of Infrastructure "RBK" (Rijkswaterstaat, 2013) should

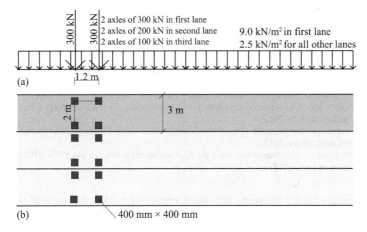

Figure 1.2 Load model 1 from NEN-EN 1991–2:2003 (CEN, 2003): (a) side view; (b) top view. The loads need to be multiplied with a factor α_{Qi} for the tandem and α_{qi} for the lane loads, with *i* the number of the lane. In the Netherlands, all α_{Qi} equal 1, $\alpha_{q1} = 1.15$, and $\alpha_{qi} = 1.4$ for $i > 1$. Conversions: 1 kN = 0.225 kip, 1 m = 3.3 ft, 1 mm = 0.04 in., 1 kN/m² = 0.021 kip/ft²

be consulted. All existing structures in the Netherlands should fulfill the requirements of the NEN 8700 series of codes, with NEN 8700:2011 (Code Committee 351001, 2011a) outlining the general safety philosophy, which is in line with NEN-EN 1990:2002 for new structures (CEN, 2002), and NEN 8701:2011 (Code Committee 351001, 2011b) outlining the loads and load factors, in line with NEN-EN 1991–2:2003 (CEN, 2003). The RBK guidelines define different safety levels at which an assessment of a highway bridge can be carried out, each with associated load factors, reliability index β, and the reference period (see Table 1.1). The defined load factors are the load factors for the permanent load γ_{perm}, which depend on whether Equation 6.10a or b from NEN-EN 1990:2002 (CEN, 2002) is governing, resulting in $\gamma_{perm,6.10a}$ or $\gamma_{perm,6.10b}$. For Equation 6.10a, a factor for the combination value of a variable action is used, whereas Equation 6.10b applies a reduction factor for unfavorable permanent loads. The remaining load factors are γ_{ll} for the live loads, γ_{wind} for the wind loads, and γ_{var} for other variable loads.

In a proof load test, the applied load should correspond to these prescribed safety levels. The load combination that is considered is the combination of the self-weight, superimposed load, and live load. The load factors are taken based on Table 1.1 at the different safety levels, with the exception that the load factor for the self-weight γ_{sw} is taken as 1.10 for all load levels. The reasoning behind this reduction is that the dimensions of an existing bridge are not a random variable anymore but instead a deterministic value. As such, only the model factor remains, which is taken as 1.10. An overview of the resulting load factors at the different load levels is then shown in Table 1.2. For the proof load, no load factor is used so that $\gamma_{proof} = 1.00$ (the load factor on the proof load) for all considered safety levels.

In the finite element model, first the design tandems are moved in their respective lanes until the position is found for which the largest sectional moment results (see Fig. 1.3). For the bending moment in concrete decks, a transverse distribution of $2d_l$ (with d_l the effective

Table 1.1 Overview of different safety levels and load factors prescribed by the Dutch guidelines for the assessment of existing bridges (RBK) (Rijkswaterstaat, 2013)

Safety level	β	Reference period	$\gamma_{perm,6.10a}$	$\gamma_{perm,6.10b}$	γ_{ll}	γ_{wind}	γ_{var}
RBK Design[1]	4.3	100 years	1.40	1.25	1.50	1.65	1.65
RBK Reconstruction[2]	3.6	30 years	1.30	1.15	1.30	1.60	1.50
RBK Usage[3]	3.3	30 years	1.25	1.15	1.25	1.50	1.30
RBK Disapproval[4]	3.1	15 years	1.25	1.10	1.25	1.50	1.30

[1]These values correspond to Consequences Class 3 from NEN-EN 1990:2002.
[2]These values correspond to the reconstruction level for Consequences Class 3 from NEN 8700:2011, taking the values for structures built before 2012.
[3]These values correspond to the disapproval level for Consequences Class 3 from NEN 8700:2011, not taking the values for structures built before 2012.
[4]These values correspond to the disapproval level for Consequences Class 3 from NEN 8700:2011, taking the values for structures built before 2012.

Table 1.2 Load factors for the different safety levels recommended for the use with proof load tests

Safety level	γ_{sw}	γ_{sd}	γ_{ll}	γ_{proof}
RBK Design	1.10	1.25	1.50	1.00
RBK Reconstruction	1.10	1.15	1.30	1.00
RBK Usage	1.10	1.15	1.25	1.00
RBK Disapproval	1.10	1.10	1.25	1.00

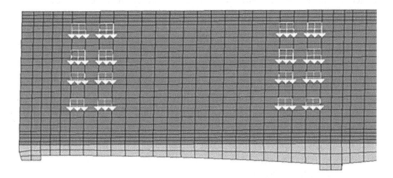

Figure 1.3 Wheel print locations for shifting the design tandem along the longitudinal axis of a viaduct that is to be proof load tested (Koekkoek et al., 2016)

depth to the longitudinal reinforcement) or sometimes 3 m (9.8 ft, notional lane width) is used – the transverse distribution width for bending moment is a topic of discussion in the Netherlands. For other bridge types, no transverse distribution is used. The position that results in the largest sectional moment or force is called the critical position. For shear in concrete slab bridges, the critical position is taken at a face-to-face distance between the load and the support of $2.5d_l$ (Lantsoght et al., 2013b), with d_l the effective depth to the longitudinal reinforcement. A transverse distribution of $4d_l$ is used for shear (Lantsoght et al., 2013a) for concrete bridge decks. For other bridge types, no transverse distribution or a distribution over $2d_l$, depending on the bridge type, is used and the critical position for shear should be determined based on the finite element model, keeping in mind that direct load transfer occurs for loads applied close to the support. For skewed bridges, the largest stress concentrations are found in the obtuse corner (Cope, 1985), so that the proof load tandem should be applied in the obtuse corner.

Once the critical position is determined, all live loads are removed from the finite element model. Then a single proof load tandem is applied in the first lane. This method is suitable for bridges with a small width (maximum three lanes in total). For wider bridges, more than one lane should be loaded in the field. Once the proof load tandem (or tandems for wider bridges) is applied in the model at the critical position, the load on the tandem is increased until the same sectional moment or shear is found as with the live load combination. On the proof load tandem, no load factors are used ($\gamma_{proof} = 1.00$). The load for which the same sectional moment or shear is found as for the live load combination is the target proof load. With this load, it can be proven that the considered span can carry the prescribed loads of the code, if the structure does not show any signs of distress during the load test. In the Netherlands, this procedure is followed for the different safety levels shown in Table 1.1, with the highest target proof load usually corresponding to the RBK Usage level.

Since the applied loads during a proof load test can be rather large, extensive preparation is required to guarantee the structural safety and the safety of the personnel and traveling public. For structural safety, the preparations include the following calculations:

• If not available from previous reports, an analytical assessment of the structure needs to be prepared. In the Netherlands, this assessment is carried out based on a Unity Check

UC, the ratio of the occurring load effect to the capacity of the cross section, taking into account the prescribed load and resistance factors. In North America, the assessment is expressed in terms of the rating factor *RF*.

- Determination of the expected bending moment capacity, based on the average and measured values of the material properties.
- Determination of the expected moment-curvature and load-displacement diagram for comparison to the measured results during the experiment.
- Determination of the expected shear capacity, based on the average and measured values of the material properties.
- Determination of the expected punching shear capacity, based on the average and measured values of the material properties.
- Verification of possible problems for the substructure.
- If during the technical inspection prior to the load test, other critical situations were observed, these should be verified analytically and monitored during the test.

Depending on the structure type, it may be valuable to carry out a nonlinear finite element simulation of the structure and proof load test for the preparations. Depending on the type of structure as well, specific methods can be used for predicting the maximum load. For reinforced concrete slab bridges, the maximum expected load that causes failure can be predicted with the Extended Strip Model (Lantsoght et al., 2017b).

If the margin between the expected capacity and the target proof load is small, the risk involved with the proof load test increases. The bridge owner then should decide if more preparations are required for the proof load test, and if the risk of testing is acceptable or not. If the risk is considered acceptable, the instrumentation during the proof load test as well as the prepared loading protocol will be crucial to decide during the test if loading to the next load step can be attempted. The sensor plan should then be developed to meet this requirement.

The current approach is fully based on the requirements of the Eurocodes. Since no rating vehicles for assessment are prescribed in the Eurocodes, another possible approach, which could significantly reduce the required target proof load and involved risk, consists of the use of weigh-in-motion data (Casas and Gómez, 2013). With this approach, the required load factors to determine the target proof load can be calibrated to the actual traffic for the site considered.

1.2.2   *AASHTO* Manual for Bridge Evaluation *method*

The method to determine the target proof load that is presented in the AASHTO *Manual for Bridge Evaluation* (MBE) (AASHTO, 2016) is based on the procedures described in the *Manual for Bridge Rating through Load Testing* (NCHRP, 1998). The target proof load should be representative of the vehicles used for rating the viaduct, taking into account the dynamic load allowance, and a load factor for the required margins of safety. The approach used in the United States is thus much more closely related to a vehicle that is used for rating, whereas the approach used in the Netherlands is linked to an entire load model. The reason why in Europe entire load models need to be proven to be equivalent is that no Eurocodes for assessment are currently available. Some countries have national codes for assessment that can prescribe a rating vehicle, which then could be linked to the target proof load.

Table 1.3 Adjustments to X_p, as shown in Table 8.8.3.3.1–1 from MBE (AASHTO, 2016)

Consideration	Adjustment
One lane load controls	+15%
Nonredundant structure	+10%
Fracture-critical details present	+10%
Bridge in poor condition	+10%
In-depth inspection performed	−5%
Ratable, existing $RF \geq 1.0$	−5%
ADTT ≤ 1000	−10%
ADTT ≤ 100	−15%

The factor that needs to be used to account for the required margin of safety is X_{pA}, and the total load that needs to be applied is the weight of the rating vehicle multiplied with the dynamic load allowance and X_{pA}. The factor X_{pA} is determined by multiplying the factor $X_p = 1.4$ with adjustment factors that take into account properties of the bridge under study. The adjustments to X_p are as given in Table 1.3, so that X_{pA} can be determined as:

$$X_{pA} = X_p \left(1 + \frac{\Sigma\%}{100} \right) \tag{1.1}$$

with $X_p = 1.4$. The value of $\Sigma\%$ is the sum of the adjustment percentages from Table 1.3 for the considered bridge. In Table 1.3, RF is used for the rating factor and ADTT is used for the average daily truck traffic.

According to the MBE, the strength based on the test is R_n:

$$R_n = X_p(L + I) + D \tag{1.2}$$

with L the live load, I the dynamic allowance, and D the dead load. The strength based on a calculation equals:

$$R_n = \gamma_{ll}(L + I) + \gamma_d D \tag{1.3}$$

with γ_{ll} the load factor for the live load and γ_d the load factor for the dead load. The dead load in the test is deterministic and does not need a load factor since the load cannot change anymore over time.

The target proof load L_T can then be determined as a function of the load of the rating vehicle L_R:

$$L_T = X_{pA}L_R(1 + IM) \tag{1.4}$$

with IM the dynamic load factor. In practice, dump trucks (Saraf et al., 1996) can be used to achieve the target proof load as well as military vehicles (Varela-Ortiz et al., 2013).

The value of $X_p = 1.4$ was based on a first-order reliability calculation, assuming all loads are normally distributed. The values assumed for the bias (mean to design value) and coefficient of variation are given in Table 1.4. The bias for the live load is based on measurements and then extrapolated to 75 years to find the mean maximum load as 1.79 HS20 vehicles. The coefficient of variation on the live load is either 18% when both the uncertainties on the heavy truck occurrences and the uncertainties of the effect of these trucks on the members of the structure are considered, and 14% when only the uncertainties of the

Table 1.4 Factors used to calibrate the first-order reliability model that was used to determine $X_p = 1.4$

	COV – prior to test	COV – after test	BIAS
Resistance	10%	0%	1.12
			1.0 distress in test
Dead load	10%	0%	1.0
Live load	18% truck + members	18%	1.79 one lane
	14% truck only	14%	1.52 two lanes
Impact	80%	80%*	1.00

*Unless a moving load test is performed to investigate the impact.

truck occurrences are considered. The target values of the reliability index β were defined as 3.5 for the inventory design levels and 2.3 for the operating rating levels. The magnitude of future live loads and possible future deterioration are the main remaining uncertainties after the proof load test. The value of X_p was derived based on an example with an 18 m (59 ft) span with $D = L$. The factor was found to be conservative for another example with a shorter span and one with a longer span with $D/L = 3.0$.

Other approaches that have been used to determine the target proof load in the past include using twice the maximum allowable load of the rating vehicle (Saraf et al., 1996). If proof load testing has as its goal to approve the passing of legal loads, a factor of 1.8 can be applied to the maximum legal load.

1.3   Procedures for proof load testing

1.3.1   *Loading methods*

Several methods are possible for applying the required target proof load during a proof load test. The following methods have been used in the past:

- Application of dead weights, where it is important to verify if no arching in the loads occurs when the structure deflects
- Using a loading vehicle
- Using a system with hydraulic jacks and counterweights.

For proof load testing of buildings, dead weights can be suitable to represent the distributed loads on the floor. The disadvantage of dead weights is that the positioning of these weights is time-consuming and thus makes using a cyclic loading protocol difficult. Similarly, if dead weights are used and signs that failure is imminent are seen, it is difficult to react quickly and offload all applied loads to prevent damage or collapse. For bridges, the use of dead weights has the additional disadvantage that this type of distributed load cannot correctly represent the concentrated wheel loads of a design truck or tandem, or rating vehicle.

When using a loading vehicle for applying a proof load, the vehicle should be able to apply the large load required for a proof load test. An option here is to use an army tank (Varela-Ortiz et al., 2013) or similar heavy weight carried by dump trucks (Saraf et al., 1996).

Figure 1.4 Load application with a loading vehicle (Lantsoght et al., 2017c)
Source: Photograph by D. A. Hordijk. Reprinted with permission.

In Germany, a special proof load testing vehicle was built, the BELFA (see Fig. 1.4) (Bretschneider et al., 2012). The advantage of using a loading vehicle is that several positions can be tested consecutively without needing much time for rearranging the test setup. Additionally, dynamic testing can be carried out.

An example of a test setup for a proof load test with a loading frame and hydraulic jacks is shown in Figure 1.5. The advantage of this system is that the load application can be carried out in a controlled manner, using the jacks for applying the load in a displacement-controlled manner and following the prescribed loading protocol. The advantage of using a displacement-controlled system is that when large deformations occur, the applied load will be reduced and irreversible damage or collapse can be avoided. The disadvantage of this system is that building up the loading frame can be time-consuming and the number of positions that can be tested is limited as every new position requires changing the setup.

For a loading system to be suitable for proof load testing, it has to fulfill the following requirements:

• The system should be suitable for applying the proof load in a safe manner and it must be able to be removed quickly when signs are observed that irreversible damage or failure is imminent.

• The system should be suitable for combination with a cyclic loading protocol: the load has to be applied in increments, loading and unloading must be done within a reasonable amount of time, keeping the load constant needs to be possible, and repeating the same load levels has to be possible.

Figure 1.5 Load application with a system of jacks and counterweights

- For bridge proof load testing, the system should be able to apply the wheel prints of the considered design truck, tandem, or rating vehicle.

The loading protocol that is required during a proof load test depends on the code that is used, and is discussed per code in Chapter 3 of Vol. 12. Most codes prescribe a loading protocol in which different load levels are applied. The trend in the current codes and guidelines is to move towards cyclic loading protocols. Using different load levels in cycles allows for checking the measurements after each load cycle to see if signs of distress are observed. The highest load level is typically the target proof load, and one of the intermediate load levels corresponds to the serviceability limit state.

When vehicles are used for the loading protocol, the different load levels can correspond to different levels of loading of the vehicles or a different total number of vehicles. Typically, different load paths are tested to study the critical structural response in different members. When hydraulic jacks are used, the load levels correspond to the load applied on the wheel prints, which should be equal for each wheel print or should correspond to the distribution of load over different axles if the loading represents a certain truck. When hydraulic jacks are used, complete unloading after the cycles is not recommended. Instead, loading to a low load level is recommended to make sure that all sensors and the jacks remain activated during the entire load test. An example of a cyclic loading protocol can be seen in Figure 1.6.

Based on field tests on reinforced concrete slab bridges (Lantsoght et al., 2017c) and laboratory testing (Lantsoght et al., 2017d), a recommended loading protocol for proof load testing of reinforced concrete (slab) bridges was developed. The loading speed during the test should be a constant value, which can be chosen between 3 kN/s and 10 kN/s

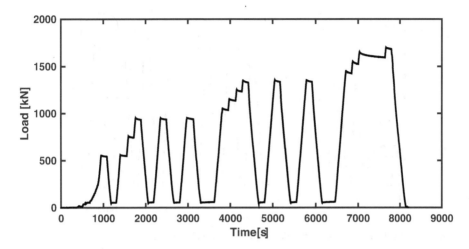

Figure 1.6 Example of loading protocol, used for bending moment position of proof load test of viaduct De Beek (Koekkoek et al., 2016)

Source: Modified from Lantsoght et al. (2017a). Reprinted with permission from ASCE.

(0.7 kip/s and 2.2 kip/s). A cyclic loading protocol is prescribed, with at least four load levels:

1   Level 1: a low load level (20%–30% of Level 4) to check if all sensors and the data acquisition and visualization system function properly.
2   Level 2: the load level corresponding to the serviceability limit state.
3   Level 3: an intermediate level between Levels 2 and 4.
4   Level 4: the highest load level that has to be demonstrated by the proof load test. In the Netherlands, this load level for existing highway bridges is typically RBK Usage, plus 5% to cover the uncertainties of proof load testing. Further research is needed to determine which percentage of additional load is necessary to cover the uncertainties. The currently used value of 5% is a first estimation.

Per load level, at least three load cycles should be carried out. Each level is kept for 2 minutes and the time of application of the baseline load is also 2 minutes. For load levels 3 and 4, an extra cycle is carried out prior to the three load cycles. In this cycle, the load is increased in small steps, and at each intermediate load level during this cycle the load is kept constant during 3 minutes to verify all measurements. Between the load cycles, a low load level should be maintained to make sure all sensors and loading equipment remains activated. A sketch of the resulting loading protocol is shown in Figure 1.7.

1.3.2   Monitoring bridge behavior during the test

Since proof load testing requires large loads, monitoring the bridge behavior during the test is important to avoid irreversible damage to the structure or collapse. This monitoring should be carried out in real time during the proof load test (see Fig. 1.8). As mentioned before, it is good practice to load to a low load level at the beginning of the proof load test to verify if all sensors are functioning properly. The effect of temperature and humidity

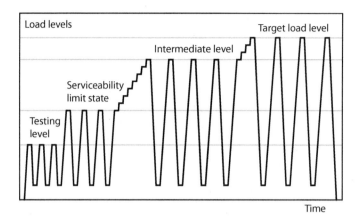

Figure 1.7 Cyclic loading protocol developed in the Netherlands.
Source: From Lantsoght et al. (2017d). Reprinted with permission from ACI.

Figure 1.8 Monitoring measurement output during a proof load test
Source: Photograph by S. Ensink. Printed with permission.

on the sensors should be known prior to the load test. A compensation sensor, which is only subjected to the effect of temperature and humidity and which is applied outside of the loaded area, should be used to find the net effect of the applied loading for the sensors that are subjected to both the structural response and the environmental influences.

According to the MBE (AASHTO, 2016), different load levels should be used during a proof load test to verify the response of the bridge under each load increment. This verification should include checking if no nonlinear behavior occurs and to limit distress due to cracking or other physical damage. A visual verification is required as well, and when visible signs of distress are observed, such as buckling patterns in compressive zones in steel elements or cracking in the concrete, the proof load test should be stopped. For this purpose, the following measurements are recommended:

• Strains (stresses) in bridge components
• Relative or absolute displacements of bridge components
• Relative or absolute rotation of bridge components
• Dynamic characteristics of the bridge.

In the Netherlands, for the proof load tests that were carried out as part of a research project, the structural response during a proof load test is followed in real time to verify if no irreversible damage occurs. It is recommended to monitor and evaluate the following structural response during a bridge load test:

• The load-displacement diagram.
• The strains in the members, to verify if strain limits are not exceeded. For girders, strains can be monitored at different locations over the height of the members, and the strain profile can be followed during the test.
• In concrete bridges, the width of existing cracks can be monitored to see if cracks are activated. Crack widths smaller than 0.05 mm (0.002 in.) can be considered to be equal to 0 mm.
• The output of all displacement sensors should be followed to see if the results are in line with the analytical predictions. If there are significant differences between the predictions and the measurements, the results should be analyzed further and a reason for the difference should be sought. Deflection profiles should be plotted to see if no nonlinearity occurs.

For shear-critical concrete bridges, fracture- and fatigue-critical steel bridges, and other bridges that can fail in a brittle manner, the instrumentation becomes important to avoid a brittle failure during the test. For these cases, special considerations are required.

To develop a sensor plan, the following guidelines can be used. It is good practice to measure the following responses:

• Deflection or deformation for at least five positions in the longitudinal direction.
• Deflection or deformation for at least three positions in the transverse direction.
• Deformations at the supports on both sides of the tested span if elastomeric bearings are used.
• Strains at the bottom for at least one position for slab bridges. For girder bridges, one girder should be instrumented over the height to follow the change in position of the neutral axis.

- Reference strain measurement to assess the influence of temperature and humidity.
- For concrete bridges, the crack width for at least one existing bending crack should be monitored if the bridge is already cracked in bending.
- The applied load, for each wheel print separately if jacks are used, or the position of the load if loading vehicles are used.
- Other parameters can be measured, for example, critical elements identified during the technical inspection.

During the proof load test, the engineers analyzing the measurements have to be in direct communication with the load operator. The responsible measurement engineer will tell the load operator when to apply or remove the load. After each load cycle, all structural responses are checked, and the measurement engineer will give permission to continue the test if no signs of distress are observed.

1.3.3   Stop criteria

The stop criteria are criteria based on the measurements obtained during the proof load test that are used to evaluate if irreversible damage is imminent. The stop criteria are an integral part of the safety philosophy for proof load testing (see Fig. 1.1), but not all codes and guidelines define the stop criteria with the same level of detail. Some codes that encompass numerous structure types give only general guidance. Other codes with a more limited scope give a quantitative description of the stop criteria. A proof load test can have two outcomes:

1   The target proof load is achieved without exceedance of a stop criterion, $F_{target} \leq F_{lim}$, and it has been shown experimentally that the structure fulfills the requirements of the code.
2   A stop criterion is exceeded before the target proof load is achieved, $F_{lim} > F_{target}$, and the test is terminated at a lower load level. The structure can then be shown to fulfill a lower load level of the code, or lower demands, if a load corresponding to this lower load level or these lower demands was achieved during the proof load test.

The stop criteria that are defined in the MBE (AASHTO, 2016) and the *Manual for Bridge Rating through Load Testing* (NCHRP, 1998) are the onset of nonlinear behavior or the occurrence of damage such as cracking in concrete or compressive distress in steel. Some codes use acceptance criteria, which are verified after a proof load test to see if the behavior of the structure is acceptable. Acceptance criteria cannot be used to warn for the onset of nonlinear behavior.

For concrete bridges, detailed stop criteria are the subject of current research (Lantsoght et al., 2018c). A proposal for stop criteria, based on pilot proof load tests, a collapse test in the field, additional testing in the laboratory, and analytical research, encompasses bridges that are flexure- and shear-critical, both for the case of a structure that has been previously cracked in bending and for the case of a structure that has no significant flexural cracking. Further experimental work is required for the verification of the proposed stop criteria for shear. Two types of stop criteria should be distinguished:

1   Stop criteria that indicate that further loading can cause irreversible damage.
2   Stop criteria that indicate that further loading can cause collapse.

For the first type, in exceptional cases the bridge owner can give permission to continue loading and will take responsibility for repairing the damage after the proof load test. For the second type, further loading can never be permitted. Further research is necessary to develop these two categories of stop criteria.

Analyzing the results of the experiments in the field and in the laboratory in the light of the existing stop and acceptance criteria resulted in a first proposal for stop criteria (Lantsoght et al., 2017d). This proposal was then refined based on theoretical considerations. The stop criteria for flexure are adequately supported by field and laboratory testing. For shear, stop criteria are proposed but further experimental validation is necessary. For flexure, flexural beam theory was used to derive a stop criterion for the limiting concrete strain and crack width (Lantsoght et al., 2018b). For shear, a limiting crack width was derived based on aggregate interlock theory (Lantsoght, 2017; Yang et al., 2017). The analysis of experiments on beams with plain bars cast in the laboratory showed that the stop criteria should be different for the failure modes of shear and flexure, and for members previously cracked in bending or not. The resulting four possible situations and their governing stop criteria are given in Table 1.5. The stop criterion for the limiting concrete strain ε_{stop} is for flexure:

$$\varepsilon_c \leq \varepsilon_{stop} \tag{1.5}$$

with ε_c the measured strain, and:

$$\varepsilon_{stop} = \varepsilon_{c,bot,max} - \varepsilon_{c0} \tag{1.6}$$

$$\varepsilon_{c,bot,max} = \frac{h-c}{d-c} 0.65 \frac{f_y}{E_s} \tag{1.7}$$

with ε_{c0} the strain caused by permanent loads, h the height of the member, d the effective depth of the member, c the height of the compression zone when the strain in the steel is 65% of the yield strain, f_y the yield stress of the steel, and E_s the Young's modulus of the steel. For shear, the strain is limited as

$$\varepsilon_{DAfStB} < \varepsilon_{c,lim} - \varepsilon_{c0} \tag{1.8}$$

Table 1.5 Currently proposed stop criteria (Lantsoght et al., 2018c)

Failure mode	Cracked in bending or not?	
	Not cracked in bending	*Cracked in bending*
Bending moment	Concrete strain ε_{stop}	Concrete strain ε_{stop}
	$w_{max} \leq w_{stop}$	$w_{max} \leq w_{stop}$
	$w < 0.05$ mm $\rightarrow w \approx 0$ mm	$w < 0.05$ mm $\rightarrow w \approx 0$ mm
	$w_{res} \leq 0.3\, w_{max}$	$w_{res} \leq 0.2\, w_{max}$
	25% reduction of stiffness	25% reduction of stiffness
	Deformation profiles	Deformation profiles
	Load-deflection diagram	Load-deflection diagram
Shear	Concrete strain ε_{DAfstB}	Concrete strain ε_{DAfstB}
	$w_{max} \leq 0.4\, w_{ai}$	$w_{max} \leq 0.75\, w_{ai}$
	$w < 0.05$ mm $\rightarrow w \approx 0$ mm	$w < 0.05$ mm $\rightarrow w \approx 0$ mm
	25% reduction of stiffness	25% reduction of stiffness
	Deformation profiles	Deformation profiles
	Load-deflection diagram	Load-deflection diagram

with ε_c the measured strain; ε_{c0} the strain caused by the permanent loads, and $\varepsilon_{c,lim} = 800\ \mu\varepsilon$ for concrete with a compressive strength larger than 25 MPa (3.6 ksi). A proposal for the limiting strain based on the critical shear displacement theory (Yang et al., 2016) is also available but needs further simplification to come to a closed-form solution (Benitez et al., 2018).

All crack widths smaller than 0.05 mm (0.002 in.) are neglected. For flexure, the maximum crack width w_{max} is limited to w_{stop}:

$$w_{stop} = 2\frac{0.65f_{ym} - f_{perm}}{E_s}\beta_{fr}\sqrt{d_c^2 + \left(\frac{s}{2}\right)^2} \tag{1.9}$$

with the stress caused by the permanent loads f_{perm}:

$$f_{perm} = \frac{d - c}{h - c}\varepsilon_{c0}E_s \tag{1.10}$$

and β_{fr} the strain gradient term, given as:

$$\beta_{fr} = \frac{h - c}{d - c} \tag{1.11}$$

The value of β_{fr} can be approximated as:

$$\beta_{fr} = 1 + 3.15 \times 10^{-3}d_c \text{ with } d_c \text{ in mm} \tag{1.12}$$

and with E_s the Young's modulus of the reinforcement steel, d_c the concrete cover to the centroid of the tension steel, and s the reinforcement spacing. The height of the compression zone c should be determined for a stress of $0.65f_{ym}$ in the reinforcement steel, with f_{ym} the mean yield strength of the steel. For shear, the maximum crack width is a fraction of the crack width w_{ai} at which the aggregate interlock capacity becomes smaller than the inclined cracking load:

$$w_{ai} = w_d + 0.01mm \tag{1.13}$$

$$w_d = \frac{0.03f_c^{0.56}\frac{s_{cr}}{d}(978\Delta_{cr}^2 + 85\Delta_{cr} - 0.27)R_{ai}}{v_{RBK}} \tag{1.14}$$

with f_c the concrete compressive strength and s_{cr} the crack spacing:

$$s_{cr} = \left(1 + \rho_s\alpha_e - \sqrt{2\rho_s\alpha_e + (\rho_s\alpha_e)^2}\right)d \tag{1.15}$$

with ρ_s the reinforcement ratio and $\alpha_e = E_s/E_c$, with E_c the Young's modulus of the concrete. For high-strength concrete, a correction factor R_{ai} is used:

$$R_{ai} = 0.85\sqrt{\left(\frac{7.2}{f_c - 40MPa} + 1\right)^2 - 1} + 0.34 \text{ for } f_c > 65MPa \tag{1.16}$$

The critical shear displacement (Yang et al., 2017) is determined as:

$$\Delta_{cr} = \frac{25d}{30610\phi_s} + 0.0022mm \le 0.025mm \tag{1.17}$$

The shear stress at inclined cracking is:

$$v_{RBK} = max(1.13k_{slab}k\sqrt{\frac{f_c}{f_{ym}}}; 0.15k_{slab}k(100\rho_s f_c)^{1/3}) \qquad (1.18)$$

with $k_{slab} = 1.2$ for slabs and 1.0 for other elements, and k the size effect factor:

$$k = 1 + \sqrt{\frac{200mm}{d}} \leq 2 \qquad (1.19)$$

For flexure, the limits to the residual crack width w_{res} after a cycle from the German guideline can be used (see Table 1.5).

The stiffness is determined as the tangent to the load-displacement diagram in the loading branch. For all cases, the reduction of the stiffness calculated on the load-deflection diagram is limited to 25%. The deformation profiles are the plots of the measured deformations (typically deflections) in the longitudinal and transverse directions. For all cases, the deformation profiles need to be qualitatively followed during the proof load test to signal changes to the structural behavior. The load-displacement graph should be plotted and observed during the load test to check for signs of nonlinear behavior.

Moreover, if the inspection identified other critical structural elements, their response should be followed during the load test and stop criteria should be agreed upon prior to the proof load test. Additional measurements, such as acoustic emission signals, may be required to capture changes prior to brittle failure modes. In certain cases, for shear-critical bridges, the development of cracks can be followed, and the stop criterion could be the development of a bending crack in the shear span, as such a crack could become the critical crack for a shear-flexural failure.

The stop criteria from the German guideline (Deutscher Ausschuss für Stahlbeton, 2000); the acceptance criteria from ACI 437.1–07 (ACI Committee 437, 2007) and ACI 437.2M-13 (ACI Committee 437, 2013); and the acceptance criteria from the Czech and Slovak (Frýba and Pirner, 2001), Polish (Research Institute of Roads and Bridges, 2008), and French (Cochet et al., 2004) guidelines for load testing are all discussed in Chapter 3 of Vol. 12.

For steel bridges, a stop criterion can be to limit the total steel strain to 90% of the yield strain and to pay attention to compressive distress in elements that are critical for buckling. Fracture- and fatigue-critical bridges need special considerations. For timber, masonry, and plastic composite bridges, the test engineer needs to determine the stop criteria prior to the load test. A limiting strain based on the known stress-strain diagram of the material can be suggested.

1.4 Processing of proof load testing results

1.4.1 On-site data validation of sensor output

The first step in reviewing the load test data is to verify all measurements after each load cycle. This step is required to check if none of the stop criteria are exceeded, but also to see if the output of the sensors is reasonable (i.e. repeatable across load cycles, symmetric where applicable, and linear). For this purpose, a cyclic loading protocol is required. As mentioned before, such a loading protocol can be achieved by using hydraulic jacks or by repeating the same driving path of one or more loading vehicles. During the first load level, the response of all sensors must be checked in detail to make sure that all sensors are working properly. If sensor malfunctioning is detected, the proof load test should not be attempted and the sensor should be replaced first. Then, this low load level should be applied again to confirm that all sensors are working properly.

When a cyclic loading protocol is used, the reproducibility of the sensor output for identical load cases needs to be confirmed. In addition, the linearity and symmetry of measurements should be verified after each load cycle. If the measurements do not align with the expected response, it is necessary to analyze in more detail what is happening within the structure or with the sensors, the data acquisition system, or the data visualization system.

1.4.2   *Final verification of stop criteria*

The stop criteria are verified in real time during a proof load test. The main conclusion of a proof load test is made at the end of the on-site test:

1   The bridge has been shown to fulfill the requirements of the code if the proof load test was successful (the target proof load is applied without observing signs of distress).
2   The bridge should be used for a lower load level if a stop criterion was exceeded prior to achieving the target load level but if a load that corresponds to a lower load level was achieved.

However, the last step of a load test, as discussed in Chapter 7 of Vol. 12, is to post-process all data, report the results of the proof load test to the owner, and formulate recommendations.

For a proof load test, the results of the measurements that are reported are the output of all sensors, in a graphical manner, as well as the verification of the measurements against the stop criteria. In this stage, the measurements should be corrected for the support displacements for bridges on elastomeric bearings and for the effect of temperature and humidity. From the corrected data, the following output should be developed for the report:

* The load-displacement diagram of the test as executed, with all load cycles, and the envelope of the load-displacement diagram.
* The measured loading protocol. The protocol is shown as the load versus time diagram (see Fig. 1.6) if hydraulic jacks were used, or loading paths and load versus time at a fixed position for testing with loading vehicles. If hydraulic jacks were used, the measured forces in the wheel prints separately should be plotted as well to show that the load was distributed as intended (equally or to fulfill a certain ratio of values between axles).
* Deformation profiles: for each load level, the deformation profiles in the longitudinal and transverse directions should be shown based on the net displacement of the superstructure. An example is shown in Figure 1.9.
* Crack width for reinforced concrete bridges: the opening of the measured crack(s) should be presented as a function of time and as a function of applied load.
* Strains: the results of the measured strains as a function of time and as a function of the applied load should be plotted. For girders, the strain distribution over the girder height and position of neutral axis should be shown.
* Other measurements: if a structural element was found to require additional attention during the technical inspection prior to the proof load test, this element should be

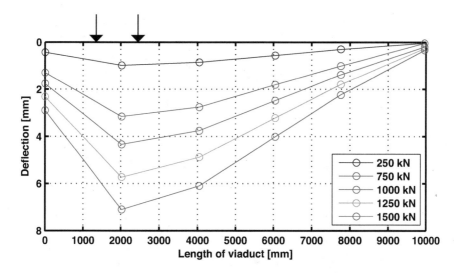

Figure 1.9 Plot of deflections in the longitudinal direction, taken from the shear-critical loading position on viaduct De Beek (Koekkoek et al., 2016). From this plot, it can be seen that the behavior over the different load levels is linear

monitored during the proof load test and the measured results should be reported and discussed.

Based on the corrected data, the final verification of the stop criteria should be reported. It is good practice to compare the strains and deflections that were measured in the experiment to the predictions made with the linear finite element model used to determine the target proof load and to prepare the proof load test.

Another element that needs to be reported is the total applied load. When dead weights are used, the number of dead weights and their individual weight should be reported. When loaded vehicles are reported, the results of weighing each vehicle without and with its cargo should be reported. When intermediate load levels are achieved by applying increments of cargo, the vehicle with its cargo should be weighed prior to each load application. When jacks are used, the weight of the additional elements that are used for the test setup should be added to the load that is applied with hydraulic pressure. For this case, for example, the weight of the jacks themselves should also be considered (see Fig. 1.10). If load cells are used to measure the applied load, the results of the calibration should be given. If the calibration has an effect on the applied load, this correction should be part of the post-processing stage. Photographs of the execution of the load test should be added to the report for future reference.

1.5   Bridge assessment based on proof load tests

If a proof load test has been carried out and the test was successful (the structure could carry the target proof load without signs of distress), then it has been demonstrated experimentally that the structure can carry the studied load combination. A further assessment is

Figure 1.10 Weight of a single jack as used during a proof load test

not necessary. This idea is followed as well by the rating procedure after a proof load test from the MBE (AASHTO, 2016). The rating factor will be at least $RF = 1$ after a proof load test when the load calculated to be equivalent to the live loads from the code is applied during the proof load test. The rating factor will be larger than 1 if loading to higher levels was carried out. The target proof load from the MBE is determined such that the rating factor becomes $RF = 1$. After the proof load test, the rating factor RF_0 is determined as:

$$RF_0 = \frac{OP}{L_R(1 + IM)} \tag{1.20}$$

with

$$OP = \frac{k_0 L_p}{X_{pA}} \tag{1.21}$$

with L_R the load rating vehicle, IM the dynamic impact factor, L_p the applied maximum load in the proof load test, and X_{pA} as defined in Equation 1.1. When a proof load test is successful, the applied maximum load L_p is the target proof load L_T. The factor k_0 takes into account if the proof load test was successful or not: $k_0 = 1$ if the target proof load was reached and $k_0 = 0.88$ if signs of distress were observed prior to reaching the target proof load. The 12% reduction is consistent with observations that show that nominal material properties used in calculations are typically 12% below observed material properties from tests. It can be seen from Equations 1.19, 1.20, and 1.4 that if $L_p = L_T$, and thus $k_0 = 1$, the value of RF_0 becomes $RF_0 = 1$.

When a Unity Check is used instead of a Rating Factor, similar procedures are followed. Whereas the Unity Check was larger than 1 prior to the proof load test or could not be determined due to the large uncertainties, it will be found to be equal to or smaller than 1 based on the results of the proof load test. A Unity Check $UC = 1$ is found when exactly the target proof load is applied during the test, and $UC < 1$ is found when a larger load is applied. When different safety levels and their corresponding load factors are considered as in the Netherlands, the assessment has to be carried out at the highest safety level required by the governing codes.

For other load combinations and/or spans that were not tested, it may be necessary to carry out an assessment by taking into account the benefit obtained from the proof load test (Lantsoght et al., 2018a). To take into account the information obtained from the proof load test, the finite element model that was used for the determination of the critical position and target proof load can be updated with the measurements from the proof load test. The principles for updating the finite element model are similar as for a diagnostic load test and have been discussed in Part III of Volume 12. The possible sources for differences (Barker, 2001) between the calculated and measured deformations and strains need to be identified, the differences between the tested structure and the finite element model should be made as small as possible, and then it should be identified which of these mechanisms are reliable for rating, resulting in a model that only includes the effect of these mechanisms. Once the finite element model is updated, it can also be used to rate the entire structure when only one span is proof load tested, or it can be used to apply other load combinations in the finite element model. Updating the finite element model also results in a better understanding of the structure's behavior and can facilitate future ratings. It is not allowed to take the ratio of the rating factor after and before the proof load test and multiply the rating factors of other spans and/or for other load combinations with this ratio. This method is too crude and does not take into account the uncertainties on the results of the proof load test or the differences between the structural elements or spans that must be assessed. For buildings, NEN 8700:2011 (Code Committee 351001, 2011a) gives a method to cover all floors of a building when only a limited number of floors are tested. This approach can only be extended to bridges when all spans are equal in geometry and material properties.

The decisions about the tested structure that are taken after a proof load test remain the full responsibility of the bridge owner. The report about the proof load test includes the measurements, the assessment of the structure, and possible advice. This advice can include the removal or application of posting to the considered bridge, or it can be a recommendation for further research and (material) testing of the considered bridge to better evaluate the material properties and state of degradation. The application of this advice is the decision and responsibility of the bridge owner.

1.6   Summary and conclusions

This chapter discussed the methodology for proof load testing. All general recommendations that apply to all types of load tests were discussed in detail in Part II of Volume 12. Here, topics that are of specific interest for proof load tests are highlighted.

The first important topic for proof load testing is the determination of the target proof load. Regardless of the considered code, the main idea for the target proof load is that this load has a magnitude that will result in a sufficient rating or assessment of the tested structure. As such, the procedures for determining the target proof load are inherently linked to the assessment practices. Where assessment codes and guidelines prescribe certain vehicles that need to be rated for, the determination of the target proof load is more straightforward, which can be seen in the provisions from the AASHTO *Manual for Bridge Evaluation* for the determination of the target proof load. When more general live load models are used, more choices need to be made to determine the target proof load, as can be seen when looking at the method used in Dutch practice.

The second topic that was discussed is related to the execution of proof load tests. For the execution, the loading method becomes important, since a relatively large load needs to be applied. Dead weights, special vehicles, and a loading frame with hydraulic jacks are options that each have their advantages and disadvantages. Nowadays, most proof load tests on bridges are carried out either by using special vehicles or by using a setup with a loading frame and hydraulic jacks. Another important element of the execution of proof load tests is the sensor plan. The applied sensors are used to monitor the structural response in real time during the bridge load test and to verify if no signs of irreversible damage occur. To verify the linearity, reproducibility, and symmetry of the measurements, a cyclic loading protocol is recommended. Moreover, a cyclic loading protocol allows for the verification of all measurements after each load step to analyze if further loading can be allowed. This analysis is carried out based on the so-called stop criteria. The stop criteria are based on the measured parameters. If a stop criterion is exceeded, further loading could result in irreversible damage to the structure or even collapse, and further loading is not permitted. After a proof load test, acceptance criteria may need to be checked, depending on the governing code.

The third topic that was discussed in this chapter is the processing of the results of the proof load test. The result of a proof load test is pass/fail: either the structure can carry the target proof load and the test is successful for the considered load combination, or a stop criterion is exceeded prior to achieving the target proof load. For the latter case, it must be analyzed if the structure was able to resist a load that corresponds to lower demands during the proof load test. In that case, the conclusion of the proof load test may be that the structure is suitable for the use at a reduced demand level. In some cases, more load than the target proof load can be applied, as long as none of the stop criteria are exceeded, to learn more about the lower bound of the structural capacity. This latter method is generally not allowed by codes and only carried out for special cases.

As can be seen from the discussion about the outcome of a proof load test, the main answer is given at the end of the proof load test itself. The result of the assessment of the structure for the considered load combination is found by the outcome of the proof load test. For other load combinations, or when other structural elements or spans that were not subjected to the proof load test must be assessed, the results of the proof load test can be taken into account. For this purpose, the finite element model that was used to determine the target proof load can be updated with the results of the experiment, as done for a diagnostic load test. The updated model, with only the effect of mechanisms that are reliable long-term and at the ultimate limit state, can then be used for an improved assessment of elements or spans that were not subjected to a proof load test or to facilitate future assessments for special vehicles and other load combinations.

For the post-processing and reporting of the test, it is important to give a visual overview of the observations made during the experiment and to write down the conclusions that were drawn at the end of the test. The post-processing also allows for the correction of the data for the effect of temperature and humidity and to find the net deflection of the superstructure by correcting for the measured deflections at the supports. In the report, the total applied load must be mentioned and the verification of the stop criteria must be included. Photographs of the execution of the load test should illustrate the report for future reference.

References

AASHTO (2016) *The Manual for Bridge Evaluation with 2016 Interim Revisions*. American Association of State Highway and Transportation Officials, Washington, DC.

ACI Committee 437 (2007) *Load Tests of Concrete Structures: Methods, Magnitude, Protocols, and Acceptance Criteria (ACI 437.1R-07)*. American Concrete Institute, Farmington Hills, MI.

ACI Committee 437 (2013) *Code Requirements for Load Testing of Existing Concrete Structures (ACI 437.2M-13) and Commentary*. American Concrete Institute, Farmington Hills, MI.

Barker, M. G. (2001) Quantifying field-test behavior for rating steel girder bridges. *Journal of Bridge Engineering*, 6, 254–261.

Benitez, K., Lantsoght, E. O. L. & Yang, Y. (2018) Development of a stop criterion for load tests based on the critical shear displacement theory. *Sixth International Symposium on Life-Cycle Civil Engineering, IALCCE 2018*, Ghent, Belgium.

Bretschneider, N., Fiedler, L., Kapphahn, G. & Slowik, V. (2012) Technical possibilities for load tests of concrete and masonry bridges. *Bautechnik*, 89, 102–110 (in German).

Casas, J. R. & Gómez, J. D. (2013) Load rating of highway bridges by proof-loading. *KSCE Journal of Civil Engineering*, 17, 556–567.

CEN (2002) *Eurocode: Basis of Structural Design, NEN-EN 1990:2002*. Comité Européen de Normalisation, Brussels, Belgium.

CEN (2003) *Eurocode 1: Actions on Structures, Part 2: Traffic Loads on Bridges, NEN-EN 1991-2:2003*. Comité Européen de Normalisation, Brussels, Belgium.

Cochet, D., Corfdir, P., Delfosse, G., Jaffre, Y., Kretz, T., Lacoste, G., Lefaucheur, D., Khac, V. L. & Prat, M. (2004) *Load Tests on Highway Bridges and Pedestrian Bridges (in French)*. Setra – Service d'Etudes techniques des routes et autoroutes, Bagneux-Cedex, France.

Code Committee 351001 (2011a) *Assessment of Structural Safety of an Existing Structure at Repair or Unfit for Use: Basic Requirements, NEN 8700:2011 (in Dutch)*. Civil center for the execution of research and standard, Dutch Normalisation Institute, Delft, The Netherlands.

Code Committee 351001 (2011b) *Assessment of Structural Safety of an Existing Structure at Repair or Unfit for Use: Loads, NEN 8701:2011 (in Dutch)*. Civil center for the execution of research and standard, Dutch Normalisation Institute, Delft, The Netherlands.

Cope, R. J. (1985) Flexural shear failure of reinforced-concrete slab bridges. *Proceedings of the Institution of Civil Engineers Part 2-Research and Theory*, 79, 559–583.

Deutscher Ausschuss für Stahlbeton (2000) *DAfStb-Guideline: Load Tests on Concrete Structures (in German)*. Deutscher Ausschuss fur Stahlbeton, Berlin, Germany.

Frýba, L. & Pirner, M. (2001) Load tests and modal analysis of bridges. *Engineering Structures*, 23, 102–109.

Koekkoek, R. T., Lantsoght, E. O. L., Bosman, A., Yang, Y., Van Der Veen, C. & Hordijk, D. A. (2015) *Proof Loading of the Viaduct in the Zijlweg: Risk Analysis and Planning*. Stevin Report nr. 25.5-15-07. Delft University of Technology, The Netherlands.

Koekkoek, R. T., Lantsoght, E. O. L., Yang, Y. & Hordijk, D. A. (2016) *Analysis Report for the Assessment of Viaduct De Beek by Proof Loading*. Stevin Report 25.5-16-01. Delft University of Technology, Delft, The Netherlands.

Lantsoght, E. O. L. (2017) *Development of a Stop Criterion for Shear Based on Aggregate Interlock*. Stevin Report nr. 25.5-17-09. Delft University of Technology, Delft, the Netherlands.

Lantsoght, E. O. L., De Boer, A., Van Der Veen, C. & Walraven, J. C. (2013a) Peak shear stress distribution in finite element models of concrete slabs. In: Zingoni, A. (ed.) *Research and Applications in Structural Engineering, Mechanics and Computation*. Taylor and Francis, Cape Town, South Africa.

Lantsoght, E. O. L., Van Der Veen, C., De Boer, A. & Walraven, J. C. (2013b) Recommendations for the shear assessment of reinforced concrete slab bridges from experiments. *Structural Engineering International*, 23, 418–426.

Lantsoght, E. O. L., Koekkoek, R. T., Van Der Veen, C., Hordijk, D. A. & De Boer, A. (2017a) Pilot proof-load test on viaduct De Beek: Case study. *Journal of Bridge Engineering*, 22, 05017014.

Lantsoght, E. O. L., Van Der Veen, C., De Boer, A. & Alexander, S. D. B. (2017b) Extended strip model for slabs under concentrated loads. *ACI Structural Journal*, 114, 565–574.

Lantsoght, E. O. L., Van Der Veen, C., De Boer, A. & Hordijk, D. A. (2017c) Proof load testing of reinforced concrete slab bridges in the Netherlands. *Structural Concrete*, 18, 597–606.

Lantsoght, E. O. L., Yang, Y., Van Der Veen, C., De Boer, A. & Hordijk, D. A. (2017d) Beam experiments on acceptance criteria for bridge load tests. *ACI Structural Journal*, 114, 1031–1041.

Lantsoght, E. O. L., De Boer, A., Van Der Veen, C. & Hordijk, D. A. (2018a) *Modelling of the Proof Load Test on Viaduct De Beek*. Euro-C, Austria.

Lantsoght, E. O. L., Van Der Veen, C. & Hordijk, D. A. (2018b) Monitoring crack width and strain during proof load testing. *Ninth International Conference on Bridge Maintenance, Safety, and Management*, IABMAS 2018, Melbourne, Australia.

Lantsoght, E. O. L., Van Der Veen, C. & Hordijk, D. A. (2018c) Proposed stop criteria for proof load testing of concrete bridges and verification. *Sixth International Symposium on Life-Cycle Civil Engineering, IALCCE 2018*, Ghent, Belgium.

NCHRP (1998) *Manual for Bridge Rating through Load Testing*. Transportation Research Board, Washington, DC.

Research Institute of Roads and Bridges (2008) *The Rules for Road Bridges Proof Loadings (in Polish)*. RIRB, Warsaw, Poland.

Rijkswaterstaat (2013) *Guidelines Assessment Bridges: Assessment of Structural Safety of an Existing Bridge at Reconstruction, Usage and Disapproval* (in Dutch), RTD 1006:2013 1.1.

Saraf, V. K., Nowak, A. S. & Till, R. (1996) Proof load testing of bridges. In: Frangopol, D. M. & Grigoriu, M. D. (eds.) *Probabilistic Mechanics & Structural Reliability: Proceedings of the Seventh Specialty Conference*. American Society of Civil Engineers, Worcester, MA.

Varela-Ortiz, W., Cintrón, C. Y. L., Velázquez, G. I. & Stanton, T. R. (2013) Load testing and GPR assessment for concrete bridges on military installations. *Construction and Building Materials*, 38, 1255–1269.

Yang, Y., Den Uijl, J. A. & Walraven, J. (2016) The critical shear displacement theory: On the way to extending the scope of shear design and assessment for members without shear reinforcement. *Structural Concrete*, 17, 790–798.

Yang, Y., Walraven, J. & Den Uijl, J. A. (2017) Shear behavior of reinforced concrete beams without transverse reinforcement based on critical shear displacement. *Journal of Structural Engineering*, 143, 04016146-1-13.

Chapter 2

Load Rating of Prestressed Concrete Bridges without Design Plans by Nondestructive Field Testing

David V. Jauregui, Brad D. Weldon, and Carlos V. Aguilar

Abstract

Load rating of concrete bridges with no design plans is currently an issue in the United States and other countries. Missing or incomplete design documentation creates uncertainties in establishing the safe load limits of bridges to carry legal vehicles. Guidance for evaluating planless concrete bridges, in particular prestressed structures, is limited in the AASHTO *Manual for Bridge Evaluation* and very few state departments of transportation have rating procedures for such bridges. In this chapter, a multi-step load rating procedure is described for planless prestressed bridges that includes (1) estimating the material properties from past specifications and amount of prestressing steel using Magnel diagrams; (2) verifying the steel estimate by rebar scanning; (3) field testing at diagnostic and/or proof load levels based on strain measurements; and (4) rating the bridges using the proof test results. Three prestressed concrete bridges (including a double T-beam, box beam, and I-girder bridge) are evaluated using nondestructive load testing and material evaluation techniques to illustrate the procedures. Rating factors are determined for AASHTO and New Mexico legal loads using the proof test results for the serviceability limit state (i.e. concrete cracking). Using a load rating software program, rating factors are also computed for the strength limit state (i.e. shear or flexural capacity) based on the measured bridge dimensions and estimated material properties. Finally, load ratings for serviceability and strength are compared to establish the bridge capacities.

2.1 Introduction

Load rating of bridges is an important and complex activity that requires knowledge and experience in bridge design, inspection, and evaluation to properly determine the load capacity of existing bridges and ensure the safety of the traveling public. Because of the economic impact associated with posting, closing, rehabilitating, or replacing a bridge, several state departments of transportation (DOTs) have adopted load testing procedures to evaluate the safety and serviceability of questionable bridges, including those without design plans and aged structures.

In the 1980s and 1990s, several state DOTs used load testing for bridge assessment purposes including New York (Kissane et al., 1980); Florida (Shahawy and Garcia, 1990; Shahawy, 1995); Michigan (Saraf et al., 1996); and Alabama (Conner et al., 1997). A study to develop load testing procedures was conducted under Project 12–28(13) of the National Cooperative for Highway Research Program (NCHRP) by Pinjarkar et al. (1990)

DOI: https://doi.org/10.1201/9780429265969

and Lichtenstein (1993). These procedures have been integrated into the AASHTO (American Association of State Highway and Transportation Officials) *Manual for Bridge Evaluation* (2011), specifically Section 8: Nondestructive Load Testing. The manual is referred to as the AASHTO Manual for the remainder of this chapter.

Numerous load tests were conducted by the aforementioned states prior to the year 2000. Furthermore, the majority of the proof tests were performed on reinforced concrete and steel bridges, with only a few on prestressed concrete bridges. More recently, an extensive project was conducted by the Georgia Institute of Technology for the Georgia DOT which involved several tasks, including an appraisal of the start-of-the-art of bridge condition assessment (Wang et al., 2009); bridge evaluation through load testing and advanced analysis (O'Malley et al., 2009; Wang et al., 2011a); and the development of guidelines for bridge condition assessment, evaluation, and load rating (Ellingwood et al., 2009; Wang et al., 2011b). The Iowa DOT (Wipf et al., 2003), Delaware DOT (Chajes and Shenton, 2006), and Vermont Agency of Transportation (Jeffrey et al., 2009) have also sponsored major research projects to develop bridge testing capabilities for load rating purposes. The testing approach most widely used in these four states has been diagnostic rather than proof testing.

To further evaluate the state load rating procedures, specifically those for bridges without plans, a questionnaire was developed by the authors and submitted to the 50 state DOTs. The survey included the following questions:

- How many planless bridges do you have in your inventory, and what are the main load rating challenges?
- Does your agency have an established procedure for the load rating of planless bridges?
- For concrete bridges, how are the steel reinforcement layout and the material strengths estimated? Is nondestructive testing employed?
- What computer software program does your agency use to perform load ratings?
- Does your state conduct any form of load testing to aid in rating bridges, and if so, what kind of tests are conducted and how many bridges are tested per year?

Thirty-three of the 50 state DOTs responded, and each survey was completed by the state's bridge rating engineer or responsible member of the bridge management team. As shown in Figure 2.1, the majority of the state DOTs that answered reported more than 100 concrete bridges without design plans in their inventory, and only eight states had specific load rating policies and/or procedures (FL, ID, MD, MI, MO, OR, TX, and UT). Furthermore, the procedures for these states generally apply to reinforced concrete bridges and not prestressed concrete bridges. The percentages of concrete bridges without design plans reported by the different states were as follows, from largest to smallest: 34% (NH), 28% (UT), 23% (MT), 21% (ND), 20% (KY), 19% (NM), 16% (NV), 9% (OR), 8% (ID), 7% (TX), 6% (WI), 4% (DE), 3% (WY), 3% (TN), 2% (IL), 1% (CO), 1% (WA), and 0.5% (MN). In addition, 85% of the respondents stated that load rating of bridges without design plans is a concern in their state and better procedures are needed. Most states indicated that the main challenge in evaluating these bridges is the unknown material properties and reinforcement details. Approximately 79% follow the AASHTO Manual to estimate the strength of the steel reinforcement, and about 33% conduct a physical inspection and use advanced testing such as rebar scanning to determine the steel layout. To estimate the

| NONE |
| 1-20 |
| 20-60 |
| 60-100 |
| >100 |
| UNKNOWN |
| NO RESPONSE |

Figure 2.1 Concrete bridges without plans in the United States

concrete compressive strength, about 76% use the AASHTO Manual values while 10% use a Schmidt hammer or Windsor probe. The load rating software program used by 58% of the states is the AASHTOWare BrR program; 24% use BRASS (Bridge Rating and Analysis of Structural Systems). About 50% conduct load tests on bridges without plans (diagnostic testing being the most common); 65% of these contract with a consultant for load testing, 12% conduct in-house testing, and 23% use a combination of consultant and in-house testing.

Apart from the United States, a major project was recently conducted in Europe titled Assessment and Rehabilitation of Central European Highway Structures (ARCHES), which focused on the use of soft, diagnostic, and proof load testing for bridge evaluation (Casas et al., 2009). Outside of the ARCHES project, just a few other studies were discovered related to diagnostic and proof load testing of concrete bridges without design plans (Bernhardt and DeKolb, 2003; Bernhardt et al., 2004; Hag-Elsafi and Kunin, 2006; Shenton et al., 2007; Commander et al., 2009). Again, the bridge type consisted mainly of reinforced concrete members (i.e. non-prestressed), and with the exception of Bernhardt and DeKolb (2003) and Bernhardt et al. (2004), the evaluation was based on diagnostic testing.

2.1.1 *Load rating of bridges*

In the AASHTO Manual, three methods for load rating are provided including the Allowable Stress Rating (ASR), Load Factor Rating (LFR), and Load and Resistance Factor Rating (LRFR) method. The LFR method evaluates bridge capacity at two rating levels (inventory and operating). The inventory rating is the smaller of the two rating levels and corresponds to the heaviest live load that can safely utilize a bridge on a continual basis, while the operating rating represents the maximum permissible live load that a bridge may be subjected to but less frequently, since continuous loading at this level could shorten the service life of the bridge. The rating equation used in the LFR method is (AASHTO, 2011):

$$RF_{LFR} = \frac{\varphi R_n - \gamma_D D}{\gamma_L L (1 + I)} \tag{2.1}$$

where RF_{LFR} is the LFR rating factor; φR_n is the design capacity of the member; γ_D and γ_L are the load factors for dead load (equal to 1.3) and live load (equal to 2.17 for inventory and 1.3 for operating), respectively; D and L are the dead load and live load effects, respectively; and I is the live load impact factor. The inventory and operating ratings are reported in terms of the vehicle used to compute the live load effects. For instance, inventory and operating rating factors of 1.05 and 1.76 for an HS-20 vehicle would be reported as HS-21.0 and HS-35.2, respectively.

The LRFR method follows the LRFD design philosophy to provide a uniform level of safety for different bridge types. In general, there are more uncertainties in the live load applied to a bridge in the design stage, whereas with evaluation (i.e. load rating) there are greater uncertainties with resistance due to material deterioration and, in the case of concrete bridges with no design plans, missing information. The load rating equation for the LRFR method is (AASHTO, 2011):

$$RF_{LRFR} = \frac{\varphi_c \varphi_s \varphi R_n - \gamma_{DC} DC - \gamma_{DW} DW}{\gamma_L LL (1 + IM)} \tag{2.2}$$

where RF_{LRFR} is the LRFR rating factor; φR_n is the design capacity of the member; φ_C and φ_S are the condition and system factor, respectively ($\varphi_c \varphi_s \geq 0.85$); γ_{DC}, γ_{DW}, and γ_L are the load factors for structural components (DC), wearing surfaces and utilities (DW), and live load (LL), respectively; DC, DW, and LL are the dead load and live load effects, respectively; and IM is the dynamic load allowance. The condition factor accounts for increased uncertainty in the resistance of deteriorated members and the probability of future deterioration. This factor depends on the member condition and ranges from 0.85 (poor) to 1.0 (good). The system factor relates to structural redundancy which is the capability of a structure to redistribute loads upon damage or failure of one or more of the load-carrying members. This factor ranges from 0.85 for nonredundant systems to 1.0 for redundant multi-girder bridges. Note that the factored dead load and live load components of the LRFR rating equation are not the same as those used in the LFR method due to differences in the design philosophy.

Regardless of the load rating method, the dead and live load effects can be determined from the bridge geometry and member dimensions using load distribution formulas; however, the major challenge for planless concrete bridges is the evaluation of φR_n. For instance, the nominal moment capacity of a reinforced concrete T-beam is given as:

$$M_n = A_s f_y \left(d - \frac{A_s f_y}{1.7 f_c' b} \right) \tag{2.3}$$

where A_s is the area of reinforcing steel, f_y is the yield stress of reinforcing steel, d is the distance from top of beam to centroid of reinforcing steel, f_c' is the concrete compressive strength, and b is the effective slab width. With the exception of b, the remaining variables are all unknown for a planless concrete bridge. Material strengths can be found in the

AASHTO Manual based on date of construction; however, A_s and d are more difficult to determine. Hence, advanced methods are needed to determine the capacity either through quantification of A_s and d or direct assessment of M_n.

In LRFR, load ratings are divided into three stages: design load rating, legal load rating, and permit load rating. The results of each rating stage serve specific uses and also determine the need for further evaluations of bridge safety and/or serviceability. Design load rating serves as the first-level assessment of bridges and is based on the HL-93 loading. Bridges that pass the design rating check at the inventory level are considered satisfactory "for routinely permitted loads on highways of various states under grandfather exclusions to federal weight laws" (AASHTO, 2011). If a bridge is rated less than 1.0 at inventory but greater than 1.0 at operating, the bridge is considered satisfactory "for legal loads that comply with federal weight limits and Formula B" (AASHTO, 2011). In both cases, no restrictive posting is required and the bridge may be evaluated for permit vehicles. If a bridge is rated below 1.0 at operating for the LRFR design load rating, a legal load rating is required under the AASHTO legal loads (Type 3, Type 3S2, and Type 3–3) and/or state legal loads. Should the legal load rating factor be less than one based on AASHTO design assumptions, a higher level evaluation may be performed that may include one or a combination of the following approaches: refined analysis, load testing, site-specific load factors, and/or direct safety assessment. Hence, the LRFR method provides a more direct approach than the LFR method for integrating nondestructive load testing into the bridge rating process.

2.1.2   Load testing of bridges

The two types of load tests used to evaluate the capacity of an existing bridge are diagnostic and proof tests (AASHTO, 2011). These two methods differ in terms of the level of load applied to the bridge, the quantity and significance of measurements taken, and the manner in which the experimental findings are used to determine the load rating. Diagnostic testing is usually employed when the original design plans of a bridge are available to create a representative analytical model (AASHTO, 2011). In a diagnostic test, a bridge is subjected to a known load below its elastic load limit, as shown in Figure 2.2. Ordinarily, the load applied is a single dump truck filled with asphalt or roadway base material. Strain and/or deflection measurements are recorded at strategic locations as the test truck is driven longitudinally across the bridge at a slow speed along different transverse paths to determine the load distribution and stiffness characteristics of the bridge. Following the test, the field measurements are compared with the results of an analytical model to understand and verify the behavior of the bridge and estimate the load capacity. Because of the low load level applied during a diagnostic test compared to a proof test (see Fig. 2.2), caution must be taken in extrapolating the measured response of the bridge to a higher rating load. Unlike a proof test approach, where the maximum load applied in the test represents the factored bridge capacity, diagnostic testing requires a more rigorous post-test analysis of the data.

A proof test is an attempt to experimentally prove that a bridge can safely carry a certain design or legal load. In situations where an analytical model cannot be developed because of the lack of design drawings or when the bridge has suffered severe deterioration difficult to quantify, proof testing is useful (AASHTO, 2011). In a proof test, increasing loads are applied to the bridge until the target proof load is reached, as shown in Figure 2.2, or some form of distress (e.g. nonlinear behavior, excessive cracking) is observed. To achieve this

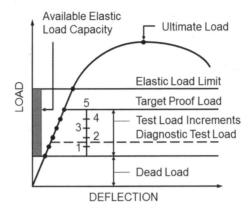

Figure 2.2 Load-deflection response for bridge testing
Source: Adapted from Pinjarkar et al. (1990).

high load level, a variety of loading systems have been employed such as multiple dump trucks, tractor-trailers loaded with concrete blocks or heavy machinery, and military tanks. In general, the target proof load represents the load necessary to produce force effects equal to those caused by the factored design or legal load plus impact. Measurements are usually taken only at a few critical locations to monitor the bridge response during a proof test and determine the maximum load carried by the bridge (considered the factored capacity which must be adjusted to determine the bridge rating).

2.2 Inspection and evaluation procedures

The load rating procedures for prestressed concrete bridges with no design plans presented herein were developed in accordance with the applicable sections of the AASHTO Manual, in particular Section 4: Inspection, Section 6: Load Rating, and Section 8: Nondestructive Load Testing. In the latter section of the AASHTO Manual, it is stated that the nondestructive load testing procedures were prescribed "to ensure consistency with the load and resistance factor load rating procedures" (AASHTO, 2011). Consequently, the LRFR method is adopted. The AASHTO Manual also states that "Many older reinforced concrete and prestressed concrete beam and slab bridges whose construction plans, design plans, or both are not available need proof testing to determine a realistic live load capacity" (AASHTO, 2011). Thus, the final procedure is based primarily on the proof test method.

The proposed procedure for load rating prestressed concrete bridges consists of various steps, as shown in Figure 2.3. The procedure will vary from bridge to bridge and depends on the confidence level with the prestressing strand estimate and the bridge condition. The following sections discuss each step of the load rating procedure in detail.

2.2.1 In-depth inspection and field measurements

The first step of the procedure is to visit the bridge site to conduct an in-depth field inspection and obtain field measurements of the structure dimensions. This inspection should be performed carefully with guidance from the latest inspection report on file (if such

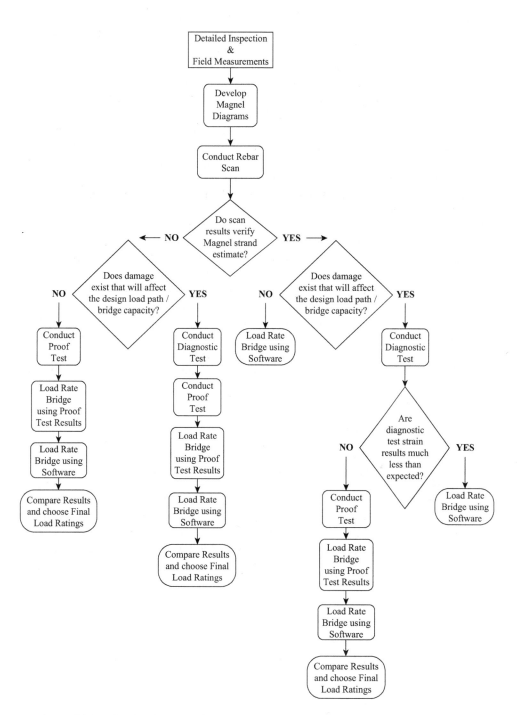

Figure 2.3 Load rating flowchart for planless concrete bridges

documentation exists). Locations with previous deterioration and/or damage should be checked closely for signs of advancement. Any damage that could possibly reduce the bridge capacity (e.g. section loss) should be noted. In double T-beam and box beam bridges as well as other bridge types without a deck, particular attention should be given to the shear connections between adjacent members. If the shear keys are damaged, the design-based distribution factor may underestimate the actual live load forces applied to the affected beams. Bridges with damaged shear keys are actually good candidates for proof load testing due to the uncertainties in live load distribution and difficulty in creating an accurate structural model, but the tests must be planned and executed with care. Shear and flexural cracks should be documented along with any section loss of the steel reinforcement in spalled areas. Section loss should be accounted for to compute the reduced capacity of the bridge and planning the load tests. As seen in Figure 2.3, damage is critical in selecting the right procedure for load rating.

During the inspection, it is also important to measure the overall bridge dimensions and the geometry of the individual girders. Bridge length and width will be used in planning the load tests and calculating the final load ratings from the test results. Girder dimensions are needed to calculate the cross-sectional properties (e.g. area, moment of inertia, section modulus) and the self-weight of the members. This information is used to estimate the amount of prestressing strands via Magnel diagrams (discussed in next section). Overall, the measurements required to create as-built drawings including a plan view, elevation view, and transverse view (i.e. cross section) of the bridge should be taken. The as-built drawings serve as an important record of the bridge configuration and properties and also will facilitate the development of analytical computer models for load rating the bridge at the strength limit state.

2.2.2 Magnel diagrams

Following the in-depth inspection, the field measurements and estimated material properties are used to develop Magnel diagrams. These diagrams provide the prestress force and eccentricity combinations that satisfy the serviceability criteria (at transfer and service) and thus serve as the initial estimate of the number and locations of the prestressing strands in the bridge beams (interior and exterior). For bridges with non-composite members, such as those with no deck where the top flanges act as the riding surface (e.g. double T-beams and box beams), the following Magnel equations are used:

$$e_1 \leq \frac{1}{P_i}(S_b * f_{bi} + M_{SW}) - \frac{S_b}{A} \tag{2.4}$$

$$e_2 \leq \frac{1}{P_i}(S_t * f_{ti} + M_{SW}) + \frac{S_t}{A} \tag{2.5}$$

$$e_3 \geq \frac{1}{k * P_i}[(-S_b * f_{bf}) + M_{TOT}] - \frac{S_b}{A} \tag{2.6}$$

$$e_4 \geq \frac{1}{k * P_i}[(-S_t * f_{tf}) + M_{TOT}] + \frac{S_t}{A} \tag{2.7}$$

where e_1, e_2, e_3, and e_4 are the range of solutions that satisfy the serviceability criteria at the beam midspan based on the bottom stress at transfer (f_{bi}), top stress at transfer (f_{ti}), bottom stress at service (f_{bf}), and top stress at service (f_{tf}), respectively. In addition, $1/P_i$ is the inverse of the initial prestress force; S_b and S_t are the section moduli at the bottom and top of the beam; A is the cross-sectional area of the beam; k is the factor for prestress losses; and M_{SW} and M_{TOT} are the self-weight moment and total moment (i.e. dead load plus live load moments), respectively, for the beam.

For bridges with composite members, such as a prestressed I-girder bridge with a cast-in-place deck, the following Magnel equations are used:

$$e_5 \leq \frac{1}{P_i}(S_{b,NC} * f_{bi} + M_{SW}) - \frac{S_{b,NC}}{A} \tag{2.8}$$

$$e_6 \leq \frac{1}{P_i}(S_{t,NC} * f_{ti} + M_{SW}) + \frac{S_{t,NC}}{A} \tag{2.9}$$

$$e_7 \geq \frac{1}{k * P_i}\left[(-S_{b,NC} * f_{bf}) + M_{DL,NC} + \frac{S_{b,NC}}{S_{b,COMP}} * (M_{DL,COMP} + M_{LL})\right] - \frac{S_{b,NC}}{A} \tag{2.10}$$

$$e_8 \geq \frac{1}{k * P_i}\left[(-S_{t,NC} * f_{tf}) + M_{DL,NC} + \frac{S_{t,NC}}{S_{t,COMP}} * (M_{DL,COMP} + M_{LL})\right] + \frac{S_{t,NC}}{A} \tag{2.11}$$

where e_5, e_6, e_7, and e_8 are the range of solutions that satisfy the serviceability criteria at midspan. In equations (2.8) through (2.11), e is the eccentricity of the prestressing steel relative to the non-composite girder centroid; $S_{b,NC}$, $S_{t,NC}$, $S_{b,COMP}$, and $S_{t,COMP}$ are the section moduli at the bottom and top of the section for both the non-composite (NC) and composite (COMP) sections, respectively; A is the cross-sectional area of the non-composite girder; k is the factor for prestress losses; M_{SW} is the moment due to the girder self-weight at transfer; $M_{DL,NC}$ and $M_{DL,COMP}$ are the dead load moments resisted by the non-composite and composite sections, respectively; and M_{LL} is the live load moment at midspan of the girder.

Note that the equations limiting the concrete stress at transfer remain the same for both non-composite (e_1 and e_2) and composite (e_5 and e_6) bridges, since only the beam self-weight is considered. In a fully non-composite bridge, the total dead and live load moment is resisted solely by the girder with no participation from a separate deck. In a composite bridge, the weights of the girder, deck, and diaphragms are resisted by the non-composite section while the superimposed dead load and live load are resisted by the composite section.

An example Magnel diagram is shown in Figure 2.4 next to a typical double T-beam section. Lines 1 through 4 represent prestressing force versus eccentricity solutions to the serviceability limit states evaluated at the bottom and top of the beam at transfer (f_{bi}, f_{ti}) and at service (f_{bf}, f_{tf}). The diagram is scaled to match the section to give a visual representation of the allowable eccentricity range for a typical beam (parallel to the area enclosed by the four equation lines). Note that the example is for a double T-beam but can be used for any typical prestressed concrete beam section (including non-composite and composite sections).

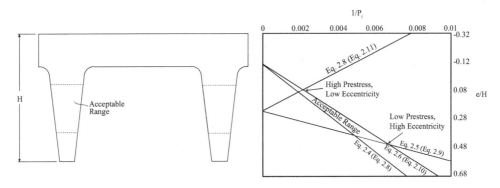

Figure 2.4 Example Magnel diagram for typical double T-beam section

Magnel diagrams should be developed for different design truck loads for both interior and exterior girders since the number of strands may vary considerably between girders, especially in cases where the bridge was widened. In the diagrams, the region enclosed by the inequalities provide acceptable eccentricity and prestress combinations that satisfy the serviceability criteria. However, two particular points of intersection should be noted: the high prestress/low eccentricity combination and the low prestress/high eccentricity combination. These two values represent the maximum and minimum amounts of prestressing steel (and corresponding eccentricities) that can be used. The initial estimate of the number of strands and eccentricity should be based on the low prestress/high eccentricity combination since it results in the least number of strands and largest eccentricity. This configuration is most commonly used by designers. Using the estimated prestress losses (k), the inverse of the initial prestress ($1/P_i$) is converted into effective prestress (P_e). The effective prestress is then divided by the effective stress of a single strand (f_{pe}), computed from the design information collected earlier to determine the total area of prestressing steel (A_s). Finally, the total area is divided by the area of a single strand to estimate the number of strands. Using the measured beam geometry, year of construction, and practical strand patterns, the diagram for the design truck that provides the best fit based on engineering judgment should be selected and used to determine the final strand estimate. Once an initial strand estimate is determined, the strands should be arranged into a practical pattern. The prestress force and eccentricity of this pattern should be checked within the Magnel diagram to confirm that these values satisfy all serviceability limits.

For bridges where it is uncertain if composite behavior was considered in the design, both non-composite and composite behavior should be investigated. Results can then be compared and the most appropriate diagram chosen for the strand estimate. Once a feasible strand estimate is obtained, nondestructive testing should be performed for confirmation (discussed in next section).

2.2.3   Rebar scan

Nondestructive testing equipment (e.g. rebar scanner) should be used to verify the prestressing strand estimate from the Magnel diagrams and to determine the bar size, spacing, and concrete cover of the transverse shear reinforcement. Scans should be taken at various locations along the bridge length to provide different areas where the strands can be viewed for purposes of confirmation. Apart from the prestressing strands, the bar sizes and spacings of the shear reinforcement need to be accurately measured so that analytical computer models can be generated to determine load rating factors. Note that an accurate measurement of the shear reinforcement is critical since the model will be used exclusively to determine the load rating factors for shear based on the strength limit state. The shear reinforcement will also be used to perform a shear check prior to load testing. Confirmation of the prestressing strands is not as critical since the proof test results are used to determine the rating factors for the serviceability limit state (e.g. concrete cracking). The following sections will provide guidance on scanning procedures for different bridge types.

2.2.3.1   Double T-beam bridges

Rebar scans should be taken along the height and length of the beam stems of double T-beam bridges near midspan and as close as possible to the abutments. An interior and exterior beam should be scanned to determine the reinforcement configurations. Differing configurations in the two beams provides strong evidence of bridge widening. Comparison of the scans taken at the abutment ends and at midspan is necessary to find out if the strands are harped and also to determine the changes (if any) in stirrup spacing. Scans should also be taken on both the inside and outside of the beam stems as shown in Figure 2.5 to determine if the stirrups are double or single legged. Note that due to close strand spacing it may not be possible to distinguish individual strands near midspan of the beams; however, strand deviation (if used) will result in larger spacing between the strand groups near the beam ends. In general, strands are harped at 0.4L and are tighter spaced in the middle 20% of the beam length, which complicates the size estimate. The seven-wire strand arrangement can also lead to inaccurate reinforcement size estimates due to its spiral pattern.

2.2.3.2   Box beam bridges

Rebar scans are generally more effective in detecting mild reinforcement than longitudinal prestressing strands of adjacent box beam bridges. Scans should be taken along the bottom flanges, primarily to locate the prestressing strands, and webs (exterior beams only), primarily to locate the shear stirrups, near the abutments and midspan. These locations are

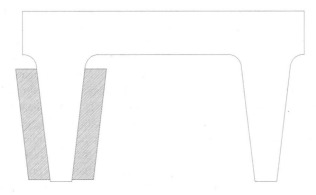

Figure 2.5 Double T-beam scan locations

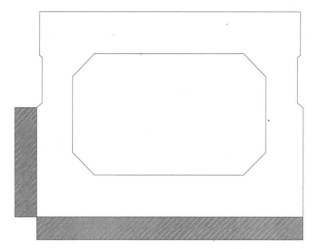

Figure 2.6 Box beam scan locations

illustrated in Figure 2.6. Similar to T-beam bridges, both interior and exterior beams should be scanned to determine if the box beams have different reinforcement, which may be due to widening. Prestressing strands typically have 9.53 mm (0.375 in.) or 12.7 mm (0.5 in.) diameters and are approximately spaced at 51 mm (2 in.) on center. Obtaining a reliable strand approximation from the rebar scan can therefore be difficult due to this narrow spacing. Strands with large spacing and small concrete cover will be easier to identify. In general, box beam bridges are more difficult to scan and obtain meaningful results since only the webs of the exterior beams are exposed and the prestressing strands are layered on the bottom flange with a large concrete cover.

2.2.3.3   *I-girder bridges*

Rebar scanning typically provides a good estimate of bar size, spacing, and concrete cover for the transverse shear stirrups in prestressed I-girder bridges but has limited usefulness in determining the prestressing strand configuration. The challenge is that I-girders have multiple layers of strands placed in a large bottom flange, which makes it difficult to detect the reinforcement past the first layer (near the exterior of the flange) with the scanner, similar to box girders. Nevertheless, scans should be taken along the bottom and side of the flange and along the I-girder web. Figure 2.7 shows the scanning locations for a typical I-girder. Both interior and exterior girders should be investigated for reasons discussed earlier. Ultimately, the results for an I-girder may provide limited information of the prestressing strands but should provide reliable results of the shear reinforcement.

For all bridge types, a rebar scanner may not authenticate the Magnel prestressing strand estimates for various reasons including close strand spacing, multiple strand layers, large concrete cover, and inaccessible areas. In such cases the prestressing strand estimate from the Magnel diagrams must be solely relied upon for planning, preparing, and performing the load testing. Typically, shear reinforcement should be easily confirmed by a rebar scanner. In summary, the rebar scanner should be used to confirm the longitudinal girder

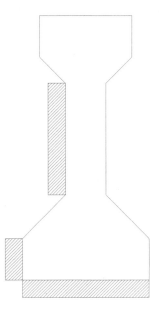

Figure 2.7 I-girder scan locations

reinforcement to the highest degree possible and if not successful, load tests are necessary as shown in the flowchart of Figure 2.3.

2.2.4 *Load testing*

When the in-depth inspection has been completed, strand estimates have been made, and the bridge has been scanned to confirm the reinforcement, load testing may then be conducted to determine the largest load the bridge can safely resist based on the serviceability limit state. Load testing may be bypassed only if the Magnel diagram estimates are confirmed by the rebar scan and the bridge has no signs of damage that could affect its capacity. In most cases involving bridges without plans, it is very difficult to confirm the reinforcement pattern from the rebar scan measurements and thus, load testing is necessary.

A diagnostic test is recommended to gain a better understanding of the bridge behavior when deterioration and/or damage exists that can significantly affect the live load distribution between the primary members. The results of the diagnostic test will provide critical truck positions that can be used to plan the loading configuration for the proof test. The magnitude of the truck load applied in a diagnostic test should be less than the target proof load due to the uncertainties in the bridge response. In a proof test, trucks are loaded to their maximum capacity, while in a diagnostic test trucks are typically loaded to two-thirds of their capacity. If no damage is present, a diagnostic load test may not be necessary. Prior to all testing, preparatory calculations are needed to determine the available moment capacity for each bridge beam and the corresponding available strain capacity. Ultimately, the available strain will be used to control the test and ensure that the load applied does not exceed the cracking limit of the prestressed beams. The beam moment and strain equations are discussed next. For non-composite behavior,

$$M_{available} = S_b * \left[P_e * \left(\frac{1}{A} + \frac{e}{S_b} \right) + f_{cr} \right] - M_{DL} \tag{2.12}$$

where f_{cr} is the modulus of rupture, M_{DL} is the dead load moment, and P_e is the effective prestress force determined from the Magnel estimate. The remaining variables were defined earlier for equations (2.4) through (2.11). For composite behavior,

$$M_{available} = S_c * \left[P_e * \left(\frac{1}{A} + \frac{e}{S_{nc}} \right) + f_{cr} \right] - M_{dnc} * \left(\frac{S_c}{S_{nc}} \right) - M_{dc} \tag{2.13}$$

where S_c, S_{nc}, M_{dc}, and M_{dnc} are the composite and non-composite section moduli and the composite and non-composite dead load moments, respectively. For non-composite and composite behavior, the available strain capacity for the beam is computed as

$$\varepsilon_{available} = \frac{M_{available}}{S_{gage} * E_{design}} \tag{2.14}$$

where S_{gage} is the section modulus at the gage location and E_{design} is the concrete elastic modulus.

To compare with the available moment and strain capacities calculated from equations (2.12) or (2.13) and (2.14), the total midspan moment should be calculated under the test truck loading. The total moment from the test truck is then multiplied by the design-based distribution factor and an "upper bound" distribution factor that represents the worst-case scenario for live load distribution between the girders (e.g. distribution factor of one for box beam bridges). The specifications that were in effect at the time of bridge construction should be used to determine the design distribution factor. These calculations provide the range of possible beam moments under the test truck load. Thus, for diagnostic testing, an appropriate load should be selected that will produce load effects well below the available moment computed by equation (2.12) or (2.13).

For all tests, the bridge should be instrumented with strain gages at various locations with the majority of gages placed at midspan of simple-span bridges where the bending moment is maximum. At least one gage should be placed towards the bottom of each bridge member where the strain is largest; however, more are desirable for redundancy purposes. A pair of gages should be placed on either side of each stem of double T-beams and on the bottom flanges of box beams or I-girders. Strain magnitudes measured at these gages will be compared to the available strain capacities and theoretical predictions (based on the design-based and "upper bound" distribution factors) from before. Thus, the extra gage is important in case a gage malfunctions during the test.

A diagnostic load test should start with a single truck placed in different transverse paths. Generally, the truck should be centered along a particular beam or in between two beams and also placed as close as possible to the curb. Throughout the test, the strains at midspan should be monitored so the loading may be halted well before the available strain capacity is reached. After the single-truck test, the paths with the highest measured strains should be loaded simultaneously with two or more trucks placed side-by-side. In the multiple truck tests, the midspan strains should again be monitored closely. At the completion of the diagnostic test, a better understanding of the live load distribution should be achieved. Strain magnitudes from midspan gages placed near the bottom of each beam will be compared to the available strain capacities and theoretical predictions made prior to testing. These results will help determine what load patterns should be used in the proof test. As shown in Figure 2.3, when the live load distribution factor estimated from the diagnostic test is equal to or less than the design-based value, a proof test may not be needed if the prestressing strand estimate was confirmed.

The proof test should be planned and conducted with the goal of applying the largest load possible to the bridge without exceeding the available moment. The test vehicles for the proof test should be loaded to their maximum rated weight and positioned along the paths deemed critical during the diagnostic test (paths producing the largest beam strains). Starting at the abutments, the trucks should be backed up and stopped at different longitudinal locations until the final positions for maximum moment are reached or until the measured beam strain approaches the available strain capacity. In general, more gages should be installed on the beams with the highest strains from the diagnostic test since these beams will dictate the outcome of the proof test. If no diagnostic test was conducted, a single truck can initially be used to load various transverse paths. This initial phase will help determine the critical paths and guide the remainder of the proof test. If no true critical path is found (i.e. similar strains are measured from path to path), an effort should be made to maximize the midspan moments of both an interior and exterior girder.

Before the proof test is executed, a shear check should be conducted to ensure that the test loads will not crack the bridge beams in shear. The shear capacity may be evaluated using the following equation (Hawkins et al., 2005):

$$V_{cw} = f_t \sqrt{1 + \frac{f_{pc}}{f_t}} b_w d + V_p \tag{2.15}$$

where f_t is the tensile concrete strength, f_{pc} is the compressive concrete stress after all losses at beam centroid, b_w is the beam web width, d is the distance from the compression face to the centroid of the tension reinforcement (not less than 80% of overall beam depth), and V_p is the vertical component of the effective prestressing force. The tensile concrete strength varies between two and four times the square root of f_c' (Hawkins et al., 2005). The shear capacity of the beam from Equation 2.15 should exceed the maximum shear produced by the dead load of the bridge and live load of the test trucks. If the computations show that cracking of the bridge due to shear is a concern, advanced methods may be used to monitor the shear regions during proof testing by acoustic emission such as in the study performed by Anay et al. (2016).

For all proof tests, an effort should be made to maximize the moment for three loading conditions: single lane loaded (one truck or two trucks, back-to-back); two lanes loaded (four trucks, back-to-back and side-by-side); and multiple lanes loaded (two, three and/or four trucks, side-by-side). In all cases, the final positions of the trucks, whether stopped early based on strain measurements or stopped after reaching the maximum positions, should be noted. These final positions will be used to determine load rating factors for the serviceability limit state using the AASHTO Manual.

2.2.5   *Serviceability ratings using proof test results*

The maximum midspan moments achieved during the proof test are used to determine the rating factors for all legal loads for the serviceability limit state. The equations used in the load rating process can be found in Section 8.8.3.3 of the AASHTO Manual. The adjusted target live load factor (X_{pA}) and the target proof load (L_T) are first determined using the following equations:

$$X_{pA} = X_p \left(1 + \frac{\Sigma\%}{100} \right)$$ (2.16)

$$L_T = X_{pA} L_R (1 + IM)$$ (2.17)

where X_p is the initial target load factor, $\Sigma\%$ is the sum of the X_p adjustment factors given in Table 2.1, L_R is the unfactored live load due to the rating vehicle, and IM is the impact factor (equal to 1.33).

The AASHTO Manual recommends a minimum value of 1.4 for X_p; however, since proof testing procedures proposed herein for prestressed concrete bridge are based on serviceability (i.e. limiting concrete tensile stress at service load) rather than strength, a factor of 1.0 is used. The L_T and L_R magnitudes are determined in terms of maximum midspan moments rather than load, since the rating vehicles and proof test trucks do not have the same axle configuration.

The target proof moments are calculated and compared to the actual moments applied by the test trucks during the proof test. The operating level (*OP*) capacity for the bridge is then computed using the following equation:

$$OP = \frac{k_0 L_p}{X_{pA}}$$ (2.18)

where k_0 is a factor that depends on how the proof test was terminated (1.0 if the target proof moment is reached and 0.88 if the test is stopped due to signs of distress) and L_P is the applied moment. In cases where the test is stopped due to excess strain, k_0 should still be taken as 1.0 since the concrete design strength is used rather than the actual strength to determine the available strain capacity.

Table 2.1 X_p adjustment factors for proof load testing (AASHTO, 2011)

Feature	Adjustment
One-lane load controls	+15%
Nonredundant structure	+10%
Fracture-critical details present	+10%
Bridges in poor condition	+10%
In-depth inspection performed	−5%
Ratable, Existing RF ≥ 1	−5%
ADTT ≤ 1000	−10%
ADTT ≤ 100	−15%

The final load ratings are then computed for all three load conditions: single lane loaded, two lanes loaded, and multiple lanes loaded. The rating factor, RF_0, for each legal load is computed using the following equation:

$$RF_0 = \frac{OP}{L_R(1 + IM)} \tag{2.19}$$

Depending on how the proof test was stopped, the final load rating will depend on the phase that produced the highest rating factors or the lowest rating factors. If the test is stopped because the available strain capacity was reached, and the test trucks did not reach the final positions for maximum moment, the phase that produced the lowest rating factors controls. In this case, loading to a level higher than these factors risks cracking the bridge. If the trucks do reach their final positions, and the available strain capacity was not met, the phase producing the highest rating factors controls since the bridge can sustain larger loads. If the serviceability rating factors are less than 1.0, safe posting loads need to be computed as follows:

$$\text{Safe Posting Load} = \frac{W}{0.7}(RF_0 - 0.3) \tag{2.20}$$

where W is the weight of the load rating vehicle. Once rating factors and safe posting loads are determined for each phase, the final step of the load rating process is to compare the serviceability-based ratings from the proof tests to the strength-based ratings from a software program.

2.2.6 Strength ratings using load rating software

Analytical load ratings from a software program for the strength limit state should be computed and compared with the serviceability-based load ratings from the proof tests to arrive at the final ratings (and possibly, load postings) for the bridge. It is important to note that the experimental load ratings are based on serviceability (i.e. concrete cracking), while the analytical load ratings from the software are based on strength (i.e. flexural and shear capacity). The AASHTO Manual indicates that legal load ratings may be based on serviceability, but strength should always be checked.

The analytical load ratings from a software program should be developed using the material properties from the applicable specifications, the field measurements of the bridge dimensions, the number and locations of the strands from the Magnel diagrams, and the shear stirrup size and spacing from the rebar scan results. Recall that for bridges without damage and/or deterioration that will affect the load distribution and/or capacity of the beams, testing may not be necessary and the bridge may be load rated solely using the software program if the rebar scan results provide unquestionable confirmation of the strand estimate from the Magnel diagrams. For bridges in suspect condition and/or with low quality scan results, proof tests are needed to determine load ratings for the serviceability limit state. In these cases, engineering judgment should be used to compare the experimental ratings from the proof tests with the analytical ratings from the software model in setting the final load ratings for the bridge.

For bridges with damaged shear connections between beams, it is very likely that the design-based distribution factor for live load does not represent the true bridge behavior. In such cases, two models should be generated for the bridge. In the first model, the original distribution factor should be used (i.e. design distribution factor). In the second model, the distribution factor should be based on the assumption that the beams act independently (i.e. wheel load distribution factor equal to 1.0). The results from these two analyses provides the capacity range for the bridge in an undamaged and damaged condition.

Finally, when the nondestructive testing equipment fails to generate results that verify the Magnel estimate, the prestressing strands should be arranged in a logical manner that satisfies the appropriate design guidelines. If available, plans from similar bridges built in the same time period should be obtained and used as a reference to arrange the strands. If limited information is available regarding the strand arrangement, different models may be created with different strand configurations. The resulting rating factors from the load rating software program for the different models may then be compared to the ratings from the proof test, and the final load ratings may be established using engineering judgment.

2.2.7   *Final load ratings*

A final check must be made that the serviceability-based rating factors from the proof tests are smaller than the strength-based rating factors from the software program. This comparison is particularly important when the proof test is stopped early and the load ratings indicate that the bridge should be load posted for legal loads. For cases where the proof tests are not stopped and all the serviceability-based rating factors are greater than 1.0, the strength-based ratings should still be checked to ensure that all factors are larger than 1.0. If the ratings for strength are less than 1.0 but the ratings for service are greater than 1.0, a decision must be made using engineering judgment on whether the bridge should be load posted. For instance, if the strand estimate and arrangement was verified by the rebar scan, the analytical strength rating factors should be used. On the other hand, if the strands were not verified, the analytical strength rating factors may be underestimating the capacity of the bridge, and the proof test ratings should be used. In cases where testing is not conducted, only analytical strength ratings are computed, eliminating the need to compare the serviceability-based and strength-based rating factors.

The next sections illustrate the use of the proposed procedures to evaluate three prestressed concrete bridges. Case studies of a T-beam, box beam, and I-girder bridge are presented.

2.3   Case studies

In this section, three prestressed concrete bridges (including a double T-beam, box beam, and I-girder bridge) are evaluated to illustrate the load rating procedures. The procedures for estimating the number and layout of steel strands in the bridge beams using Magnel diagrams are first covered. Measurements are then taken with a rebar scanner to check the Magnel strand estimates and also to determine the size and location of the mild shear reinforcement. Next, the load testing activities for each bridge including the preparatory calculations, instrumentation plans, truck configurations, and strain measurements are discussed. Legal load ratings are then computed for serviceability (i.e. concrete cracking) in accordance with the AASHTO Manual using the results of the proof tests. In addition, ratings based on flexure and shear strength are computed for each bridge using a software program. The serviceability-based and strength-based legal load ratings are then judged to arrive at final recommendations regarding the capacity of each bridge.

2.3.1   Bridge 8761 (double T-beam)

Bridge 8761 has five prestressed concrete double T-beams and a 28° skew. The bridge was originally built in 1989 with three beams and subsequently widened in 2005 with two additional beams. The bridge has a total width (out to out) of 12.2 m (40 ft) and total length of 11.0 m (36 ft). The beams are 2.44 m (96 in.) wide by 0.69 m (27 in.) deep by 11 m (36 ft) long and the beam flanges serve as the riding surface. The flange thickness is 152 mm (6 in.) and the stem thickness varies from 146 mm (5.75 in.) at the top to 95 mm (3.75 in.) at the bottom. The beams are numbered 1 through 5 from south to north and the abutments are numbered 1 and 2 from east to west. Overall, the concrete beams were in fair condition based on the 2012 New Mexico Department of Transportation (NMDOT) bridge inspection report. See pictures and bridge cross-sections given in Figure 2.8.

Based on the construction years, Magnel diagrams for the interior and exterior beams were developed based on the 13th (1983) and 17th (2002) editions, respectively, of the AASHTO *Standard Specifications*. The grade of prestressing strands is normally determined from the AASHTO Manual; however, historical documentation was found for the bridge that specified the grade as 1862 MPa (270 ksi). Furthermore, the nominal strand diameters were given as 15.2 mm (0.6 in.) for the exterior beam and 12.7 mm (0.5 in.) for the interior beam. Concrete strengths of $f_c' = 37.9$ MPa (5500 psi) and $f_{ci}' = 31.0$ MPa (4500 psi) were selected for the interior beams using the AASHTO Specifications (1983), while f_c' and f_{ci}' were taken as 41.4 MPa (6000 psi) and 31.0 MPa (4500 psi) for the exterior beams, respectively, as given in the rehabilitation plans. Prestress losses of 20% were assumed to estimate the number of strands. During the latest field inspection, the shear keys for Bridge 8761 appeared to be functioning. Thus, the wheel distribution factor for the interior beams was calculated as $S/6$, where S is the beam spacing in feet (AASHTO, 1983). For $S = 2.44$ m (8 ft), the distribution factor equaled 1.333 for wheel loading. For the exterior beam, the distribution factor was computed using the lever rule (AASHTO, 2002), which amounted to 1.290 for wheel loading. The dead load moment was computed based on self-weight only for the interior beam and self-weight and the bridge railing for the exterior beam.

From the Magnel diagrams, the number of strands and corresponding eccentricities for the three AASHTO trucks (H-15, H-20, and HS-20) were estimated. The eccentricities computed for the H-15 and H-20 trucks at the low-prestressing points exceeded the available eccentricity range, which suggested that the H-15 or H-20 was not the design truck. Historical documentation for the original bridge and the rehabilitation plans for the widened bridge both confirmed the HS-20 vehicle as the design truck. For the HS-20 truck, the strand estimate was 7.0 for the exterior beam, which was rounded up to 8 (four strands per stem), and 10.1 for the interior beam, which was rounded up to 12 (six strands per stem). The corresponding strand eccentricities for an exterior and interior beam were 434 mm (16.69 in.) and 463 mm (18.23 in.), respectively. Fewer strands were needed for the exterior beam due to the smaller distribution factor and also the larger strand diameter. Assuming the strands are bundled with 51 mm (2 in.) cover (from the stem bottom to the center of the lowest strands), the maximum eccentricities for the exterior and interior beams amounted to 469 mm (18.46 in.).

Rebar scans were generated for the exterior beams (widened bridge) and interior beams (original bridge). Figure 2.9 shows the beam elevation given in the rehabilitation plans along with the results from the rebar scanner for the south exterior beam. The scans showed two longitudinal bars (one small diameter and one large diameter) and a series

Figure 2.8 Initial inspection pictures, typical cross-sections, instrumentation plan and critical load paths for Bridge 8761

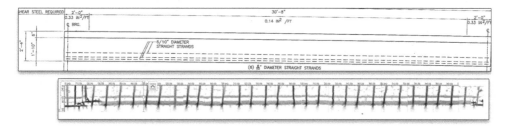

Figure 2.9 Elevation view of the exterior beam of Bridge 8761 from rehabilitation plans with rebar scan

Figure 2.10 Rebar scan of the interior beam of Bridge 8761 from abutment to midspan

of vertical bars spaced approximately 305 mm (12 in.) on center. The rehabilitation plans specify a 102 mm (4 in.) and 51 mm (2 in.) spacing between the three strands from top to bottom which agreed with Figure 2.9. For the top strand, the rebar scanner gave a bar size ranging between #10 (#3) and #16 (#5). Note that the cross-sectional area of a 15.2 mm (0.6 in.) diameter strand is 140 mm^2 (0.217 in.^{2}), which equates to a 13.5 mm (0.530 in.) diameter solid bar (i.e. #13 [#4]), which is the average of the scan results. Due to the small spacing, the middle and bottom strands were more difficult to distinguish in the scans but the two strands could still be reasonably identified. This amounted to three strands per stem, which is less than the number estimated from the Magnel chart (four strands per stem). For the vertical steel, the scan results consistently gave a #10 (#3) bar size. To determine the number of stirrup legs in each beam stem, scans were taken on the inside and outside faces of a single stem. The two scans showed a stirrup leg at the face of each stem and thus, signified a two-legged stirrup. The rehabilitation plans specified the required shear steel area as 698.5 mm^2/m (0.33 in.^{2}/ft) for a distance of 0.61 m (2 ft) from the beam ends and 296.3 mm^2/m (0.14 in.^{2}/ft) for the remaining 9.35 m (30.67 ft) of the beam length. As mentioned previously, the scan detected #10 (#3) vertical bars spaced at 305 mm (12 in.) (equal to 232.8 mm^2/m [0.11 in.^{2}/ft] × 4 = 931.3 mm^2/m [0.44 in.^{2}/ft]), which exceeded the specified area of 698.5 mm^2/m (0.33 in.^{2}/ft).

The south interior and center beams of the original bridge were also scanned their full lengths and both beams showed similar results. Figure 2.10 shows half the scan of an interior beam. As shown in the figure, the steel reinforcement appears as a clear grid pattern near the beam ends and a hazy grid towards midspan. The original 1989 design calculations denoted two layers of welded wire fabric as the shear reinforcement over a distance of 3.05 m (10 ft) from the beam ends and a single layer for the remainder of the beam which matches the scans. However, the bar size estimate was inconclusive due to the mesh. At the end of the wire mesh, the prestressing strands appeared in the scan which

Table 2.2 Preparatory calculations for proof load testing of Bridge 8761

Variable[1]	Exterior	Interior
Concrete compressive strength, f_c' (MPa)	41.37	37.92
Modulus of elasticity, $E \times 10^3$ (MPa)	30.44	29.14
Section modulus at beam bottom, $S_b \times 10^6$ (mm^3)	27.20	28.14
Number of strands	6	12
Area per strand (mm^2)	140.0	98.71
Effective prestress force, P_e (kN)	821.1	1157.9
Eccentricity, e (mm)	423.9	463.0
Cracking moment, M_{cr} (kN-m)[2]	481.5	686.6
Dead load moment, M_{dl} (kN-m)	214.8	199.2
Available moment, $M_{available}$ (kN-m)	266.7	487.4
Estimated diagnostic test truck moment, M_{truck} (kN-m)[4]	205.9	216.0
Type 3 legal load moment, M_{Type3} (kN-m)[4]	321.1	336.6
Available strain, $\varepsilon_{available}$ ($\mu\varepsilon$)[3]	303	562
Estimated diagnostic test truck strain, ε_{truck} ($\mu\varepsilon$)[3]	232	249
Type 3 legal load strain, ε_{Type3} ($\mu\varepsilon$)[3]	362	388

Notes:
[1] 1 MPa = 145.04 psi; 1 mm = 39.370 × 10^{-3} in.; 1 kN-m = 737.56 lb-ft; 1 kN = 224.81 lb.
[2] Computed based on allowable tension of six times square root of f_c'.
[3] Computed one inch from bottom of beam section (location of strain gage).
[4] Computed based on load distribution factor of 0.667 for interior beam and 0.645 for exterior beam, and impact factor of 1.33 (for Type 3 legal load only).

according to the 1989 calculations are bundled with 12.7 mm (0.5 in.) spacing and harped at 40% of the beam length. Since the strands are bundled, there is no clear distance between the strands and thus the scan could not provide an accurate estimate of the size and number of strands. Thus, six bundled strands in the interior beam stem were assumed.

The evaluation of Bridge 8761 served as the pilot study in developing the load rating procedures given in this chapter, and thus diagnostic and proof tests were performed. Preliminary calculations were done in preparation for the diagnostic testing to safeguard against cracking the bridge beams. The cracking moments were computed as 481.5 kN-m (355.1 kip-ft) for the exterior beam and 686.6 kN-m (506.4 kip-ft) for the interior beam using Equation 2.12 given earlier assuming prestress losses of 30% and an allowable tension of six times the square root of f_c'. The available moments for live load were then calculated, which amounted to 266.7 kN-m (196.7 kip-ft) for an exterior beam and 487.4 kN-m (359.5 kip-ft) for an interior beam. Recall that the available moment equals the cracking moment minus the moment due to dead load (214.8 kN-m [158.4 kip-ft] for an exterior beam and 199.2 kN-m [146.9 kip-ft] for an interior beam). From the available moments, available strains were calculated at 303 $\mu\varepsilon$ for an exterior beam and 562 $\mu\varepsilon$ for an interior beam using Equation 2.14. Note that the available strain was calculated using the section modulus corresponding to the location of the strain gage during load testing. In addition, expected diagnostic truck moments were calculated at 205.9 kN-m (151.9 kip-ft) and 216.0 kN-m (159.3 kip-ft) for an exterior and interior beam, respectively. These moments were calculated using the distribution factors mentioned earlier. The corresponding expected strains were 232 $\mu\varepsilon$ and 249 $\mu\varepsilon$ for an exterior and interior beam, respectively. Table 2.2 summarizes the preparatory calculations for Bridge 8761. The diagnostic test of Bridge 8761 provided the opportunity to compare the analytical and measured strains at a lower load level and to determine the critical truck locations and axle weights for proof testing. Single-truck tests and double-truck tests (side-by-side and back-to-back) were conducted. In the first phase, a single truck was positioned along the following paths: center of each interior beam (paths 3, 5, and 7); center of each shear key between beams (paths 2, 4, 6, and 8); and 0.92 m (3 ft) from the edge of deck (paths 1 and 9). Along each path, the front axle of the dump truck was stopped 3.66 m (12 ft) and 4.88 m (16 ft) from the east abutment. In the second phase, two dump trucks were placed side-by-side along paths 1 and 3, 4 and 6, and 7 and 9, and the back axles were stopped at midspan of the bridge. Figure 2.8 shows the second phase of testing, with trucks placed along paths 1 and 3. In the third phase, two dump trucks were placed back-to-back along paths 1, 5, and 9.

The maximum measured strains for the first phase of diagnostic testing were comparable for symmetrical truck paths and equaled 51 µε for paths 4 and 6, 58 µε for paths 3 and 7, and 62 µε for paths 1 and 9. Only path 2 and path 8 showed a small difference in maximum strains (50 µε for path 2 and 60 µε for path 8). Based on these comparisons, it was concluded that the shear keys for Bridge 8761 were functioning. The first phase results showed paths 1, 5, and 9 to be critical since the measured strains were largest for these paths. As a result, the trucks for the double-truck tests were positioned to maximize the strains in the two exterior beams (i.e. beams 1 and 5) and the middle beam (i.e. beam 3). The maximum strains for the interior and exterior beams were 172 µε and 160 µε, respectively, which were approximately 70% of the estimated strains of 249 µε and 232 µε for the second phase of testing. This means that the AASHTO distribution factor was larger than the measured value for Bridge 8761. The maximum strains during the third phase of testing (i.e. back-to-back truck loading) were larger than the second phase (i.e. side-by-side truck loading) only at the exterior beams. Thus, the diagnostic test results for Bridge 8761 provided confidence for the proof load test and showed that single lane loading controls for the exterior beams and multiple lane loading controls for the interior beam.

The proof test consisted of three phases and in each phase, the trucks were transversely positioned to maximize the load effects on beams 1, 3, and 5 (i.e. two exterior beams and center interior beam) as these were the critical beams from the diagnostic test. On average, the test trucks weighed 249.1 kN (56 kips) with axle spacings of 4.42 m (14.5 ft) between the first and second axles, and 1.37 m (4.5 ft) between the second and third axles. To monitor the bridge response, 36 strain gages were installed as shown in Figure 2.8, with 24 gages placed at midspan. Before testing, the beams were evaluated for shear to check the possibility of web-shear cracking under proof loading. The shear analysis of the bridge resulted in an ultimate shear under dead and live load of 225.1 kN (50.6 kips) and 227.3 kN (51.1 kips) for the interior and exterior beams, respectively. Using Equation 14.15, V_{cw} ranged from 252.7 kN (56.8 kips) to 408.8 kN (91.9 kips) for the interior beam and from 236.2 kN (53.1 kips) to 392.8 kN (88.3 kips) for the exterior beam. Thus, concrete cracking due to shear was not expected.

Two trucks were placed side-by-side in phase one with the front of the back axles at various distances from the support bearing for each path. The maximum strain for the interior beam was 264 µε, which is less than the available strain of 562 µε, while the maximum strain for the exterior beam was 213 µε, which is less than the available strain of 303 µε. The weights of the rear axles in the proof test were about 43% larger than the diagnostic test, while the ratio of the maximum strains between the proof and diagnostic tests ranged between 1.36 and 1.51. This shows that the bridge behavior was linear. In phase two, three trucks were placed side-by-side. Compared to phase one, the strains increased in phase two (due to the larger distribution factor) but remained significantly below the

Table 2.3 Maximum measured strains for all phases of proof testing (Bridge 8761)

Transducer Location[1]	Strain(με)								
	Phase 1			Phase 2			Phase 3		
	Paths 1 & 3	Paths 4 & 6	Paths 7 & 9	Paths 1, 3, & 5	Paths 3, 5, & 7	Paths 5, 7, & 9	Paths 1 & 3	Paths 4 & 6	Paths 7 & 9
B1S1	203	22	9	212	36	32	268	−14	7
B1S2	**213**	46	5	**218**	159	66	**323**	48	6
B2S1	**220**	93	11	230	175	136	280	135	17
B2S2	206	186	35	240	231	196	307	230	53
B3S1	201	258	115	**277**	307	**310**	**341**	**355**	154
B3S2	80	**264**	196	258	260	252	163	326	283
B4S1	25	235	**277**	227	**313**	286	67	335	**399**
B4S2	4	139	236	142	211	249	21	195	348
B5S1	−10	66	**209**	66	193	**217**	−2	81	**308**
B5S2	−7	41	141	22	57	160	3	6	209

Note: [1] B#S# = Beam #, Stem #. Beams and stems are numbered from south to north.

Figure 2.11 Instrumentation and proof testing of Bridge 8761

available strains for the exterior and interior beams. Table 2.3 shows the strain results for all phases of testing. Another important finding from phases one and two was that the test moment (per lane) produced by the test trucks was approximately 454.2 kN-m (335.0 kip-ft) which is 89% of the AASHTO Type 3 moment (509.8 kN-m [376.0 kip-ft]) and 71% of the AASHTO HS-20 moment (638.6 kN-m [471.0 kip-ft]). This comparison shows that the target moment could not be achieved with side-by-side trucks only, and thus a larger load was required since the trucks were already loaded as heavily as possible. Consequently, four trucks were placed side-by-side and back-to-back in phase three to reach the target moments. The trucks were placed in the same paths as phase one. Pictures of the instrumentation and proof test of Bridge 8761 are shown in Figure 2.11.

As shown in Table 2.3, the maximum exterior beam strain was 323 με which is slightly larger than the available strain of 303 με. However, the available strain was estimated conservatively using a concrete strength of 41.4 MPa (6000 psi) and an allowable tension equal to 80% of the modulus of rupture. The exterior beam was also inspected carefully after the proof test and no flexural cracks were found. The maximum interior beam strain was 399 µε which is significantly less than the available strain of 562 µε. Note that the moment produced by the test trucks in phase three was approximately 650.8 kN-m (480 kip-ft). This moment is 102% of the HS-20 truck moment (638.6 kN-m [471.0 kip-ft]) and 127% of the Type 3 truck moment (509.8 kN-m [376.0 kip-ft]). Thus, the proof test showed the bridge supported the target load without cracking.

Based on the proof test results reported in Table 2.3, load rating factors were determined for Bridge 8761 under the AASHTO and New Mexico legal loads. The adjusted target live load factor (X_{pA}) and the target proof load (L_T) were determined using Equations 2.16 and 2.17 given earlier. Recall that the proof test was performed based on the limiting concrete tensile stress at service load rather than strength to determine the legal load rating. From Table 2.1, the adjustment factors that applied for Bridge 8761 included +15% for one-lane loaded, −10% for an ADTT less than 1000, and −5% for an in-depth inspection. Thus, the Σ% was 0% for one-lane loading and −15% for multiple-lane loading resulting in X_{pA} values of 1.0 and 0.85, respectively. To load rate the bridge, only multiple-lane loading was considered and an X_{pA} of 0.95 was used rather than 0.85 to limit the reductions. Single-lane loading was ignored since the measured strains from the proof test were well below the available strain for this load case.

Table 2.4 shows the final load ratings for Bridge 8761. Based on the final positions of the test trucks, the operating level (*OP*) moment capacities were computed using Equation 2.18 and equaled 474.5 kN-m (350 kip-ft) for phases one and two, and 684.7 kN-m (505 kip-ft) for phase three. Final load ratings from the proof test are based on phase three which produced the largest moment on the bridge without exceeding the available capacity. The ratings from the proof test all exceeded 1.0 by more than 30% which proved that the bridge was safe under legal loading and posting was not necessary. A legal load rating analysis of Bridge 8761 was also performed for the strength limit state using the AASHTOWare BrR load rating program according to the LFR and LRFR methods. As shown in Table 2.4, the LFR and LRFR ratings are comparable for the interior beam due to similarities in the load distribution factors; however, the LRFR ratings are smaller for the exterior beam since the load distribution factors are larger. Overall, the rating factors for strength (controlled by ultimate flexure) exceeded the values for service (based on concrete cracking) determined from the proof test with the exception of the LRFR ratings for the exterior beam which were slightly smaller. Since the service-based rating factors from the proof test were about equal to the LRFR strength-based rating factors for the exterior beam and smaller

Table 2.4 Proof test (service) and BrR program (strength) rating factors for Bridge 8761

Legal Load	Proof Test	LFR		LRFR	
		Interior Beam	Exterior Beam	Interior Beam	Exterior Beam
HS-20	1.09	1.64	1.24	1.63	1.02
Type 3	1.34	2.14	1.57	2.13	1.28
Type 3S2	1.41	2.35	1.62	2.21	1.33
Type 3–3	1.64	2.57	1.92	2.55	1.57
NM 2	1.79	2.85	2.08	2.82	1.70
NM 3A	1.36	2.20	1.58	2.15	1.29
NM 5A	1.31	2.15	1.51	2.06	1.23

than the LFR ratings, the decision was made to use the service limit state to establish the legal load capacity of the bridge.

2.3.2   Bridge 8825 (box beam)

Bridge 8825 consists of 10 prestressed concrete box beams (section type 5B20) which are 1.52 m (60 in.) wide and 0.51 m (20 in.) deep. Originally built in 1990, the bridge has a span length of 12.8 m (42 ft) and curb-to curb width of 13.41 m (44 ft). The bridge has no concrete deck, a 76 mm (3 in.) asphalt overlay, and concrete sidewalks (254 mm [10 in.] thick, 152 mm [6 in.] wide) that extend 0.91 m (3 ft) past the exterior beams. Overall, the superstructure was in good condition based on the 2011 NMDOT bridge inspection report. Bridge pictures and cross-sections are presented in Figure 2.12.

The AASHTO Manual was used to obtain the material properties of the concrete and prestressing strands along with the 14th edition of the AASHTO *Standard Specifications* (1989) since the bridge was constructed in 1990. For bridges constructed in 1963 and later, the ultimate tensile strength of the prestressing strand, f_{pu}, is specified as 1724 MPa (250 ksi). Accordingly, the prestressing steel was assumed to be Grade 250, seven-wire, low-relaxation strands with 12.7 mm (0.5 in.) diameter. Prestress losses were initially taken as 20% to estimate the number of prestressing strands and then as 25% to calculate the cracking moment (discussed later). Concrete strengths were taken as $f_c' = 34.5$ MPa (5000 psi) and $f_{ci}' = 31.0$ MPa (4500 psi).

From AASHTO Section 9.15.2 (1989), the allowable stresses were equal to the following magnitudes: $f_{bi} = 0.6f_{ci}' = 18.6$ MPa (2700 psi); $f_{ti} = 3(f_{ci}')^{1/2} = 1.39$ MPa (201.2 psi); $f_{bf} = 6$ $(f_c')^{1/2} = 2.93$ MPa (424.3 psi); and $f_{tf} = 0.4f_c' = 13.8$ MPa (2000 psi). The lateral distribution factors for interior concrete box beams are specified as $S/8$ and $S/7$ for one lane and two or more lanes, respectively, where S is the average beam spacing in feet (i.e. 0.625 and 0.714 for $S = 1.52$ m [5 ft]). For exterior beams, the distribution factor is $w_e/8$, where w_e is the exterior beam width in feet which amounted to 0.625. Thus, the multiple lanes distribution factor for an interior beam controlled and was used to compute the live load moments of 305.2 kN-m (225.1 kip-ft), 229.7 kN-m (169.4 kip-ft), and 173.6 kN-m (127.3 kip-ft) for HS-20, H-20, and H-15 truck loading, respectively. The dead load moment equaled 323.8 kN-m (238.8 kip-ft) including the beam self-weight, asphalt overlay, pedestrian sidewalks, and steel guardrails. The sidewalk and guardrail loads were evenly distributed among all beams.

The Magnel diagram under HS-20 loading provided the best fit in terms of prestress and eccentricity resulting in 16 strands with an eccentricity of 200.2 mm (7.88 in.) for the interior and exterior beams. An attempt was made to verify the strand estimate using a rebar scanner. However, due to the large concrete cover and problems with scanning box beams (i.e. adjacent beams allow for only the undersides of interior beams to be scanned), the scans were of limited use. Longitudinal prestressing strands could not be distinguished but the spacing of shear reinforcement was found to be 305 mm (12 in.). The rebar scanner could not detect the stirrup size and thus, #10 (#3) bars were conservatively assumed. Although the strand estimates were not confirmed, only a proof test was conducted for Bridge 8825 due to the good condition of the bridge. Using the prestressing strand estimate, the following bending moments were computed: cracking (610.0 kN-m [449.9 kip-ft]), available (286.1 kN-m [211.0 kip-ft]), target (235.5 kN-m [173.7 kip-ft]), and test truck (211.1 kN-m [155.7 kip-ft] for a single truck in a single lane, 300.4 kN-m

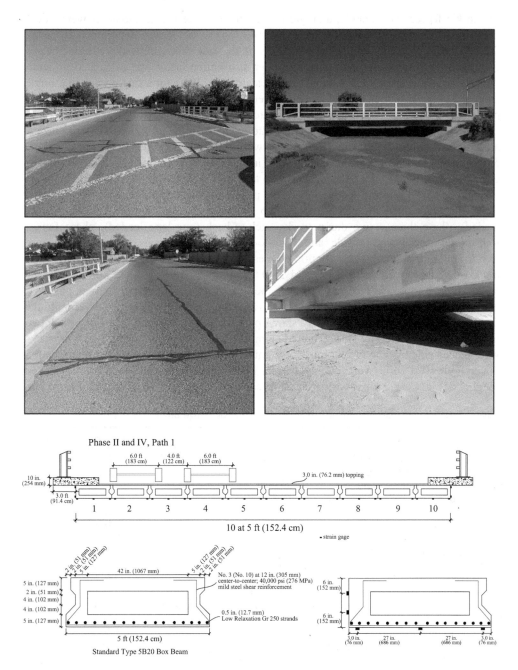

Figure 2.12 Initial inspection pictures, typical cross-sections, instrumentation plan, and critical load paths for Bridge 8825

[221.6 kip-ft] for two trucks in a single lane). Target and test truck moments were calculated using a design distribution factor of 0.429. This factor was determined by using the lever rule on an exterior beam; the size of the sidewalks limits the exterior beams from being directly loaded and thus, a smaller distribution factor was calculated. In addition, the available strain (177 $\mu\varepsilon$) and test truck strains (130 $\mu\varepsilon$ for a single truck in a single lane, 185 $\mu\varepsilon$ for two trucks in a single lane) were computed to monitor the beams during the proof test. Table 2.5 summarizes the preparatory calculations for Bridge 8825.

Strain transducers were mounted on the bottom flanges of the interior and exterior beams at midspan (see Fig. 2.12). Three transducers were installed on the soffit of six beams (1, 3, 5, 6, 8, and 10), while two transducers were installed on four beams (2, 4, 7, and 9). The gages were centered under the beam webs and also centered on the bottom flange of selected beams. The proof test included five phases of loading: phase one – single dump truck (one lane loaded); phase two – two dump trucks side-by-side (multiple lanes loaded); phase three – two dump trucks back-to-back (one lane loaded); phase four – four dump trucks side-by-side and back-to-back (multiple lanes loaded); and phase five – four trucks side-by-side (multiple lanes loaded). For phases one and three, single trucks were placed in paths 1, 2, and 3: 0.61 m (2 ft) from the east curb, along the center of the bridge, and 0.61 m (2 ft) from the west curb, respectively. The truck positions were the same for phases two and four; however, a second truck was placed adjacent to the first truck in paths 1 and 3. Figure 2.12 shows path 1 in phases two and four. For path 2, two trucks were placed side-by-side, centered along the bridge width. In phase five, four trucks were placed side-by-side in two transverse paths: path 1 towards the east curb and path 3 towards the west curb (with roughly 1.22 m [4 ft] between trucks). Each truck weighed roughly 258 kN (58.0 kips) with front and rear axle weights of 66.7 kN (15.0 kips) and

Table 2.5 Preparatory calculations for proof load testing of Bridge 8825

Variable[1]	Interior and Exterior
Concrete compressive strength, f_c' (MPa)	34.47
Modulus of elasticity, $E \times 10^3$ (MPa)	27.79
Section modulus at beam bottom, $S_b \times 10^6$ (mm³)	58.44
Number of strands	16
Area per strand (mm²)	92.90
Effective prestress force, P_e (kN)	1345.1
Eccentricity, e (mm)	200.2
Cracking moment, M_{cr} (kN-m)[2]	610.0
Dead load moment, M_{dl} (kN-m)	323.8
Available moment, $M_{available}$ (kN-m)	286.1
Estimated test truck moment, M_{truck} (one truck)(kN-m)[4]	211.1
Estimated test truck moment, M_{truck} (two trucks, back-to-back)(kN-m)[4]	300.4
Type 3 legal load moment, M_{Type3} (kN-m)[4]	235.5
Available strain, $\varepsilon_{available}$ ($\mu\varepsilon$)[3]	177
Estimated test truck strain, ε_{truck} (single truck) ($\mu\varepsilon$)[3]	130
Estimated test truck strain, ε_{truck} (two trucks, back-to-back) ($\mu\varepsilon$)[3]	185
Type 3 legal load strain, ε_{Type3} ($\mu\varepsilon$)[3]	145

Notes:
[1] 1 MPa = 145.04 psi; 1 mm = 39.370 × 10⁻³ in.; 1 kN-m = 737.56 lb-ft; 1 kN = 224.81 lb.
[2] Computed based on allowable tension of six times square root of f_c'.
[3] Computed at bottom of beam section.
[4] Computed based on load distribution factor of 0.429 for the interior and exterior beams, and impact factor of 1.33 (for Type 3 legal load only).

Figure 2.13 Instrumentation and proof testing of Bridge 8825

Table 2.6 Maximum measured strains for all phases of proof testing (Bridge 8825)

Gage Location[1]	Strain (με)												
	Phase 1			Phase 2			Phase 3		Phase 4			Phase 5	
	Path			Path			Path		Path			Path	
	1	2	3	1	2	3	1	3	1	2	3	1	3
B1G1	69	–	–	107	–	–	114	–	175	–	–	118	–
B1G3	44	–	–	82	–	–	70	–	104	–	–	75	–
B2G1	39	–	–	75	–	–	67	–	98	–	–	74	–
B2G3	40	–	–	79	–	–	47	–	83	–	–	75	–
B3G1	38	–	–	76	–	–	46	–	94	–	–	80	–
B3G3	22	30	–	79	66	–	30	–	91	82	–	96	–
B4G1	50	20	–	105	60	–	63	–	127	85	–	109	–
B4G3	35	29	–	85	66	–	55	–	95	87	–	96	–
B5G1	13	49	–	62	87	–	19	–	76	109	–	114	–
B5G3	10	45	–	41	74	–	10	–	53	102	–	100	–
B6G1	–	48	20	–	85	56	–	24	–	109	56	–	116
B6G3	–	39	24	–	75	66	–	29	–	96	83	–	103
B7G1	–	33	24	–	71	76	–	35	–	94	99	–	108
B7G3	–	16	37	–	63	82	–	60	–	78	116	–	99
B8G1	–	33	29	–	80	69	–	27	–	86	81	–	106
B8G3	–	–	45	–	–	88	–	63	–	–	119	–	109
B9G1	–	–	31	–	–	61	–	47	–	–	83	–	73
B9G3	–	–	59	–	–	87	–	81	–	–	125	–	98
B10G1	–	–	64	–	–	98	–	92	–	–	150	–	114
B10G3	–	–	73	–	–	106	–	110	–	–	176	–	118

Note: [1]B#G# = Beam #, gage #. Beams and gages are numbered from west to east, with three gages per beam.

191.3 kN (43.0 kips), respectively. Pictures of the instrumentation and proof test of Bridge 8825 are shown in Figure 2.13.

In all phases, the test trucks reached their final position, with their rear axles centered at midspan. Table 2.6 shows the maximum bottom flange strains recorded at midspan for all phases. The transducers are labeled B#G# from west to east according to the beam number and gage position (e.g. B3G1 through B3G3 denotes the west, center, and east gages of beam 3). The maximum strains for phase two were 107 $\mu\varepsilon$ (beam 1), 87 $\mu\varepsilon$ (beam 5), and 106 $\mu\varepsilon$ (beam 10). The exterior beam strains were about 85% of the estimated strain (130 $\mu\varepsilon$), which shows the AASHTO live load distribution prediction reasonably matched the measured behavior. The maximum moment per lane produced by the test vehicles in phases one and two was 211.1 kN-m (155.7 kip-ft), which is lower than the target moment of 235.5 kN-m (173.7 kip-ft). This shows that the target moment was not achieved and therefore, a larger load is required. Consequently, two trucks back-to-back and four trucks (side-by-side and back-to-back) were used in the third and fourth phases to reach the target moments. Strains as high as 114 $\mu\varepsilon$ were recorded for phase three (one lane loaded) and 176 $\mu\varepsilon$ for phase four (two lanes loaded). The maximum strain for the latter test phase was very close to the expected strain of 185 $\mu\varepsilon$ and available strain of 177 $\mu\varepsilon$ which again confirms the AASHTO live load distribution prediction and also that the bridge reached the critical load based on concrete tension. Since the bridge was wide enough to accommodate four trucks, the fifth and final loading consisted of four trucks side-by-side. As shown in Table 2.6, the maximum strains under four trucks (phase five) were about equal to the maximum strains under two trucks (phase three). However, both the interior and exterior beams were heavily loaded in phase five, whereas only the exterior beams were heavily loaded in phase three.

Load rating factors were determined for Bridge 8825 from the proof test. The adjusted target load factor, X_{pA}, was 1.0 for one lane loaded and 0.85 for multiple lanes loaded. The distress factor, k_0, was also taken as 1.0 since the proof test was completed without incident. For the different test phases, the applied maximum moments at midspan, L_p, were 603.3 kN-m (445 kip-ft) (phases one and two), 852.8 kN-m (629 kip-ft) (phases three and four), and 593.8 kN-m (438 kip-ft) (phase five). Phases one and three represent one lane loading, and phases two, four, and five represent multiple-lane loading. The final operating rating factors were based on phase four. This test phase had four trucks placed side-by-side and back-to-back, which controlled since the target moment and available strain were achieved simultaneously. Load rating results from the proof test are presented in Table 2.7.

Bridge 8825 was evaluated at the strength limit state for legal loads using the BrR program, results of which are also given in Table 2.7. In both the LFR and LRFR analyses, the rating factors for the interior and exterior beams were the same since there were no differences in the steel reinforcement configurations or load distribution percentages. Strength rating factors were computed under legal load plus pedestrian loading as required for

Table 2.7 Rating factors from proof test (service) and BrR program (strength) for Bridge 8825

Legal Load	Proof Test	LFR		LRFR	
		Interior Beam	Exterior Beam	Interior Beam	Exterior Beam
HS-20	1.17	1.40	1.40	1.44	1.44
Type 3	1.50	1.78	1.78	1.82	1.82
Type 3S2	1.61	1.90	1.90	1.95	1.95
Type 3–3	1.82	2.13	2.13	2.18	2.18
NM 2	2.07	2.40	2.40	2.45	2.45
NM 3A	1.54	1.82	1.82	1.87	1.87
NM 5A	1.46	1.76	1.76	1.81	1.81

bridges with sidewalks. The rating results show that the strength-based rating factors (based on ultimate flexure) exceeded the service-based rating factors (based on concrete cracking) determined from the proof test. This comparison confirmed the optional check at the service limit state controlled the bridge capacity.

2.3.3 Bridge 8588 (I-girder)

Bridge 8588 consists of six prestressed PCI AASHTO Type III I-girders spaced at 1.98 m (6 ft 6 in.) on center with a 15° skew. Built in 1973, the bridge has a span length of 22.6 m (74.1 ft) with a 191 mm (7.5 in) reinforced concrete deck and a 1.75 m (5 ft 2 in.) wide, 0.31 m (12 in.) thick sidewalk on the east side. The bridge width is 11.1 m (36 ft 4 in.) from rail to rail, and 9.94 m (32 ft 7 in.) from west side rail to east side sidewalk. Overall, the bridge is in fair condition based on the 2011 NMDOT bridge inspection report. Bridge pictures and cross-sections are presented in Figure 2.14.

Since the bridge was constructed in 1973, the 10th edition of the AASHO Standard Specifications (1969) was used. Concrete strengths at release (f_{ci}') and 28 days (f_c') were taken as 31.0 MPa (4500 psi) and 34.5 MPa (5000 psi), respectively. The prestressing steel was assumed to be 12.7 mm (0.5 in.) diameter, Grade 270, seven-wire, stress-relieved strands. The value for k, the amount of prestress loss, was estimated from AASHO (1969) Section 1.6.8B as 241.3 MPa (35,000 psi) or 18.5% to determine the amount of prestressing strands and then as 30% to calculate the cracking moment (discussed later). The AASHO allowable stresses equaled $f_{bi} = 0.6f_{ci}' = 18.6$ MPa (2700 psi); $f_{ti} = 3(f_{ci}')^{1/2} = 1.39$ MPa (201.2 psi); $f_{bf} = 3(f_c')^{1/2} = 1.46$ MPa (212.1 psi); and $f_{tf} = 0.4f_c' = 13.8$ MPa (2000 psi). Live load moments were determined for HS-20, H-20, and H-15 truck loading. Impact and wheel distribution factors for live load were computed as 1.25 and 1.18, respectively, according to AASHO (1969) Sections 1.2.12 and 1.3.1.

A typical interior girder of Bridge 8588 was first evaluated by Magnel diagrams assuming both non-composite and composite behavior. Figure 2.15 shows the diagrams for HS-20 loading. The diagram on top assumes non-composite behavior while the diagram on bottom assumes composite behavior. In the top diagram, the area less than (i.e. above) the Equation 2.4 line and greater than (i.e. below) the Equation 2.6 line do not overlap; thus, there is no design region that satisfies the top and bottom girder stresses at transfer and service. As shown in Figure 2.15, however, consideration of composite action resulted in a design region of feasible prestress/eccentricity combinations in the Magnel diagram that satisfy the girder stresses at transfer and service.

The centroid of a standard AASHTO Type III I-girder section is 516 mm (20.3 in.) from the bottom flange. Assuming a 51 mm (2 in.) cover to the center of the bottom strand layer, the maximum eccentricity is 465 mm (18.3 in.). From the Magnel analysis, the girder geometry and span length of Bridge 8825 resulted in a 22-strand estimate for HS-20 loading. Arranging strands into acceptable patterns, the strand estimate and eccentricity calculations were repeated which converged to 24 strands with a 427 mm (16.8 in.) eccentricity.

A rebar scanner was used to measure the size, spacing, and location of the steel reinforcement (prestressed and non-prestressed) in the girders of Bridge 8588. The scanner limitations, in particular the minimum required bar spacing and incapability to detect the steel past the first layer of reinforcement, complicated the measurement of strands. Scans were taken over the bottom flange height of 178 mm (7 in.) to estimate the number of layers of prestressing. The scans clearly showed the presence of strands over a vertical bandwidth of

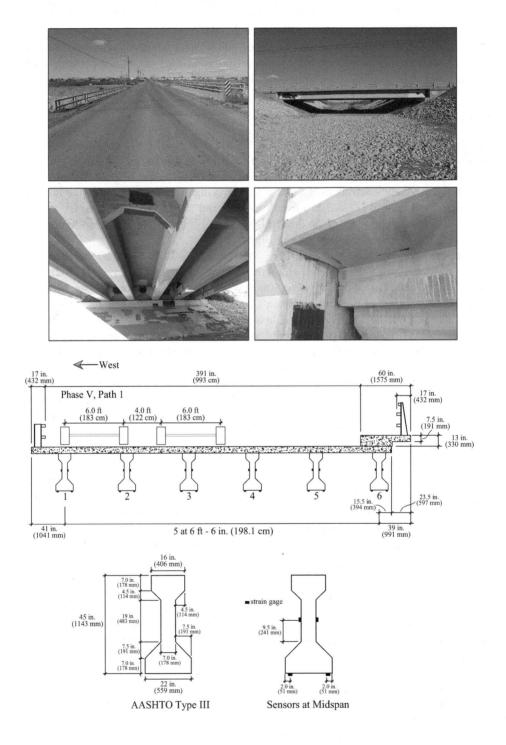

Figure 2.14 Initial inspection pictures, typical cross-sections, instrumentation plan, and critical load paths for Bridge 8588

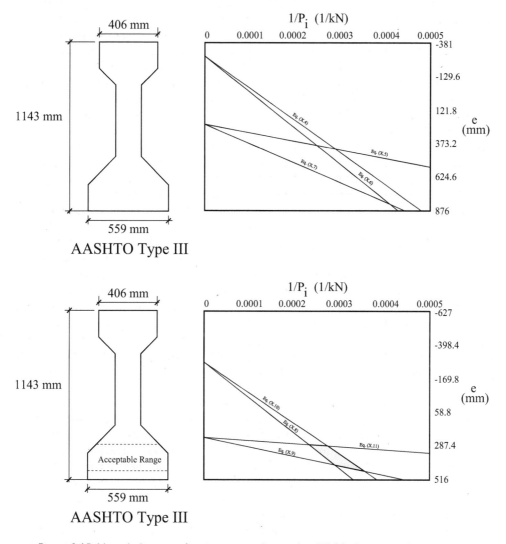

Figure 2.15 Magnel diagrams for interior girder under HS-20 design truck assuming non-composite and composite behavior (Bridge **8588**)

152 mm (6 in.), a possible indication that the bottom three layers have prestressing. For a standard AASHTO Type III girder, the maximum amount of strands in each layer is 10; thus, the Magnel estimate of 24 would require three layers. This partially verified the Magnel estimate but the amount in each layer could not be confirmed. Scans were also taken along the sides of the web near the abutment to check for harped strands and mild shear reinforcement. The scans showed the shear stirrups but no evidence of harping, meaning that some strands were likely debonded. The shear reinforcement was measured to be double-legged #10 (#3) bars spaced at 533 mm (21 in.) near midspan and 229 mm (9 in.) near the abutment. The spacing transitions occurred at a distance of 3.66 m (12 ft) from the abutment.

Due to the good condition of Bridge 8588 and composite girder construction, it was anticipated that the live load distribution factor would be similar or smaller than the design value from the AASHO specifications and therefore, no diagnostic load test was conducted. The available moments and strains were determined for four cases and amounted to: average girder (999.5 kN-m [737.2 kip-ft] and 219 με), interior girder (1110.4 kN-m [819.0 kip-ft] and 244 με), east exterior girder (671.0 kN-m [494.9 kip-ft] and 155 με), and west exterior girder (1061.9 kN-m [783.2 kip-ft] and 232 με). For the average girder case, the composite dead loads (including the wearing surface, sidewalk, and guardrails) were equally distributed to all six girders. In the remaining three cases, loads were applied based on tributary area. For the interior girder case, one-sixth of the wearing surface dead load was applied. Apart from the wearing surface, the east exterior girder was assigned the sidewalk and east guardrail weights and the west exterior girder was assigned the west guardrail weight. As a result, the east exterior girder had the lowest available strain of 155 με, while the interior and west exterior girders both had an available strain of approximately 240 με (note that the average girder case resulted in an available strain of 219 με for all girders). Under truck loading, the expected strains were 272 με for back-to-back trucks and 160 με for one truck in a single lane for the interior and west exterior girder since the live load distribution factors were about the same. Expected strains of 57 με and 96 με were calculated for the east exterior girder for one truck in a single lane and back-to-back trucks, respectively. The lower available and expected strain for the east exterior girder is due to the large sidewalk; the sidewalk provides additional dead load moment to the girder which decreases its available moment, yet it also restricts the girder from direct truck loading and therefore lowers its distribution factor and expected strain. Table 2.8 summarizes the preparatory calculations for Bridge 8588.

A shear evaluation was done to check the I-girders for potential diagonal cracking. Assuming an average truck weight of 267 kN (60 kips), the maximum undistributed shear for a single girder caused by back-to-back truck loading is 133.5 kN (30 kips). Live load distribution factors for an interior and exterior girder equaled 1.38 and 1.29, respectively, using the lever rule. Therefore, ultimate live load shear forces were 185.0 kN (41.6 kips) and 172.1 kN (38.7 kips). Using Equation 2.15, the web-shear cracking force V_{cw} ranged from 541.8 kN (121.8 kips) to 786 kN (176.7 kips) for concrete tensile strengths of two to four times f_c'. The strands were assumed to be straight and debonded, therefore V_p was zero. Under dead and live loads, the maximum shear forces were 387.9 kN (87.2 kips) (west exterior girder), 392.8 kN (88.3 kips) (interior girder), and 518.2 kN (113.5 kips) (east exterior girder) which all were less than 541.8 kN (121.8 kips) and therefore, cracking due to shear was not expected.

Bridge 8588 was instrumented with the majority of strain gages placed at midspan. Each girder had four gages, two on the bottom flange and two on each side of the web. Figure 2.14 shows the overall instrumentation layout and gage placement. The gages were placed in line with the skew. The proof test consisted of five phases: phase one – single dump truck (one lane loaded); phase two – two dump trucks side-by-side (multiple lanes loaded); phase three – three dump trucks side-by-side (multiple lanes loaded); phase four – two dump trucks back-to-back (one lane loaded); and phase five – four dump trucks side-by-side and back-to-back (multiple lanes loaded). In paths 1, 2, and 3, the load was situated on the east edge, center, and west edge of the bridge, respectively, similar to Bridge 8825. Phases one and two were used to determine the critical paths for phases

Table 2.8 Preparatory calculations for proof load testing of Bridge 8825

Variable[1]	Average	Interior	Exterior (East)	Exterior (West)
Girder compressive concrete strength, f_c' (girder) (MPa)	34.47	34.47	34.47	34.47
Deck compressive concrete strength, f_c' (deck) (MPa)	20.68	20.68	20.68	20.68
Girder modulus of elasticity, E (girder) $\times 10^3$ (MPa)	27.79	27.79	27.79	27.79
Deck modulus of elasticity, E (deck) $\times 10^3$ (MPa)	21.53	21.53	21.53	21.53
Composite section modulus at beam bottom, $S_c \times 10^6$ (mm³)	164.1	164.1	156.0	164.6
Non-composite section modulus at beam bottom, $S_{nc} \times 10^6$ (mm³)	101.4	101.4	101.4	101.4
Area of non-composite section, $A \times 10^3$ (mm²)	361.0	361.0	361.0	361.0
Number of strands	24	24	24	24
Area per strand (mm²)	98.71	98.71	98.71	98.71
Effective prestress force, P_e (kN)	2160.9	2160.9	2160.9	2160.9
Eccentricity, e (mm)	426.7	426.7	426.7	426.7
Girder modulus of rupture, f_{cr} (kPa)[2]	2923.4	2923.4	2923.4	2923.4
Non-composite dead load moment, M_{dnc} (kN-m)	1092.5	1092.5	1092.5	1092.5
Composite dead load moment, M_{dc} (kN-m)	187.0	63.99	758.2	107.8
Available moment, $M_{available}$ (kN-m)	999.5	1110.4	671.0	1061.9
Estimated test truck moment, M_{truck} (one truck) (kN-m)[4]	–	730.4	245.5	665.4
Estimated test truck moment, M_{truck} (two trucks, back-to-back) (kN-m)[4]	–	1239.4	416.8	1129.4
Available strain, $\varepsilon_{available}$ (µε)[3]	219	244	155	232
Estimated test truck strain, ε_{truck} (one truck) (µε)[4]	–	160	57	146
Estimated test truck strain, ε_{truck} (two trucks, back-to-back) (µε)[4]	–	272	96	247

Notes:
[1] 1 MPa = 145.04 psi; 1 mm = 39.370 $\times 10^{-3}$ in; 1 kN-m = 737.56 lb-ft; 1 kN = 224.81 lb.
[2] Computed based on allowable tension of 6 times square root of f_c'.
[3] Computed at the bottom of beam section (gage location).
[4] Computed based on a live load distribution factor of 1.18 (interior beam), 0.40 (east exterior beam), and 1.03 (west exterior beam), and a 60 kip dump truck.

four and five, respectively, while phase three was used to maximize the total midspan moment for a multiple lanes loaded condition. Figure 2.14 shows the trucks in path one for phase five. Table 2.9 summarizes the maximum measured strains at the gages located on the bottom flange at midspan. The gages are labeled G#G# representing the girder number and gage position (1 for west, 2 for east). Pictures of the instrumentation and proof test for Bridge 8588 are shown in Figure 2.16.

In the first phase of the proof test, the largest strains were measured at the exterior girder in path 1. This was expected since the truck was positioned 0.61 m (2 ft) from the west rail directly above girder 1. Path 2 showed a symmetrical response between the interior and exterior girders on either side of the bridge centerline (i.e. average strain for girders 1 and 6 = 5 µε, girders 2 and 5 = 29 µε, and girders 3 and 4 = 64 µε). The exterior girders were considered critical since girder 6 (i.e. east exterior girder) had the lowest available strain (155 µε), and girder 1 (i.e. west exterior girder) showed the highest measured strains in phase one (87 µε). In the remaining test phases, path 2 (i.e. dump trucks centered on bridge) was omitted since the exterior girders were more critical. Both paths in phase three resulted in about the same level of strain at the interior girders since the bridge

Table 2.9 Maximum measured strains for all phases of proof testing (Bridge 8588)

Gage Location[1]	Strain (με)								
	Phase 1			Phase 2		Phase 3		Phase 4	Phase 5
	Path 1	Path 2	Path 3	Path 1	Path 3	Path 1	Path 3	Path 1	Path 1
G1G1	68	**14**	0	89	14	86	84	117	141
G1G2	**87**	−4	−4	**113**	−6	**105**	92	**160**	**174**
G2G1	61	31	8	115	44	122	116	107	178
G2G2	**69**	26	6	**118**	35	122	113	**117**	**181**
G3G1	30	64	22	90	79	109	109	53	152
G3G2	43	64	13	106	78	**128**	**128**	61	171
G4G1	15	60	59	49	**102**	101	112	18	98
G4G2	11	**66**	52	44	100	101	112	22	100
G5G1	3	24	57	15	71	64	77	−6	29
G5G2	5	34	**64**	31	87	82	90	6	65
G6G1	−6	−1	**44**	−6	**35**	27	43	−9	−14
G6G2	4	11	34	9	34	34	48	5	24

Note: [1]G#G# = Girder #, gage #. Beams and gages are numbered from west to east, with two gages per beam.

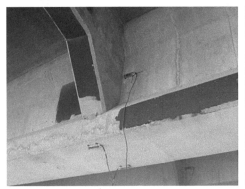

Figure 2.16 Instrumentation and proof testing of Bridge 8588

was fully loaded in the transverse direction with three trucks. In summary, the measured strains in the first three phases of testing were about 50% or more below the available strains, and 16% to 23% smaller than the expected strains.

Based on the measured strains from the first three phases, path 1 was considered the most critical for phases four and five since the west exterior girder and several adjacent interior girders were most heavily loaded simultaneously for this path. Since the truck load could not be positioned directly over girder 6 due to the presence of the sidewalk, the possibility of exceeding the available strain was low, and therefore path 3 was not considered in the final two phases. In phase four, the largest measured strains for the one lane loaded condition were 160 με at the west exterior girder and 117 με at the adjacent interior girder. For the multiple lanes loaded condition (i.e. phase five), the maximum measured strains increased to 174 and 181 με for the exterior and interior girders, respectively, which are roughly 75% of the available strains and 70% of the expected strains. These percentages indicated that the AASHTO distribution factors are conservative. In all five test phases,

Table 2.10 Proof test (service) and BrR program (strength) rating factors for Bridge 8588

Legal Load	Proof Test	LFR		LRFR	
		Interior Girder	Exterior Girder	Int. Girder	Ext. Girder
HS-20	1.17	1.64	1.77	1.80*	1.68**
Type 3	1.60	2.27	2.44	2.46**	2.26**
Type 3S2	1.44	1.93	2.07	2.21*	2.06**
Type 3–3	1.51	1.93	2.07	2.29*	2.15**
NM 2	2.29	3.20	3.43	3.46**	3.19**
NM 3A	1.68	2.38	2.58	2.58**	2.38**
NM 5A	1.62	1.91	2.05	2.27*	2.17**

Note: LRFR factors are controlled by shear strength (*) or flexural strength (**.)

the available strains were well above the measured strains, and thus the loading was not stopped until the trucks reached the final positions.

Based on the proof test results, load rating factors were determined for Bridge 8588. The adjustment factors for X_p were +15% for one lane loaded and −5% for an in-depth inspection. Thus, the final adjusted factors were 1.1 and 0.95 for one lane and multiples lane loaded, respectively. Phase five (i.e. four trucks, side-by-side and back-to-back) was used to determine the final load ratings since all trucks reached their final positions and measured strains were less than available strains by 25%. The applied moment (L_p) corresponding to the final loading in phase five was 2077 kN-m (1532 kip-ft). Using a k_0 factor of 1.0, OP amounted to 2186 kN-m (1612 kip-ft). Table 2.10 lists the final rating factors for Bridge 8588. For all legal loads, the rating factors exceed 1.0 by a significant margin indicating the bridge is safe under legal loading and no posting is necessary. Table 2.10 provides the LFR and LRFR legal load rating results for the strength limit state. The rating factors were different for the interior and exterior girders primarily due to differences in load distribution, and were controlled by shear strength for LFR and shear or flexural strength for LRFR (see note given in Table 2.10). The service-based rating factors from the proof test were all significantly lower than the strength-based rating factors from the BrR program. Consequently, the bridge capacity for legal loads was conservatively set based on the service limit state.

2.4 Conclusions

This chapter focused on the development of load rating procedures for prestressed concrete bridges with no design plans. The procedures involve the following steps: estimation of steel reinforcement using Magnel diagrams in accordance with the applicable design specifications and the AASHTO Manual; detection and confirmation of steel reinforcement estimate with a rebar scanner; diagnostic and/or proof load testing and strain measurements; load rating for serviceability using the proof test results according to the AASHTO Manual; and load rating for strength using a load rating program based on the LFR and LRFR methods. Three case studies were presented including a double T-beam, box beam, and I-girder bridge to illustrate the load rating procedures. Attention was given to prestressed concrete bridges since the evaluation procedures for reinforced concrete and steel bridges without design plans are much better established.

It is important to note that even though the AASHTO Manual was generally used in this study, there are no specific guidelines in the Manual that address the load rating of prestressed concrete bridges through proof load testing. The Manual provides guidance for proof testing based on strength but not serviceability. Although common testing procedures were employed herein, the procedures were adapted to the service limit state for concrete tension, which required a unique approach not given in the AASHTO Manual. In addition, the literature review done prior to this evaluation uncovered just a few publications (mainly reports from state DOTs or consultants) that gave minimal direction for the load rating of bridges via proof testing, in particular prestressed concrete bridges without design plans.

References

American Association of State Highway and Transportation Officials (AASHTO) (1983) *Standard Specifications for Highway Bridges*, 13th edition. AASHTO, Washington, DC.

American Association of State Highway and Transportation Officials (AASHTO) (1989) *Standard Specifications for Highway Bridges*, 14th edition. AASHTO, Washington, DC.

American Association of State Highway and Transportation Officials (AASHTO) (2002) *Standard Specifications for Highway Bridges*, 17th edition. AASHTO, Washington, DC.

American Association of State Highway and Transportation Officials (AASHTO) (2011) *Manual for Bridge Evaluation*, 2nd edition. AASHTO, Washington, DC.

American Association of State Highway Officials (AASHO) (1969) *Standard Specifications for Highway Bridges*, 10th edition. AASHTO, Washington, DC.

Anay, R., Cortez, T. M., Jáuregui, D. V., ElBatanouny, M. K. & Ziehl, P. (2016) On-site acoustic-emission monitoring for assessment of a prestressed concrete double-tee-beam bridge without plans. *Journal of Performance of Constructed Facilities*, 30(4), 1–9 (04015062).

Bernhardt, R. P. & DeKolb, S. P. (2003) *Inspection and Load Testing of Three Bridges*. Final Report, Project No. DE-01739-2P. US Army Signal Center, Fort Gordon, GA.

Bernhardt, R. P., DeKolb, S. P. & Kwiatkowski, T. L. (2004) *Proof Testing of Reinforced Concrete Bridges: A Useful Way to Provide Load Rating*. TECHBrief 2004 No. 3. Burns & McDonnell. pp. 3–5.

Casas, J. R., Olaszek, P., Sajna, A., Znidarie, A. & Lavric, I. (2009) *Sustainable Development, Global Change and Ecosystems (Sustainable Surface Transport): Deliverable D16: Recommendations on the Use of Soft, Diagnostic, and Proof Load Testing*. Document No. ARCHES-02-DE16. Assessment and Rehabilitation of Central European Highway Structures (ARCHES) Management Group.

Chajes, M. J. & Shenton, H. W. (2006) Using diagnostic load tests for accurate load rating of typical bridges. *Journal of Bridge Structures*, 2(1), 13–23.

Commander, B., Valera-Ortiz, W., Stanton, T. R. & Diaz-Alvarez, H. (2009) *Field Testing and Load Rating Report, Bridge FSBR-514, Fort Shafter, Hawaii*. Final Report No. ERDC/GSL TR-09-9. US Army Corp of Engineers, Arlington, VA.

Conner, G. H., Stallings, J. M., McDuffie, T. L., Campbell, J. R., Fulton, R. Y., Shelton, B. A. & Mullins, R. B. (1997) *Steel Bridge Testing in Alabama*. Transportation Research Record, No. 1594. pp. 134–139.

Ellingwood, B. R., Zureick, A. H., Wang, N. & O'Malley, C. (2009) *Condition Assessment of Existing Bridge Structures: Report of Task 4: Development of Guidelines for Condition Assessment, Evaluation and Rating of Bridges in Georgia*. Report of GTRC Project No. E-20-K90 and GDOT Project No. RP05-01. Georgia Department of Transportation and Georgia Institute of Technology, Atlanta, GA.

Hag-Elsafi, O. & Kunin, J. (2006) *Load Testing for Bridge Rating: Dean's Mill over Hannacrois Creek*. Report FHWA/NY/SR-06/147, Transportation Research and Development Bureau, New York State Department of Transportation.

Hawkins, N. M., Kuchma, D. A., Mast, R. F., Marsh, M. L. & Reineck, K.-H. (2005) *Simplified Shear Design of Structural Concrete Members*. Final Report, NCHRP Report 549. National Cooperative Highway Research Program, Washington, DC.

Jeffrey, A., Breña, S. F. & Civjan, S. A. (2009) *Evaluation of Bridge Performance and Rating through Non-Destructive Load Testing*. Final Report, Research Report 2009-1. Vermont Agency of Transportation.

Kissane, R. J., Beal, D. B. & Sanford, J. A. (1980) *Load Rating of Short-Span Highway Bridges*. Interim Report on Research Project 156–1, Research Report 79 (US DOT/FHWA). Engineering Research and Development Bureau, New York State Department of Transportation.

Lichtenstein, A. G. (1993) *Bridge Rating through Nondestructive Load Testing*. Final Draft Report for NCHRP Project 12-28(13)A. Transportation Research Board, National Research Council, Washington, DC.

O'Malley, C., Wang, N., Ellingwood, B. R. & Zureick, A.-H. (2009) *Condition Assessment of Existing Bridge Structures: Report of Tasks 2 and 3: Bridge Testing Program*. Report of GTRC Project No. E-20-K90 and GDOT Project No. RP05-01. Georgia Department of Transportation and Georgia Institute of Technology, Atlanta, GA.

Pinjarkar, S. G., Guedelhoefer, O. C., Smith, B. J. & Kritzler, R. W. (1990) *Nondestructive Load Testing for Bridge Evaluation and Rating*. Final Report for NCHRP Project 12-28(13). Transportation Research Board, National Research Council, Washington, DC.

Saraf, V., Sokolik, A. F. & Nowak, A. S. (1996) *Proof Load Testing of Highway Bridges*. Transportation Research Record, No. 1541. pp. 51–57.

Shahawy, M. A. (1995) Nondestructive Strength Evaluation of Florida Bridges. *Proceedings of SPIE: The International Society of Optical Engineering (Nondestructive Evaluation of Aging Bridges and Highways), Society of Photo-Optical Instrumentation Engineers*, Volume 2456. pp. 101–123.

Shahawy, M. E. & Garcia, A. M. (1990) *Structural Research and Testing in Florida*. Transportation Research Record, No. 1275. pp. 76–80.

Shenton, III, H. W., Chajes, M. J. & Huang, J. (2007) *Load Rating of Bridges without Plans*. Final Report, Research Report DCT 195. Delaware Center for Transportation.

Wang, N., Ellingwood, B. R., Zureick, A.-H. & O'Malley, C. (2009) *Condition Assessment of Existing Bridge Structures: Report of Task 1: Appraisal of State-of-the-Art of Bridge Condition Assessment*. Report of GDOT Project No. RP05–01. Georgia Department of Transportation, Atlanta, GA.

Wang, N., O'Malley, C., Ellingwood, B. R. & Zureick, A.-H. (2011a) Bridge rating using system reliability assessment, I: Assessment and verification by load testing. *ASCE Journal of Bridge Engineering*, 16(6), 854–862.

Wang, N., Ellingwood, B. R. & Zureick, A.-H. (2011b) Bridge rating using system reliability assessment, II: Improvements to bridge rating practices. *ASCE Journal of Bridge Engineering*, 16(6), 863–871.

Wipf, T. J., Phares, B. M., Klaiber, F. W., Wood, D. L., Mellingen, E. & Samuelson, A. (2003) *Development of Bridge Load Testing Process for Load Evaluation*. Final Report on Iowa DOT Project TR-445 / CTRE Project 00-65, Research Report TR-445. Iowa Highway Research Board, Iowa Department of Transportation.

Chapter 3

Example of Proof Load Testing from Europe

Eva O. L. Lantsoght, Dick A. Hordijk, Rutger T. Koekkoek, and Cor van der Veen

Abstract

This chapter describes the proof load testing of viaduct Zijlweg at a position that is critical for bending moment and at a position that is critical for shear. The viaduct Zijlweg has cracking caused by alkali-silica reaction, and the effect of material degradation on the capacity is uncertain. Therefore, the assessment of this viaduct was carried out with a proof load test. This chapter details the preparation, execution, and evaluation of viaduct Zijlweg. The outcome of the proof load test is, according to the currently used methods for proof load testing, that the viaduct fulfills the code requirements, and that strengthening or posting is not required.

3.1 Introduction to viaduct Zijlweg

3.1.1 Existing bridges in the Netherlands

Many of the existing bridges in the Netherlands were built in the decades of the European reconstruction after the Second World War. These structures are now approaching the end of the service life they were originally designed for. Additionally, the currently governing codes are prescribing larger live loads (representing the large traffic volumes and intensities as compared to the codes that were used at the time of design of these structures) and some-times lower capacities (e.g. the shear capacity according to the Dutch NEN 6720:1995 [Code Committee 351001, 1995] is higher than according to the NEN-EN 1992–1–1:2005 [CEN, 2005]). As a result, the capacity of some existing bridges is subject of discussion (Walraven, 2010). Other sources of uncertainty are the effect of material degradation or uncertainty about the available reinforcement when as-built plans are missing.

For the assessment of a bridge by calculation, different Levels of Approximation can be used. Each higher level requires more computational time but also gives a value that is expected to be closer to the experimental (real) value, although still keeping a margin of safety into account (fib, 2012). For reinforced concrete slab bridges in the Netherlands, the assessment on the first Level of Approximation is a hand calculation (Lantsoght et al., 2013; Vergoossen et al., 2013). If this calculation shows that the capacity of the bridge is sufficient, then it is not necessary to go to higher Levels of Approximation. If the calculation shows that the capacity is not sufficient, there is no indication that the real capacity is insufficient. A more refined calculation technique should then be used. For the assessment of bridges at Level of Approximation 2, a linear finite element calculation is

DOI: https://doi.org/10.1201/9780429265969

used (Lantsoght et al., 2017a). If an even higher Level of Approximation is necessary, non-linear finite element calculations need to be made. Further Levels of Approximation include probabilistic analyses and coupling of nonlinear finite element techniques with spatial variability of materials (Li et al., 2004) and the measured distribution of loads.

If there are still uncertainties to determine if a given structure can carry the code-prescribed loads (e.g. when the effect of concrete deterioration over time is unknown), then proof loading can be applied. In such a test, sufficient structural capacity of the bridge is proven by demonstrating that it can withstand a calculated proof load without causing irreversible (additional) damage. The proof loading can be applied with respect to different safety levels: disapproval level, reconstruction level, and the design level, as prescribed in the Dutch code NEN 8700:2011 (Normcommissie 351001, 2011a). This Dutch code describes different safety levels, with associated different load factors, calibrated for different values of the reliability index (Steenbergen and Vrouwenvelder, 2010). For existing bridges and their associated consequence class, these requirements and the added "usage" level are prescribed in the Guidelines for Assessment for Existing Bridges (Richtlijnen Bestaande Kunstwerken, RBK) (Rijkswaterstaat, 2013). The load factors, reliability index β, and reference period ("ref period") are shown in Table 3.1, which shows the values used in NEN-EN 1990:2002 (CEN, 2002), NEN 8700:2011 (Code Committee 351001, 2011), and the RBK (Rijkswaterstaat, 2013). The load factors that are listed in Table 3.1 are those used for the assessment of existing bridges: γ_{DC} for the self-weight, γ_{DW} for the superimposed dead load, and γ_{LL} for the live loads.

Proof load testing can be an interesting assessment method for the existing bridges in the Netherlands, and the technique can be used for different bridge types, owned by municipalities, provinces, and the Ministry of Infrastructure and the Environment. When the technique of proof load testing will be standardized and streamlined, it can become a fast and economic method for the direct evaluation of existing bridges. To develop standardized procedures, a number of open research questions need to be answered. Therefore, a number of pilot proof load tests have been carried out in the Netherlands (Lantsoght et al., 2017e). The testing of viaduct Zijlweg (Lantsoght et al., 2017c), demonstrated in this chapter, formed part of this campaign of pilot proof load tests.

The difficulty in assessing viaduct Zijlweg is caused by the uncertainty on the capacity. The superstructure of viaduct Zijlweg is affected by alkali-silica reaction (ASR). The effect of ASR on the shear capacity of concrete elements is subject to discussion. Using a conservative approach to take into account the effect of ASR resulted in an insufficient assessment for the bridge. Therefore, viaduct Zijlweg is an excellent candidate for a pilot proof load test to investigate if the code requirements are fulfilled. The failure modes that were assessed through this load test were flexure and shear. Proof load testing for shear is a novel application, as this brittle failure mechanism requires additional preparations.

Table 3.1 Considered safety levels governing in the Netherlands

Safety level	β	Ref period	γ_{DC}	γ_{DW}	γ_{LL}
Eurocode Ultimate Limit State	4.3	100 years	1.25	1.35	1.50
RBK Design	4.3	100 years	1.25	1.25	1.50
RBK Reconstruction	3.6	30 years	1.15	1.15	1.30
RBK Usage	3.3	30 years	1.15	1.15	1.25
RBK Disapproval	3.1	15 years	1.10	1.10	1.25
Eurocode Serviceability Limit State	1.5	50 years	1.00	1.00	1.00

The viaduct Zijlweg is representative of the viaducts affected by ASR in the Netherlands. In total, between 40 and 50 structures (bridges, tunnels, and locks) have material degradation caused by ASR (Nijland and Siemes, 2002). The aggregates that contain reactive silica leading to ASR are porous chert, chalcedony, and impure sandstones that often react as porous chert. Viaduct Zijlweg forms part of a series of bridges in and over highway A59, built during the 1960s and 1970s, which are all affected by ASR. A number of the viaducts with ASR damage are posted, strengthened, and/or monitored (Borsje et al., 2002). For viaduct Zijlweg, monitoring is used. Maintenance activities focus on preventing the ingress of moisture so that the swelling of the ASR gel is stopped. Monitoring activities focus on the moisture content and deformations and aim at assessing the effectiveness of the rehabilitation strategies.

3.1.2 *Viaduct Zijlweg*

3.1.2.1 *General information and history*

The Ministry of Infrastructure and the Environment commissioned TU Delft to perform a proof load test on the viaduct Zijlweg (Fig. 3.1) over highway A59 in the Zijlweg between Waspik and Raamsdonksveer (see Fig. 3.2). Viaduct Zijlweg is a reinforced concrete slab bridge with four spans. The spans are supported by concrete piers at the mid supports and by an abutment at the end supports. The forces from the deck are transferred via elastomeric bearing pads to the piers and abutments. It was built in 1965 for the province of Noord Brabant with a design life of 80 years.

Figure 3.1 Viaduct Zijlweg in the highway A59 in the province Noord Brabant

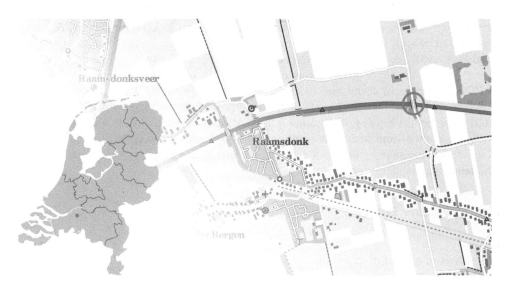

Figure 3.2 Location of viaduct Zijlweg

Viaduct Zijlweg is one of a series of viaducts in and over the A59 highway with alkali-silica reaction (ASR). This viaduct was field tested as the influence of ASR on the capacity was uncertain and complicated the analytical assessment. In 1997, ASR was detected in the viaduct Zijlweg (Projectteam RWS/TNO Bouw, 1997). Material testing at that time led to the observation that the uniaxial tensile strength is very small, which, in addition to the large number of cracks observed, resulted in concerns with regard to the shear capacity. As typical for reinforced concrete slab bridges, the structure does not contain shear reinforcement. Moreover, the bridge was designed for class 30 from VOSB 1963 (Nederlands Normalisatie Instituut, 1963) but is currently in use as class 45.

In 2002, the most recent renovation of viaduct Zijlweg was done (Rijkswaterstaat, 2002). At that time, the concrete of the main superstructure was repaired with cement-bound polymer-modified mortar, the damage caused by ASR was inspected, the concrete at the support was repaired, the asphalt layer was inspected and replaced, the joint at the end support was replaced and inspected, maintenance was given to the handrail, the concrete of the sidewalk was repaired, and all other concrete elements were inspected. According to the maintenance and management plan of viaduct Zijlweg, the main superstructure and supports should be inspected every five years. The joints are to be given maintenance every 20 years and the handrail every 12 years. The waterproofing layer of the concrete should be replaced every 12 years.

3.1.2.2   *Material properties*

The concrete compressive strength is determined based on cores drilled from the viaduct (Witteveen+Bos, 2014). The average cube concrete compressive strength was found to be $f_{c,cube,m}$ = 44.4 MPa (6.4 ksi) and the characteristic concrete cylinder compressive strength is f_{ck} = 24.5 MPa (3.6 ksi). Experimental results about the properties of the reinforcement steel are not available. The symbols on the reinforcement drawings indicate that

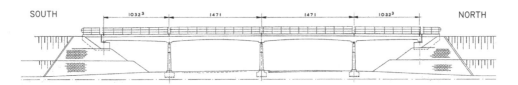

Figure 3.3 Structural system of viaduct Zijlweg. Units: cm. Conversion: I cm = 0.4 in.

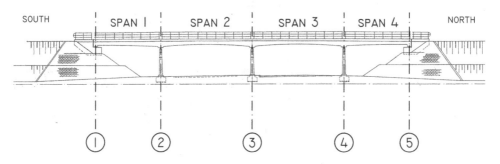

Figure 3.4 Numbering of the spans and supports

plain bars QR22 or QR24 (not specified which steel grade) were used. For QR22, the characteristic yield strength is $f_{yk} = 220$ MPa (32 ksi), and for QR24, $f_{yk} = 240$ MPa (35 ksi).

3.1.2.3 *Structural system and description of tested span*

Viaduct Zijlweg crosses the highway under an angle of 75.6°, resulting in a skew of 14.4°. The length of the end spans is 10.32 m (34 ft) and the length of the central spans is 14.71 m (49 ft), as can be seen in Figure 3.3. Figure 3.4 shows the numbering of the spans and the location of the tested span. Span 4 was chosen because it is not directly above the highway A59, so that closing of the highway is not necessary during the proof load test. The width of the viaduct is 6.60 m (22 ft) and the width of the carriageway is 4.00 m (13 ft). Only a single lane is available, and vehicles need to take turns in entering the viaduct. A sidewalk of 1.3 m (4 ft) wide is located on both sides (see Fig. 3.5).

Span 4 has a variable thickness. The thickness is 550 mm (22 in) at support 5 and increases to 850 mm (33 in) at support 4. The thickness varies parabolically with a radius of curvature $r = 150$ m (495 ft) (see Fig. 3.6). The layout of the reinforcement in span 4 is shown in Figure 3.7.

3.2 Preparation of proof load test

3.2.1 *Preliminary assessment*

In the Netherlands, an assessment is carried out based on the Unity Check, which is the ratio of the factored load effect to the factored capacity. If the Unity Check is larger than 1, the cross section is found not to fulfill the code requirements. For span 4 of

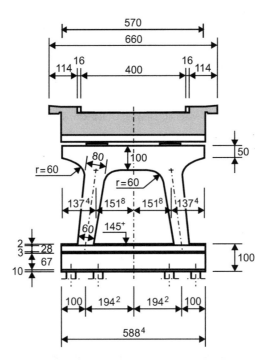

Figure 3.5 Sketch of cross section. Units: cm. Conversion: I cm = 0.4 in.

viaduct Zijlweg, an initial assessment resulted in a Unity Check of 5.4 at the end support for shear. Therefore, in accordance to the approach with different levels of assessment (Lantsoght et al., 2017b), a refined calculation was carried out. The tensile strength was determined based on 51 specimens in uniaxial tension and 10 specimens in splitting tension, and finite element calculations were used to determine the acting sectional shear. The highest Unity Check with this refinement was UC = 1.31 at support 4 (see Fig. 3.4) (Projectteam RWS/TNO Bouw, 1997). For this purpose, the viaduct Zijlweg was considered a good candidate for a proof load test.

3.2.2 Inspection

The inspection that was planned for 2007, was carried out on 18 April 2008. A visual inspection was used to fine-tune the plan for future repair and inspections. The analysis of the inspection was based on risk (see Table 3.2). Based on the calculated risk, it is decided whether a given risk is acceptable or not. For unacceptable risks, actions for management and maintenance are proposed. In the conclusions of the report, the risk associated with the insufficient capacity of the superstructure was emphasized. The viaduct was considered to be in moderate conditions. Maintenance actions were proposed for 2019 (replacement of asphalt upper layer, repair of deck, sidewalk, and support, cleaning of bearings, and conservation of steel handrails) and 2025 (replacement of full asphalt layer). From the analysis of the ASR measurements, it was concluded that the expansion in the longitudinal direction was reaching its maximum value.

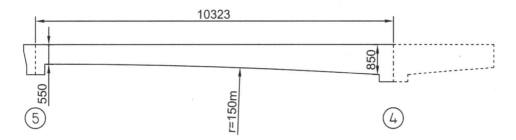

Figure 3.6 Longitudinal view of span 4

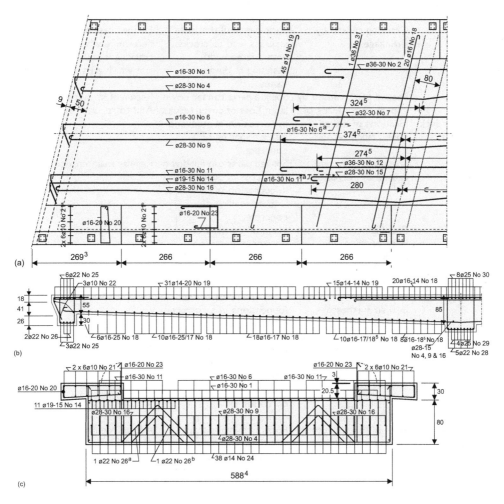

Figure 3.7 Reinforcement in span 4: (a) top view; (b) side view; (c) crossbeam. Units: bar diameters in mm and other dimensions in cm. Conversion: 1 mm = 0.04 in., 1 cm = 0.4 in.

Source: Reprinted with permission from Lantsoght et al. (2017c).

Table 3.2 Overview risks and recommendations from inspection 2008

Bridge part	Problem	Explanation	Recommendation
Entire structure	Assumptions of design do not correspond with current use	Increase of traffic intensity	Try to find out traffic load class Assessment through recalculation of structural capacity
	Assumptions of design do not correspond with current use	Change in usage	No action required
	Design errors	Design errors	No action required
Concrete general	Cracking caused by ASR	Swelling causes disintegration of concrete, resulting in reduced capacity	Monitoring is in place No further actions required
	Cracking caused by shrinkage	Shrinkage and thermal stresses lead to cracks with a large width, resulting in durability issues	Further research to determine which part of cracking is caused by ASR and which part by shrinkage
Main load-carrying structure: deck and supports	Spalled parts caused by impact, loose filling of locations of drilled cores, failed repair	Spalled parts can fall down later. Loss of cover results in corrosion of steel and further spalling of concrete.	Repair of impact damage Add to regular maintenance
	Spalling of concrete caused by corrosion	Spalled parts can fall down later. Loss of cover results in corrosion of steel and further spalling of concrete.	Regular maintenance
Handrail	Possible accidents from large openings in handrail	Handrail consists of horizontal parts that children can use to climb and fall through	Attention owner of structure
Supports	Elastomeric bearings show cracking and are in their furthest position	Support is not able to carry the loads to the foundations in a correct manner	Possibly caused by ASR Same actions as for ASR
Sidewalk	Spalling, loose filling of cores, shrinkage cracks, poor compaction of concrete	Increased deterioration from deicing salts, caused by poor maintenance of the concrete surface. Possibility for future corrosion of reinforcement.	Add to regular maintenance
Joint (steel + rubber)	Steel of joint is rusted and rubber has been damaged	Possible damage from chlorides in structure and damage of traffic by loosening of finger joint. Damage attributed to traffic loads. Water tightness and safety are compromised.	Add to regular maintenance

Figure 3.8 Top deck

Figure 3.9 Bottom of deck

A visual inspection was carried out during the preparation stage of the proof load test. Figure 3.8 shows the deterioration of the top deck, present in the pedestrian sidewalk. Figure 3.9 shows an impression of the bottom of the deck. The typical cracking pattern of ASR (map cracking) is not clearly visible from the distance at which the inspection was carried out. Closer observations on scaffolding right before the execution of the proof load test allowed for the drawing of the full crack map (see Fig. 3.10). Figure 3.11 shows the small distance remaining in the joint, resulting from the expansion in the longitudinal direction.

3.2.3   Effect of alkali-silica reaction

3.2.3.1   Effect of alkali-silica reaction on capacity

Alkali-silica reaction (ASR) is the reaction between the reactive (glasslike) silica that is present in aggregates from some sources and alkali that is present in cement paste (Neville, 2012). The result of this reaction is a gel of calcium silicate hydrate. Upon contact with water, this gel will expand, and this expansion causes stresses in the concrete. If these stresses exceed the tensile strength of the concrete, cracking occurs. The expansion and resulting internal stresses in the concrete are influenced by the geometry and structure. For example, the existence of reinforcement limits the ASR expansion in the direction of the bars. In the vertical direction where no reinforcement is present, the expansion will not be limited. As a result, cracking occurs parallel to the reinforcement.

The effect of ASR on the concrete strength is typically expressed as a reduction on the strength as a function of the amount of free expansion of the concrete (Siemes et al., 2002). In the literature, the reduction of the tensile strength due to ASR is reported to vary between 5% (Ahmed et al., 1999) and 82% (Siemes et al., 2002). Research (Siemes et al., 2002) showed that the uniaxial tensile strength is lower than the expected value based on the compressive strength determined from testing cores or derived from the splitting tensile strength. The ratio of uniaxial tensile strength to the splitting tensile strength tested on core samples varied between 0.18 and 0.77; with an average of 0.5 for an expansion of 0.5‰ to 1.0‰ (Siemes et al., 2002). These high values, however, result from uniaxial tension tests on cracked specimens, in which the stress flow is disrupted. Moreover, the existing cracking reduces the area. Therefore, the use of uniaxial tension tests on specimens affected by ASR may not be advisable. While generally the tensile strength of concrete is neglected in design calculations, it plays an important role in the bond and anchorage strength, the shear and punching strength of members without shear reinforcement, the resistance against splitting caused by dowel action, and the load transfer in plain concrete elements such as footings. Low uniaxial tensile strengths are not unique for structures with ASR damage, but have also been found in relatively old structures where no ASR is present (Siemes et al., 2002; Yang et al., 2010).

Since bending moment capacity is governed by crushing of the concrete (failure of concrete in compression), ASR is expected not to have an influence on the bending moment capacity. Experiments showed that no clear conclusions on the influence of ASR on the concrete compressive strength can be drawn (Giaccio et al., 2008). One series of experiments (Talley, 2009) confirms that no reduction in the moment capacity occurs, unless a high expansion has taken place, in which case 25% reduction of the bending moment capacity was found. Another comparison between a reinforced concrete member with and without ASR damage on a small scale (Haddad et al., 2008) found that the flexural capacity of a specimen with ASR damage was lower than the specimen without damage. The reduction of the capacity due to ASR damage was 11%. The serviceable load and flexural stiffness of the ASR-damaged element were increased. The precompression of the

Figure 3.10 Observed cracking pattern, showing cracking on bottom face of slab and side faces

Source: From Lantsoght (2017), reprinted with permission.

Figure 3.11 Expansion joint at the south side of viaduct Zijlweg

concrete on the tension side delayed the crack initiation in the ASR beam as compared to the control beam. Once the precompression was offset at the tension side, during loading, the shift of the neutral axis toward the compression zone would be faster in the ASR-damaged specimens as compared to that of the control specimens because of the lower splitting strength of the damaged concrete and the "prestress" in the compression zone. The flexural failure of the ASR-affected beams was more abrupt than that of the unaffected beams.

Since shear failure is related to the tensile capacity of the concrete, ASR is expected to reduce the shear capacity. On the other hand, the precompression in the direction of the reinforcement may reduce or even revert this reduction as it increases the aggregate inter-lock capacity of the beams when the crack width is decreased. Measurements by (Ahmed et al., 1998) showed that the introduction of compression reinforcement reduced the expansion from ASR. ASR caused an irreversible upward deflection of the beam that was largely proportional to the amount of expansion measured in the beams. In fatigue testing, a larger fatigue life was found for the specimens affected by ASR. In an overview of the literature, Schmidt et al. (2014) found that sometimes experiments indicate that ASR cracks lead to a reduction of the shear capacity and sometimes do not lead to a reduction of the shear

Figure 3.12 Testing of beams with ASR damage in the Stevin laboratory (den Uijl and Kaptijn, 2002)

Source: Photograph by J. A. Den Uijl, used with permission.

capacity. In an exceptional case, an increase of 78% of the shear capacity was observed. When carrying out shear tests on bridge deck slabs, the measured capacity was found to exceed the calculated capacity for most of the cases (depending on the assumptions for the cross-sectional height and depending on the position along the bridge). More than anything, the test results open the question whether the compressive strength from core samples provides the right basis for evaluation of the shear capacity (Schmidt et al., 2014). Most experiments on the effect of ASR on shear have been carried out on beams of small size that were cast in the laboratory. Experiments on beams sawn from two 35-year-old viaducts with ASR damage are available in the literature (see Fig. 3.12) (den Uijl and Kaptijn, 2002). Material testing on the bridges from which the beams were sawn showed a small uniaxial tensile strength. Comparing the experimental test results to the predicted value from Rafla's equation (Rafla, 1971) showed that for the beams that failed in shear (from the second viaduct), a tested to predicted ratio of 0.77 (with a coefficient of variation of 7.5%) was found. In the conclusions, the recommendation was given to reduce the shear capacity of reinforced concrete members affected by ASR by 25%. This recommendation is used in the Netherlands to determine the shear capacity of ASR-affected members. Note that this recommendation is based on a comparison to a formula derived from a shear database, and that this formula possibly does not correctly take the size effect into account. A last series in which the effect of low direct tensile strength was studied on old and new concrete showed a ratio of uniaxial tensile strength to splitting tensile strength $f_{ct}/f_{ct,spl}$ of 47% for old concrete taken from an existing viaduct and 70% for beams cast in the lab (Yang et al., 2010). Upon testing, no clear differences between the old and new beams with regard to the failure mode, cracking load and ultimate load could be observed. This result seems to confirm that the low uniaxial tensile strength of ASR-affected members does not automatically imply small shear capacities.

3.2.3.2   Load testing of ASR-affected viaducts

In the literature, reference to a number of load tests on ASR-affected viaducts are available. The first example is the proof loading of a bridge in the Hanshin expressway in Japan. Testing to 80% of the design load showed sufficient capacity of the bridge. The difference in deflection between the elements with damaged and undamaged concrete was less than 0.2 mm (0.008 in.). A second example is the bridges on the A26 highway in France, several of which are suffering from ASR damage. Diagnostic testing of new bridges is standard practice in France, so baseline information is available (Talley, 2009). Load testing can thus give a comparison to the initial state of the bridge upon completion of the construction. It was found in a reference case that the reduction in stiffness of the viaduct was less than 10%. Repair after the load test consisted of applying a coating to prevent further moisture ingress. A third example includes testing to 85% of the design capacity on a double-deck road structure in South Africa, which led to the conclusion that this structure was adequate (Talley, 2009). A final example relates to ASR-affected bridges in Denmark (Schmidt et al., 2014), tested to evaluate their performance in shear. It was found that the compressive strength and splitting tensile strength were reduced significantly compared to the design strengths while no reduction of the shear capacity was found.

3.2.3.3   Monitoring results

Since 2003, monitoring of the viaduct Zijlweg for ASR is in place (Koenders Instruments, 2015). An update of the equipment was provided in 2007. Sensors are applied in spans 1 and 4, measuring temperature, thickness of the deck, moisture in the concrete, and the longitudinal expansion of the deck. The data analysis of 2008 (Rijkswaterstaat, 2008) concluded that shrinkage was occurring over the depth of the cross section, expansion was occurring over the longitudinal direction, and the moisture content is increasing.

Prior to the proof load test, the available data are analyzed. The analysis uses the environmental data obtained from the KNMI (Royal Dutch meteorological institute) at a weather station 14 km (8.7 mi) south of the proof loading location (Royal Dutch Meteorological Institute, 2017). There is a clear correlation visible between the fluctuations of the temperature over the year and the thickness, joint size, or moisture content (see Fig. 3.13). Since the beginning of the year 2009, a clear increase in thickness is visible. Figure 3.13a shows the relation between the average thickness of the cross section and the ambient temperature. The joint size (Fig. 3.13b) is inversely correlated to the temperature. This effect is due to the expansion of the bridge at higher temperatures, which results in a smaller joint size. The joint size gradually decreases over time, which means that the bridge is expanding longitudinally. Considering the moisture content in Figure 3.13c, a clear fluctuation of the moisture content with an average of 10% can be seen. From Figure 3.13d it can be concluded that there is a good correlation between the ambient and deck temperature.

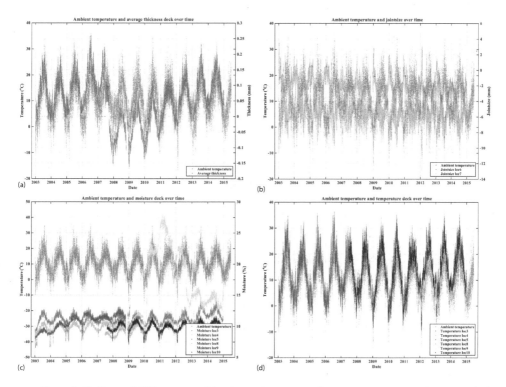

Figure 3.13 Data of ASR sensors: (a) relation between ambient temperature and average deck thickness; (b) relation between ambient temperature and joint size; (c) relation between ambient temperature and moisture in the deck; (d) relation between ambient temperature and measured temperature at the deck

3.2.4 Determination of target proof load and position

3.2.4.1 Finite element model

The target proof load was determined based on a linear finite element model, developed with TNO DIANA (TNO DIANA, 2012). The slab was modeled with a uniform thickness, and the larger depth of the sidewalk was not taken into account. Eight-node quadrilateral isoparametric flat shell elements were used. The final mesh contains 106 elements in the length direction and 12 elements in the width direction. Each crossbeam of the viaduct is supported at two locations.

The permanent loads used in the model include the self-weight of the structure and the superimposed dead load. A density of 2500 kg/m³ (150 pcff)) is used for the concrete. The equivalent load of the elements that are not modeled is added. These elements are the sidewalks, the wearing surface, and the asphalt layer. For the wearing surface, a load of 1.15 kN/m² (24 psff)) is used. The width in the model is 5.7 m (18.7 ft), which is smaller than the full width of 6.6 m (21.7 ft) shown in Figure 3.5, so that a distributed load to represent the additional width is added. For the asphalt layer, a load of 2.5 kN/m² (52 psff)) is applied, which corresponds to a layer of 100 mm (4 in.) thickness.

The variable loads used in the model are in accordance with live load model 1 from EN 1991–2:2003 (CEN, 2003), which combines a distributed lane load with a design tandem per lane. The notional lane width is 3 m (10 ft). The distributed lane load is 9 kN/m² (188 psff). To find the largest load effects in span 4, the lane load is applied in spans 2 and 4, whereas no distributed lane load is applied in spans 1 and 3. The remaining part of the actual lane of the viaduct is loaded with a load of 2.5 kN/m² (52 psf). On the sidewalks, a pedestrian load of 5 kN/m² (104 psf) is used. According to the Dutch NA, a reduction factor of $\psi = 0.4$ can be applied to the pedestrian load, which was not considered for this case study. As such, the approach is conservative. The design tandem consists of two axles of 300 kN (67 kip) spaced 1.2 m (4 ft) apart. The wheel print of the tandem is 400 mm × 400 mm (15.7 in. × 15.7 in.). For the considered model using shell elements, only the mid-depth of the cross section is modeled. It is assumed that the concentrated wheel loads are distributed under 45° to the mid-depth, resulting in a surface of 950 mm × 950 mm (37.4 in × 37.4 in.). In the model, the concentrated loads are applied as a distributed load of 155 kN/m² (3.3 ksf) on a surface of two elements by two elements. The tandem is centered in the notional lane. The Dutch National Annex NEN-EN 1991–2/NA:2011 (Normcommissie 351001, 2011b) permits the use of reduction factors for bridges and viaducts subjected to less than 250,000 vehicles per year. These factors are 0.97 for the lane load and design tandem and 0.90 for the remaining area of the lane. As the factor only relates to the number of vehicles, it does not affect the modeled pedestrian traffic.

The load factors that are used for the load combination are as described by the Dutch guidelines for the assessment of bridges ("RBK" [Rijkswaterstaat, 2013]) (see Table 3.1). Table 3.3 gives an overview of the safety factors that are used for the load combinations at different safety levels in combination with proof load testing. These load combinations are all altered for γ_{DC} with regard to the prescribed load factors in the code. For all cases, $\gamma_{DC} = 1.10$. This value is used since the self-weight of an existing structures is not a random variable anymore, but can be considered as a deterministic value. Only the model factor remains (Walraven, 2012).

Once the loads are defined, the governing sectional moment and shear have to be found. For this purpose, the position of the design truck is changed to a flexure-critical and shear-critical position, respectively. To find the flexure-critical position, the truck is moved along span 4 until the position is found that results in the largest sectional moment. To find the shear-critical position, the truck is placed at $2.5d_l$, with d_l the effective depth to the longitudinal reinforcement (Lantsoght et al., 2013). Note that this distance was developed as the critical distance for straight slab bridges, and no experimental evidence is available about its validity for skewed bridges. For a skewed viaduct, the critical position occurs in the obtuse corner (Cope, 1985). The peak shear stress is then distributed in the transverse direction over $4d_l$ to find the representative sectional shear stress (Lantsoght et al., 2017a).

Table 3.3 Load factors in combination with proof load testing at different load levels

	γ_{DC}	γ_{DW}	γ_{LL}
ULS	1.10	1.35	1.50
RBK Design	1.10	1.25	1.50
RBK Renovation	1.10	1.15	1.30
RBK Usage	1.10	1.15	1.25
RBK Disapproval	1.10	1.10	1.25
SLS	1.00	1.00	1.00

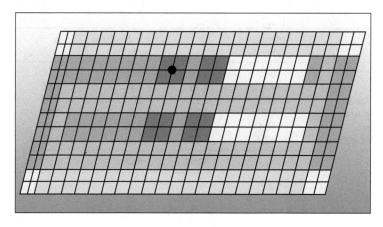

Figure 3.14 Position of wheel loads for flexure-critical case

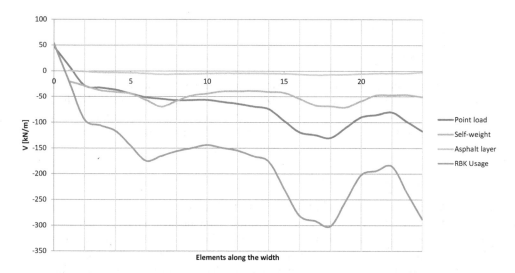

Figure 3.15 Sectional shear over width of slab. The x-axis shows the numbering of the elements in the finite element model. Conversion: 1 kN/m = 0.07 kip/ft

The flexure-critical position in span 4 is at 3382 mm (11 ft) from the face of the support (see Fig. 3.14). The shear-critical position in span 4 is at 1208 mm (4 ft) from the face of the support. The contributions of the different loads to the total sectional shear at the cross section at $0.5d_l$ from the support are shown in Figure 3.15. These contributions are then summed and averaged over $4d_l$ to find the governing sectional shear that should equal the sectional shear caused by the load combination with the proof load tandem.

An example of the results using the RBK (Rijkswaterstaat, 2013) Usage load level is shown in Table 3.4. First, the bending moment caused by the different loads is calculated

Table 3.4 Results of applied loads in linear finite element model. Conversion: 1 kNm/m = 0.225 kipft/ft

Load	Moment (kNm/m)	γ	$\gamma \times$ Moment (kNm/m)	α	M_{Ed} (kNm/m)
Self-weight	−84.1	1.10	−92.5	1.00	−92.5
Asphalt layer	−7.7	1.15	−8.8	1.00	−8.8
UDL Notional Lane	−56.7	1.25	−70.9	0.97	−68.7
UDL Remaining Area	−4.7	1.25	−5.9	0.90	−5.3
Eurocode tandem	−184	1.25	−229	0.97	−222
Pedestrian traffic sidewalk	−12.1	1.25	−15.1	1.00	−15.1
		Total:	−423		−413

Figure 3.16 Flowchart for determination of required proof loads

separately. The resulting moment for the Eurocode tandem is obtained at the flexure-critical position. Then, each load is multiplied by the load factor, γ, and subsequently by the factor that takes the reduced traffic into account, α. Then, to find the bending moment caused by the code-prescribed loads at the RBK Usage level, all contributions are summed and a bending moment M_{Ed} = 413 kNm/m (93 kipft/ft) is found.

3.2.4.2   *Resulting target proof load*

For the proof load, the wheel print of Rijkswaterstaat is used, which has a size of 230 mm × 300 mm (9 in. × 12 in.). For the application in the finite element model, which uses shell elements, vertical load distribution to the middle of the cross section is used. The wheel print in the finite element model is then 780 mm × 850 mm (31 in. × 33 in.).

The position of the design truck is moved along span 4 in order to find the position at which the largest bending moment results: the flexure-critical position. Then, the live loads from the Eurocode are removed, and the permanent loads remain in place. Subsequently, the proof load tandem is placed at the flexure-critical position. Then, the magnitude of the proof load is determined so that the maximum sectional moment caused by the factored permanent loads and the proof load tandem becomes the same as the maximum sectional moment resulting from the factored Eurocode live loads and permanent loads. Since this comparison is based on factored loads, the load factors associated with the considered safety level (see Table 3.1 for the prescribed load factors and Table 3.3 for the application to proof load testing) govern. A flowchart summarizing this procedure is given in Figure 3.16.

The same procedure is used for the shear-critical position, for which the critical position is taken at $2.5d_l$ and the peak shear stress is distributed in the transverse direction over $4d_l$. A summary of the results for the flexure-critical and shear-critical position is listed in Table 3.5.

Table 3.5 Results for required proof loads P_{tot} at different safety levels for the flexure-critical and shear-critical positions. Conversion: 1 kN = 0.225 kip, 1 metric ton = 1.1 short tons

Safety level	Bending moment		Shear	
	P_{tot} (kN)	P_{tot} (metric ton)	P_{tot} (kN)	P_{tot} (metric ton)
EC	1259	128	1228	125
RBK Design	1257	128	1228	125
RBK Reconstruction	1091	111	1066	109
RBK Usage	1050	107	1027	105
RBK Disapproval	1049	107	1025	104
SLS	815	83	791	81

Table 3.6 Bending moment M and curvature κ at cracking, yielding, and ultimate. Conversion: 1 kNm/m = 0.225 kipft/ft and 1/mm = 25.4 1/in.

	QR22			QR24		
	Cracking	Yielding	Ultimate	Cracking	Yielding	Ultimate
M (kNm/m)	204	383	413	204	415	459
κ (10^{-6} 1/mm)	0.372	2.69	68.2	0.372	2.91	61.2

Table 3.7 Bending moment capacity $M_{Rd,c}$, sectional moment M_{Ed}, and resulting Unity Checks UC. Conversion: 1 kNm/m = 0.225 kipft/ft

	QR 22		QR 24	
	RBK Usage	RBK Design	RBK Usage	RBK Design
M_{Ed} (kNm/m)	413	476	413	476
$M_{Rd,c}$ (kNm/m)	413	413	459	459
UC	1.00	1.15	0.90	1.04

3.2.5   *Expected capacity and behavior*

Part of the preparation of a proof load test includes estimating the capacity of the viaduct and relating the expected capacity to the load that can be applied to the bridge. The values of the expected maximum load at a certain position of the proof load tandem help to know when meticulous attention must be paid to the measurements and stop criteria. For the preparation of the proof load test, all failure modes need to be considered. For this purpose, the capacity of the viaduct has to be determined for flexure, shear, and punching.

First, the bending moment capacity is determined and the expected moment-curvature diagram is developed. The results for the cracking, yielding, and ultimate moment are given in Table 3.6 for the cross section with the maximum sagging moment, and the results of the Unity Checks based on a unit meter of width are given in Table 3.7 for both QR 22 and QR 24 steel. In Table 3.7, M_{Ed} is the sectional moment caused by the load combination prescribed by the code, and $M_{Rd,c}$ is the capacity as prescribed by the code. These results show that a priori the viaduct could not be deemed satisfactory for bending moment capacity at all safety levels (as defined in Table 3.1), since for the Design safety level, the Unity Check is larger than 1.

Two levels of approximation (Lantsoght et al., 2017b) were followed to check the shear capacity at the shear-critical cross section: Level of Approximation I (Quick Scan

[Vergoossen et al., 2013]) and a Level of Approximation II (linear finite element analysis [Lantsoght et al., 2017a]). The shear capacity of concrete affected by ASR is assumed to be reduced by 25% (den Uijl and Kaptijn, 2002). In the Level of Approximation I, the Quick Scan sheet is used in a load combination consisting of the self-weight, superimposed dead load, and proof load tandem. To find the maximum load that can be applied, the proof load is increased until a Unity Check of 1 is found. The results are given in Table 3.8 for the RBK (Rijkswaterstaat, 2013) Usage and Design safety levels. The column with "Mean" is the resulting maximum load when all load and resistance factors are taken as equal to one and average instead of material properties are used. The value of P_{max} is the total maximum load for which a Unity Check UC = 1 is found for the considered safety level. The value of $P_{max,ASR}$ takes the capacity of the ASR-affected cross section as 75% of an undamaged cross section. The row with "UC" gives the resulting Unity Check for the cross section, taking into account the effect of ASR damage, at the considered safety level and using the code-prescribed loads. These Unity Checks are calculated for the values of P_{tot} from Table 3.5. These calculations considered the obtuse corner, as in the obtuse corner a larger concentration of shear stresses occurs (Cope, 1985). For the reinforcement, QR 24 is assumed, since the yield strength of the steel is inversely proportional to the lower bound of the shear capacity v_{min} (Walraven, 2013). A lower steel quality would thus result in a higher value of v_{min}, so that it is conservative to assume QR 24 for the determination of the shear capacity.

Next, the Unity Check for shear was determined at a Level of Assessment II. The peak sectional shear force v_{Ed} resulting from the factored loads is distributed over $4d_l$ and then compared to the shear capacity $v_{Rd,c}$ and v_{min} as used in Level of Assessment I. The results for the Unity Check are given in Table 3.9, from which it can be seen that the calculations indicate that the shear-critical cross section does not fulfill the requirements for all safety levels. For comparison, the resulting shear stress v_{Ed} is also compared to the average shear capacity $v_{R,c}$ (calculated assuming that the resistance factors equal one and using the average material properties). As can be seen from comparing Table 3.8 to Table 3.9,

Table 3.8 Results for shear capacity at obtuse corner: maximum load to have a Unity Check = 1 based on the Quick Scan spreadsheet, and Unity Check for code-prescribed loads. Conversion: 1 kN = 0.225 kip, 1 metric ton = 1.1 short tons

	Usage	Design	Mean
P_{max} (kN)	1644	1136	4360
$P_{max,ASR}$ (kN)	1233	1002	3270
UC	0.833	1.225	–

Table 3.9 Shear capacity $v_{Rd,c}$, average shear capacity $v_{R,c}$, sectional shear v_{Ed}, and resulting Unity Checks UC. Conversion: 1 kN/m = 0.07 kip/ft, 1 MPa = 145 psi

	Usage	Design
V_{Ed} (kN/m)	241	276
v_{Ed} (MPa)	0.477	0.547
$v_{Rd,c}$ (MPa)	0.538	0.538
UC	0.887	1.017
$v_{R,c}$	0.984	0.984
$v_{Ed}/v_{R,c}$	0.485	0.556

Level of Approximation II is less conservative than Level of Approximation I (average decrease of the Unity Check with 9.8%) because it results in lower sectional shear forces, and thus follows the principles of a Level of Approximation approach (fib, 2012).

For both proof load locations, the punching capacity has been verified. At the shear-critical position, the Unity Check for punching is 0.95 and at the flexure-critical position UC = 0.93. A punching shear failure is thus not governing.

To have a better estimate of the average ultimate capacity of the slab that can be expected when loading until failure, the capacity is determined with the Extended Strip Model (Lantsoght et al., 2017d). The maximum expected load on the bending moment position is then 3297 kN (741 kip) and on the shear position 3025 kN (680 kip).

Lastly, the shear capacity of the sidewalk, which has stirrups, has been verified. It was observed that shear failure could occur in the sidewalk before it occurs in the solid slab, and therefore the behavior of the sidewalk was monitored during testing.

The analysis of the cross-sections under study for the calculated values for the proof loading for bending and shear showed that the capacity was only found to be sufficient for some safety levels (but not all required levels) prior to proof loading the viaduct. The proof load test is thus a suitable tool to show experimentally that the bridge fulfills the code requirements. Since high loads are applied, for which prior to the load testing Unity Checks larger than one are found, the interpretation of the measured responses during the proof load test becomes of the utmost importance.

3.2.6   *Sensor plan*

Prior to a proof load test, saying that loading to a certain load level is safe or unsafe, is related to the calculations that are made (as discussed in Section 3.2.5) without the additional knowledge that proof loading offers the assessing engineers. During testing, the observed behavior in terms of deformations, strains, cracks, and so forth form the first basis for determining whether loads can be further increased or not. The preparatory calculations are used as background for the decisions to further increase the load or not. As the load is increased, the importance of the measurements and interpretation thereof increases.

For viaduct Zijlweg, the following responses were monitored: deflections of the deck, deflections of the crossbeams, crack width, strain, movement in the joint, and acoustic emissions. The force in the jacks was measured separately with load cells. As the pilot proof load test was used to gain more experience in the technique of proof load testing and because stop criteria (especially for shear) are not well-defined yet, an extensive sensor plan was used. In total 16 LVDTs (linear variable differential transformers), six laser distance finders, 15 acoustic emission sensors, and four load cells were used. The acoustic emission measurements were carried out for research purposes (Yang and Hordijk, 2015). An overview of the sensor plan is given in Figure 3.17. The LVDTs that are not shown in this figure are the reference LVDT that is used outside the loaded span to check the strain resulting from the effects of temperature and humidity (see Fig. 3.18), the three LVDTs used to follow the width of existing cracks, and the four LVDTs used to monitor the expansion joint.

The lasers and LVDTs in the longitudinal direction at the middle of the notional lane are used to find the longitudinal deflection profile of the tested span. Two additional lasers are used at different positions in the transverse direction to find the transverse

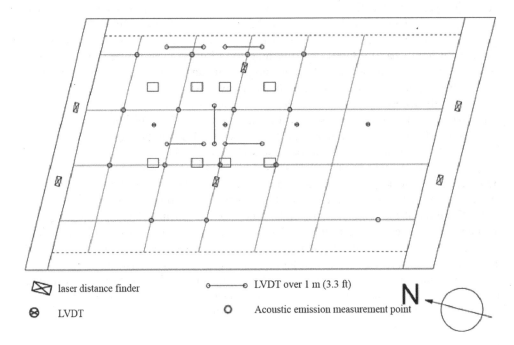

laser distance finder   o——o LVDT over 1 m (3.3 ft)

⊗ LVDT   O Acoustic emission measurement point

N

Figure 3.17 Sensor plan of viaduct Zijlweg

Source: Reprinted with permission from Lantsoght et al. (2018a).

deflection profile. The deflections of the crossbeams are measured with two lasers per support. These measurements are used to measure the compression of the elastomeric bearings.

LVDTs are placed over existing cracks to monitor crack opening during the proof load test. On the side of the deck, a possible shear crack is monitored, and on the soffit of the crack, a longitudinal and transverse crack close to the center of the proof load tandem are monitored. To measure strains, LVDTs are placed over 1 m (3.3 ft) to find the average strain over this region. Two LVDTs are placed longitudinally and one LVDT is placed transversely to measure the longitudinal and transverse bending moments, respectively. Finally, the movement in the joints is measured by using LVDTs of which one end is connected to the abutment and the other end to the slab. Two LVDTs are used to monitor the joint on the east side of the slab and two LVDTs on the west side.

3.3   Execution of proof load test

3.3.1   *Loading protocol*

Two proof load tests were carried out: one proof load test at a flexure-critical position and one proof load test at a shear-critical position. Both tests were carried out in span 4, on June 17th, 2015. The proper functioning of all sensors was verified during the afternoon of June 16th. A person jumping on the bridge resulted in small but measurable

Figure 3.18 Position of reference LVDT to measure influence of temperature and humidity. Picture showing support 5

responses (see Fig. 3.19), and driving a car over the viaduct resulted in clear responses (see Fig. 3.20).

For the proof load tests at the shear- and flexure-critical positions, a setup with a load spreader beam, counterweights, and four hydraulic jacks was used. Figure 3.21 shows a photograph of the load application system. Before the beginning of the proof load test, the jacks are not extended, and the load caused by the applied counterweight is carried off at the supports to the substructure. During the proof load test, the load on the jacks can be controlled and the superstructure (reinforced concrete slab) is loaded.

To measure the repeatability and linearity of the response and to follow the acoustic emission measurements, a cyclic loading protocol was selected. This protocol also allowed for the verification of the measurements after each cycle of loading and unloading. The loading protocol used at the flexure-critical position is shown in Figure 3.22a and the loading protocol used at the shear-critical position in Figure 3.22b. The first level of 400 kN (90 kip) was used to check the responses of all sensors. The second load level corresponds to the serviceability limit state. The third and fourth load level are intermediate load levels. The final load level corresponds to a load slightly larger than the safety level RBK design. This load level was only applied once. A baseline load level of 100 kN (23 kip) is used between the load cycles to keep all jacks and sensors activated. The maximum load in the bending moment test was 1368 kN (308 kip) when the weight of the jacks and steel plate are taken into account. The maximum load in the shear test was 1377 kN (310 kip) when all loading is considered.

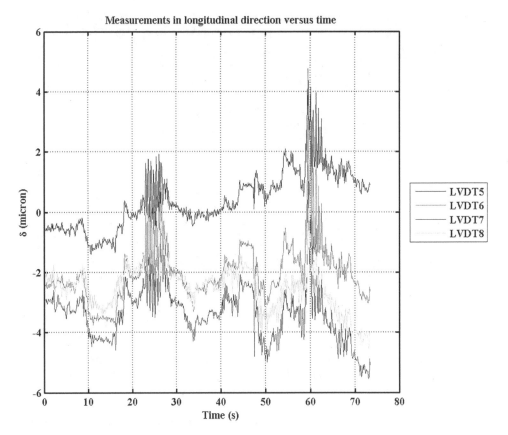

Figure 3.19 Measurement of person jumping. Conversion: 1 micron = 4×10^{-5} in.

3.3.2 *Measurements and observations*

3.3.2.1 *Load-deflection curves*

The envelope of the load-displacement curve of the bending moment test is shown in Figure 3.23. Some reduction of the stiffness can be observed for increasing loads, which may be caused by redistribution of stresses in the transverse direction. A reversal of the stiffness occurred during the unloading branch after the proof load test. The envelope of the load-deflection diagram for the shear-critical position is shown in Figure 3.23. From this figure, it can be seen that during this proof load test the behavior was fully linear. The observed stiffness in the test at a shear-critical position is larger than for the flexure-critical position, which was expected as a result of the loading position closer to the support for the shear-critical position.

3.3.2.2 *Deflection profiles*

The longitudinal deflection profile obtained during the bending moment test is shown in Figure 3.24a. From the longitudinal deflection profiles, it can be seen that no significant

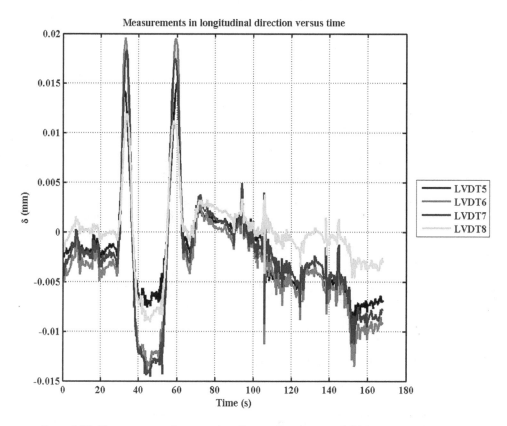

Figure 3.20 Measurement of car passing. Conversion: 1 mm = 0.04 in.

changes occur in the structural behavior. The transverse deflection profile is shown in Figure 3.24b. A change in the behavior in the transverse direction can be seen. The stiffness of the sidewalk is not the same as the stiffness of the carriageway. At low load levels, the sidewalk deflects more because of the presence of cracks. At higher load levels, the stirrups are activated and the sidewalk behaves in a stiffer way than the carriageway.

The longitudinal deflection profile resulting from loading at a shear-critical position is shown in Figure 3.25a. At two intermediate load levels, the deflection profile is disturbed. An analysis of the data showed that for this range of loads, a shift in the values of the measuring LVDT occurred, possibly caused by the sensor being unable to be further compressed. The transverse deflection profile is fully linear (see Fig. 3.25b). The structural response under the sidewalk is stiffer than under the carriageway, caused by the different reinforcement layout.

3.3.2.3 *Strains and crack width*

For strains and the increase in crack width, LVDTs were placed horizontally on the slab soffit. During the bending moment test, the strains at the bottom were measured and a linear relation between the applied load and resulting strains was observed (see Fig.

Figure 3.21 Setup of the load used on viaduct Zijlweg

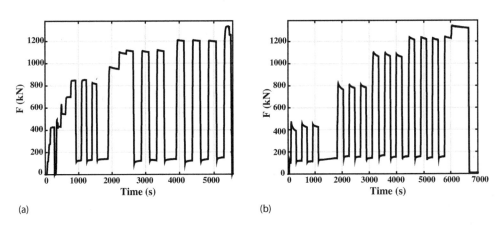

(a)   (b)

Figure 3.22 Loading protocol: (a) bending moment test; (b) shear test. Conversion: 1 kN = 0.225
kip

3.26a). The increase in crack width for the applied load is shown in Figure 3.27a. The
largest increase in crack width is less than 0.04 mm (0.0016 in), so that the effect of cracking can be considered negligible, since cracks smaller than 0.05 mm (0.0020 in) are typically not considered as structural.

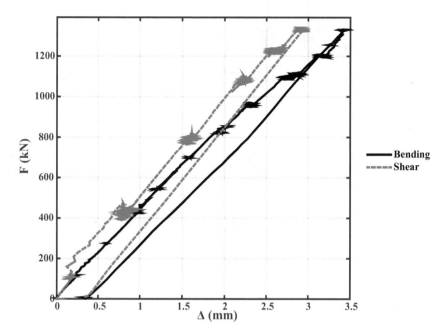

Figure 3.23 Envelope of the force-displacement diagram for the flexure-critical and shear-critical position. Conversion: 1 kN = 0.225 kip, 1 mm = 0.04 in.

For the shear test, no nonlinear behavior was observed from the measurements of the strains (see Fig. 3.26b). The increase in crack width for the applied load is shown in Figure 3.27b. The largest increase in crack width is less than 0.025 mm (0.0010 in.), so the effect of cracking can be considered negligible.

3.3.2.4 Movement in joint

The results of the measurements of the LVDTs applied over the expansion joint are given in Figure 3.28. As the measurements are started at the beginning of the proof load test, it makes sense that a negative displacement is measured during the test. The test is carried out with a system of a steel spreader beam, jacks, and counterweights. Prior to the test, all load goes to the supports and affects the joint due to small eccentricities in the setup. During the test, the jacks apply the load to the superstructure, which results in unloading of the joint. The last part of the measurements (7000 s and further in Fig. 3.28) correspond to the removal of the counterweights (per ballast block), which results in unloading of the joint.

3.3.2.5 Influence of temperature

The strains caused by changes in temperature and humidity are measured by the reference LVDT (see Fig. 3.18). The information of the ambient temperature can be taken from the Dutch meteorological institute (Royal Dutch Meteorological Institute, 2017). The measurement of the temperature of the bridge deck is part of the monitoring system for ASR, as discussed in Section 3.2.3.3. During the bending moment test (see Fig. 3.29a), the

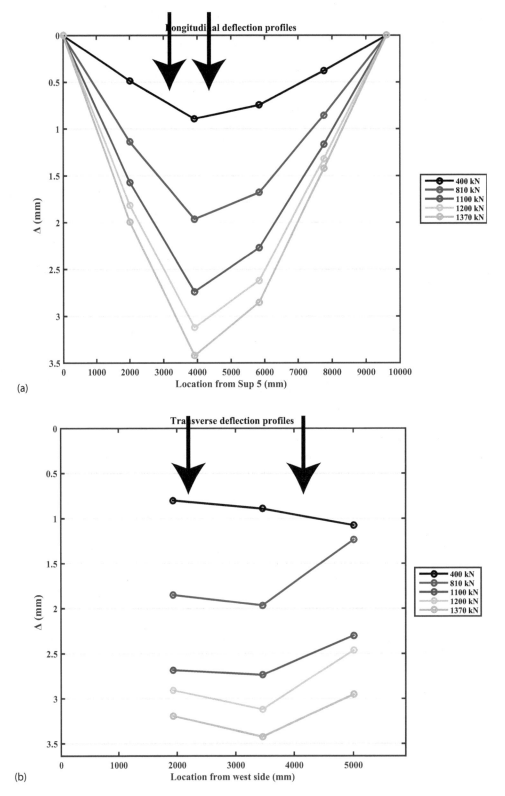

Figure 3.24 Deflection profiles for loading at flexure-critical position: (a) longitudinal profile; (b) transverse profile. Arrows indicate position of center of axles. Conversion: 1 kN = 0.225 kip, 1 mm = 0.04 in.

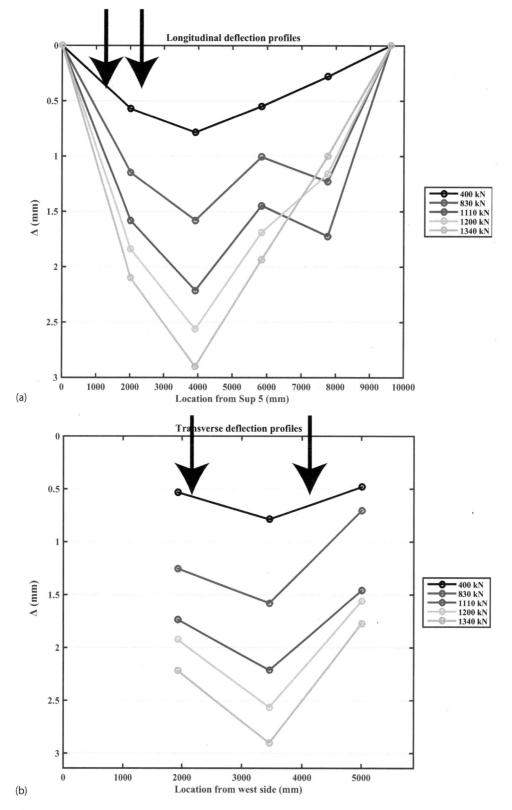

Figure 3.25 Deflection profiles for loading at shear-critical position: (a) longitudinal profile; (b) transverse profile. Arrows indicate position of center of axles. Conversion: 1 kN = 0.225 kip, 1 mm = 0.04 in.

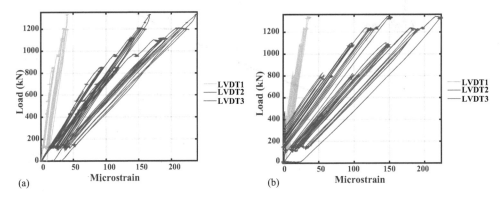

(a)   (b)

Figure 3.26 Relation between strain and load (a) during the bending moment proof load test; (b) during the shear test. Conversion: 1 kN = 0.225 kip.

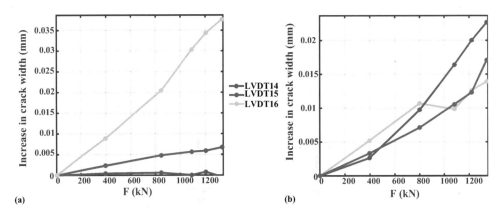

(a)   (b)

Figure 3.27 Increase in crack width (a) during the bending moment proof load test; (b) during the shear proof load test. Conversion: 1 mm = 0.04 in., 1 kN = 0.225 kip

bridge deck temperature and ambient temperature increased similarly. The measured strain is inversely proportional to the temperature. As the temperature increases, the aluminum strip expands. This expansion leads to compression of the LVDT, or negative values.

For the shear proof load test (see Fig. 3.29b), the observations are different: the ambient temperature decreased but the temperature of the bridge deck increased slightly. The weather conditions explain this behavior: the wind was blowing, reducing the ambient temperature, and at the same time the sun was shining, warming up the bridge deck. The measurements of the reference LVDT are in correspondence to the ambient temperature.

3.4   Post-processing and rating

3.4.1   Development of final graphs

After the proof load test, the final graphs are developed and reported. These graphs are similar to the graphs that are observed in real time during the experiment in the field.

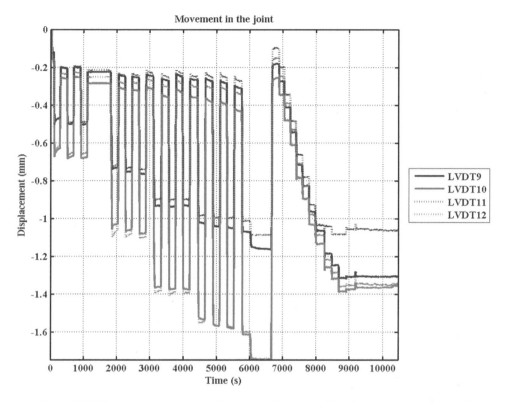

Figure 3.28 Movement in the joint during the shear proof load and during removal of the counterweight. Conversion: 1 mm = 0.04 in.

The net deflections are found by subtracting the deflections of the supports from the deflections measured over the span. The net strains result from subtracting the reference strain for environmental effects from the measured strains in the tested span. The graphs shown in Section 3.3.2 are the post-processed graphs from testing viaduct Zijlweg.

3.4.2 *Comparison with stop criteria*

3.4.2.1 *ACI 437.2M acceptance criteria*

The limit of residual deformation to maximum deformation according to ACI 437.2M-13 (ACI Committee 437, 2013) is 25%. In the bending moment test, the maximum deflection is Δ_{max} = 3.42 mm (0.14 in.) and the residual deflection is Δ_{res} 0.333 mm (0.013 in.), resulting in a ratio of 9.7%. The acceptance criterion is thus fulfilled. For the shear test, Δ_{max} = 2.9 mm (0.11 in.) and Δ_{res} = 0.28 mm (0.01 in.), so that the ratio equals 9.7%. The acceptance criterion is fulfilled.

The remaining acceptance criteria from ACI 437.2M-13 (ACI Committee 437, 2013) are the permanency ratio and deviation from linearity index. These acceptance criteria are directly related to the cyclic loading protocol described by ACI 437.2M-13. As such, they cannot be evaluated for the different loading protocol that is used for testing viaduct Zijlweg.

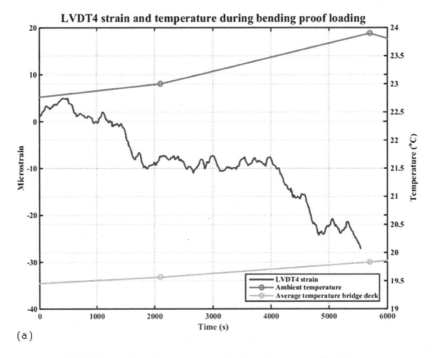

(a)

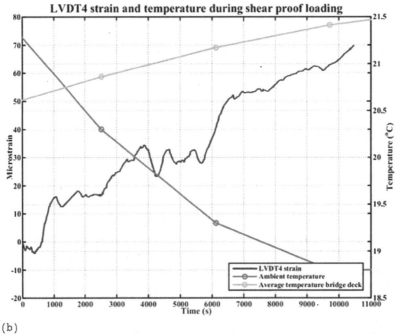

(b)

Figure 3.29 Reference strain and temperature during (a) testing at flexure-critical position; (b) testing at shear-critical position

3.4.2.2 German guideline stop criteria

The limit of residual deformation to maximum deformation according to the German guideline (Deutscher Ausschuss für Stahlbeton, 2000) is 10%. As mentioned in Section 3.4.2.1, the maximum ratio in the bending moment test was 9.7%. The stop criterion for residual deflection was thus never exceeded. Similarly, for the shear test, the stop criterion for residual deflection was not exceeded during the proof load test.

The stop criterion for concrete strains is:

$$\varepsilon_c < \varepsilon_{c,lim} - \varepsilon_{c0} \tag{3.1}$$

with ε_c the measured strain, $\varepsilon_{c,lim} = 800$ με for concrete compressive strengths above 25 MPa (3.6 ksi), and ε_{c0} the strain caused by the permanent loads present at the time of the proof load test. The value of ε_{c0} can be taken from the output of the finite element program described in Section 3.2.4.1. The maximum strain measured during the bending moment experiment is 240 με when corrected for the effect of temperature and humidity. The corresponding strain due to permanent loads at the position where the maximum strain was measured is 38 με. The maximum strain in the experiment would thus be 762 με, which is not exceeded. The stop criterion is thus not exceeded. For the shear test, $\varepsilon_c = 224$ με (corrected with the reference strain for environmental effects), and the strain due to permanent loads is $\varepsilon_{c0} = 45$ με. The limiting strain, 755 με, is not exceeded and the stop criterion is never exceeded during the experiment.

The limiting crack width and residual crack width form a stop criterion in the German guideline as well. However, since none of the crack widths measured are larger than 0.05 mm (0.002 in.), no structural cracking occurs and this stop criterion is not exceeded.

3.4.2.3 Proposed stop criteria

The proposed stop criteria (Lantsoght et al., 2018b) are compared with the measurements on viaduct Zijlweg. The stop criteria are based on the concrete strain, the limiting crack width, the stiffness, the load-deflection diagram, the longitudinal deflection profiles, and the transverse deflection profiles. Note that these stop criteria are still a topic of research, and that the stop criteria that were developed for the proof load test on the viaduct Zijlweg are an older version of stop criteria than those presented in Chapter 1 of Volume 13.

The stop criteria related to the load-deflection diagram and the longitudinal and transverse deflection profiles are qualitative in nature. As shown in Sections 3.3.2.1 and 3.3.2.2, no signs of nonlinear behavior were observed from these plots. The limiting crack width only needs to be verified for crack widths larger than 0.05 mm (0.0020 in.). The data in Section 3.3.2.3 show that none of the cracks that were monitored during the proof load tests were activated.

The stop criterion for the limiting concrete strain ε_{stop} is for flexure:

$$\varepsilon_c < \varepsilon_{stop} \tag{3.2}$$

with ε_c the measured strain, and:

$$\varepsilon_{stop} = 0.65\varepsilon_{bot,max} - \varepsilon_{c0} \tag{3.3}$$

$$\varepsilon_{bot,max} = \frac{h-c}{d-c}0.9\frac{f_y}{E_s} \tag{3.4}$$

with ε_{c0} the strain caused by permanent loads, h the height of the member, d the effective depth of the member, c the height of the compression zone when the strain in the steel is 90% of the yield strain, f_y the yield stress of the steel, and E_s the Young's modulus of the steel. The limiting strain ε_{stop} for the flexure-critical position is 751 $\mu\varepsilon$. This value is never exceeded, as can be seen in Figure 3.26a. Until a suitable stop criterion is developed based on research, the stop criterion for the limiting concrete strain for shear is kept as the same criterion used in the German guideline, Equation 3.1, which was not exceeded during the proof load test for shear.

For flexure and shear, the stiffness (measured as the tangent to the load-displacement diagram) during the proof load test cannot be less than 75% of the initial stiffness. In other words, the maximum reduction of the stiffness that is allowed is 25%. In the bending moment test, the stiffness reduction was 4%. In the shear test, the stiffness reduction was 12%. The stop criterion for the stiffness is thus never exceeded.

As shown in the previous paragraphs, none of the proposed stop criteria were exceeded during the proof load tests for flexure and shear. This observation corresponds with the conclusions that were drawn from the extensive sensor plan on the viaduct: no nonlinear behavior or signs that irreversible damage had occurred were observed during the proof load test. The bridge could thus withstand the applied target proof load, representative of the factored live loads. Therefore, it was shown experimentally that the tested element of the bridge (the superstructure) fulfills the requirements of the code.

3.4.3 *Final rating*

The proof load tests on the viaduct Zijlweg were successful. The planning and preparation facilitated the execution of the test. All sensors and the data acquisition system functioned properly and allowed for an evaluation of the data in real time. The goal of the proof load test (i.e. to show that the superstructure fulfills the code requirements) was also met based on the current knowledge on proof load testing. Further research is necessary to confirm assumptions made in the procedures. The highest safety level that was fulfilled (according to current knowledge) is the RBK Design level (with 5% extra loading). During the proof load experiment, there was no sign of distress in the measurements. If the currently followed procedures can be confirmed with further research, the field tests have shown that it is safe to keep the viaduct open to all traffic. Until then, however, it has been shown that the bridge is not unsafe, but it is not necessarily demonstrated that the bridge fulfils the safety requirements of the code, since this aspect requires further research on the probabilistic aspects of proof load testing. The result of the proof load test is important because the uniaxial tensile strength of the concrete in the viaduct is measured to be very low as a result of damage caused by alkali-silica reaction, so that the shear capacity of this viaduct was subject to discussion. The results of the proof loading test and the past experiments in the laboratory (see Section 3.2.3.1) indicate that, even though it is not possible to quantify the reduction or increase in the capacity of an element with ASR damage, the problems for existing bridges appear to be limited. When a proof load test demonstrates that an existing bridge fulfils the code requirements, the cost of replacement of the superstructure is avoided (Lantsoght, 2017). This cost includes the direct economic cost but also the environmental cost associated with the CO_2 emissions of the materials and transportation of materials, and the social cost of driver delays.

3.4.4 Lessons learned and recommendations for practice

Based on the experience of the proof load test on the viaduct Zijlweg, the following lessons learned can be identified:

• Proof load testing for shear is possible and gives valuable results, but requires further research.
• The presented case study is a good example of a bridge with large uncertainties on the capacity caused by ASR cracking. It shows that load testing can be used for the assessment of existing bridges when the uncertainties on the capacity are large. The conclusion is that for such cases, proof load testing is an interesting and economic method.

The following practical recommendations result from the pilot proof load test:

• A good preparation for the proof load test is essential.
• It is necessary to carry out a visual inspection so that the limitations on site can be identified during the preparation stage. For the case of viaduct Zijlweg in particular, marking positions on the top face of the bridge was made difficult because of the sidewalk.
• Selecting the critical crack that should be monitored during a proof load test is difficult. Either the test engineers should follow the crack width of a number of cracks during the test, or non-contact methods should be used to follow all cracks within a certain region.
• The presented loading method with the steel spreader beam and counterweights performs well in practice and can be recommended for proof load testing.
• For shear-critical bridges and bridges with large uncertainties, extensive preparation and a large number of sensors during the test are necessary to avoid a brittle failure during the test. For other bridge types, a shorter and simpler approach to proof load testing should be developed, so that with minimal preparation and equipment an answer can be found to the question of whether the bridge fulfills the code requirements or not.
• To save time in the field, non-contact sensors should be applied where feasible. The use of non-contact sensors, and stop criteria related to the output of such sensors, requires further research and pilot testing.

3.4.5 Discussion and elements for future research

Proof load testing of shear-critical reinforced concrete slab bridges requires a thorough understanding of shear-critical concrete structures, and currently a large number of uncertainties on this method still exist. As such, it is not possible yet to develop a step-by-step method for proof load testing of shear-critical concrete bridges that can be implemented in a code or guideline. The main uncertainties at this moment are the definition of safe stop criteria for shear-critical concrete bridges and the determination of the target proof load. To develop safe stop criteria for reinforced concrete slab bridges, further experimental work is necessary to study crack development in laboratory conditions. If safe stop criteria are available, it also becomes possible to load a bridge past the target proof load and until reaching the first stop criterion to learn about the full linear elastic range of the bridge and to know the maximum live load on the bridge. To verify the currently used simple approach for determining the target proof load, a probabilistic study is necessary so that the target proof load indeed represents the required reliability index for the tested bridge.

A limitation of a proof load test with only two loading positions is that in fact only the experimental results from two positions are obtained. To extrapolate this result to all positions of the bridge, which may have some local weaknesses perhaps, is an element that need further probabilistic studies. Currently, an additional few percent of load is applied to cover variability and uncertainties on the method itself, but this number should be quantified with rigorous studies.

A proof load test does not result in the experimental capacity of a tested bridge as this would require a test to failure. However, it gives a lower bound of the capacity and can express the capacity in terms of traffic that the bridge can carry.

3.5   Summary and conclusions

On June 17th, 2015, a proof load test was carried out on the viaduct Zijlweg in the province of Noord Brabant, the Netherlands. Two positions were identified for testing: a flexure-critical position and a shear-critical position. The viaduct Zijlweg was selected for proof loading, as it showed severe cracking caused by alkali-silica reaction (ASR) damage. There-fore, the capacity of the viaduct, especially its shear capacity, was subject to discussion. To evaluate the effect of ASR, monitoring of the expansion, temperature, and moisture was installed in 2003. The end span was tested, as this span is not directly located above the highway.

The literature review on ASR and the behavior of structural elements with ASR damage showed that the uniaxial tensile strength of concrete with ASR damage is very low. The effect of ASR on the shear capacity is subject to discussion. For small elements, no reduction, or even an increase in the shear capacity of an ASR-damaged beam as compared to an undamaged beam, was observed. These beams were fabricated in the laboratory and sub-jected to higher temperatures and moisture cycles to activate the swelling of the ASR gel. For beams sawn from a viaduct with ASR damage, the shear capacity was 75% of the the-oretically predicted capacity. In some experiments, an increase in the capacity was found for the ASR-affected elements. This increase was attributed to the prestressing induced in the sections by the restraint of the swelling. Load testing on ASR-deteriorated bridges in Japan, France, South Africa, and Denmark in the past did not result in a marked influence of the ASR damage on the measured response of the structures.

The position and magnitude of the target proof load were determined by using a linear elastic finite element model. For the bending moment proof load test, the position of the proof loading tandem was determined by finding the most unfavorable position of the design truck in the finite element model. The live loads were then replaced by the proof load tandem at the critical position. The magnitude of the load on the proof load tandem was then increased until it reaches the bending moment that was found by using the load combination from the code. This procedure is followed for the different safety levels, which each use a different set of safety factors. For the shear proof load test, the critical position is at $2.5d_l$. The target proof load to get the same sectional shear as for the load combination from the code was then determined. Further probabilistic studies are required to develop a better method for determining the target proof load. An assess-ment prior to the test showed that at the RBK Design Level, the viaduct does not fulfill the requirements for shear and bending.

The viaduct was heavily instrumented during both load tests to carefully monitor the structural behavior. Currently, no suitable and safe stop criteria for shear-critical reinforced concrete slab bridges exist. The large amount of sensors applied served the purpose of catching signs of possible nonlinearity or occurrence of irreversible damage, as stop criteria (especially for shear) are still a subject of research. A cyclic loading protocol was used for the analysis. The maximum load in the bending moment test corresponds with the safety level RBK Design + 9%. The maximum load in the shear test corresponds with the safety level RBK Design + 12%.

The proof load test was carried out at a flexure-critical and at a shear-critical position. For both proof load tests, it was found that the viaduct could carry loads corresponding to the safety level RBK Design (and some extra reserve) without signs of distress. Given the uncertainties on the method and need for future research, the final conclusion about the viaduct is that it has been shown that the viaduct is not unsafe, but it has not yet been shown that it is safe and that it fulfills the code requirements. Regular inspections and analysis of the ASR data is necessary, as the remaining space in the expansion joints is small. The proof load test on the viaduct in the Zijlweg showed the important conclusion that the capacity of a viaduct affected by ASR is higher than the analytically determined capacity, since the Unity Check for the applied load was analytically determined to be larger than 1. The problem of the small strength of the concrete in uniaxial tension is thus of a lower magnitude than feared.

Acknowledgments

The authors wish to express their gratitude to the Province of Noord Holland and the Dutch Ministry of Infrastructure and the Environment (Rijkswaterstaat) for financing this pilot proof load tests. This research would not have been possible without the contributions and help of our colleagues from Delft University of Technology A. Bosman, S. Fennis, P. van Hemert, and W. Vos, of F. Linthorst and D. den Boef of Witteveen + Bos, responsible for practical preparations and safety inspections on site, and of O. Illing and the late C. Huissen from Mammoet, responsible for applying the load. The many discussions with A. de Boer, S. Fennis, M. Naaktgeboren, and H. van der Ham of the Dutch Ministry of Infrastructure and the Environment have been crucial in the development of this work, and are gratefully acknowledged.

References

ACI Committee 437 (2013) *Code Requirements for Load Testing of Existing Concrete Structures (ACI 437.2M-13) and Commentary*. American Concrete Institute, Farmington Hills, MI.

Ahmed, T., Burley, E. & Rigden, S. (1998) The static and fatigue strength of reinforced concrete beams affected by alkali-silica reaction. *ACI Materials Journal*, 95, 376–388.

Ahmed, T., Burley, E. & Rigden, S. (1999) Effect of alkali-silica reaction on tensile bond strength of reinforcement in concrete tested under static and fatigue loading. *ACI Materials Journal*, 96, 419–428.

Borsje, H., Peelen, W. H. A., Postema, F. J. & Bakker, J. D. (2002) Monitoring Alkali-Silica Reaction in structures. *Heron*, 47, 96–109.

CEN (2002) *Eurocode: Basis of Structural Design, NEN-EN 1990:2002*. Comité Européen de Normalisation, Brussels, Belgium.

CEN (2003) *Eurocode 1: Actions on Structures, Part 2: Traffic Loads on Bridges, NEN-EN 1991-2:2003*. Comité Européen de Normalisation, Brussels, Belgium.

CEN (2005) *Eurocode 2: Design of Concrete Structures, Part 1–1: General Rules and Rules for Buildings. NEN-EN 1992–1–1:2005.* Comité Européen de Normalisation, Brussels, Belgium.

Code Committee 351001 (1995) *NEN 6720 Technical Foundations for Building Codes, Concrete Provisions TGB 1990: Structural Requirements and Calculation Methods (VBC 1995).* Civil Engineering Center for Research and Regulation, Dutch Normalization Institute, Delft, The Netherlands.

Code Committee 351001 (2011) *Assessment of Structural Safety of an Existing Structure at Repair or Unfit for Use: Basic Requirements, NEN 8700:2011 (in Dutch).* Civil Center for the Execution of Research and Standard, Dutch Normalisation Institute, Delft, The Netherlands.

Cope, R. J. (1985) Flexural shear failure of reinforced-concrete slab bridges. *Proceedings of the Institution of Civil Engineers Part 2-Research and Theory*, 79, 559–583.

Den Uijl, J. A. & Kaptijn, N. (2002) Structural consequences of ASR: An example on shear capacity. *Heron*, 47, 1–13.

Deutscher Ausschuss für Stahlbeton (2000) *DAfStb-Guideline: Load Tests on Concrete Structures (in German).* Deutscher Ausschuss für Stahlbeton, Berlin, Germany.

FIB (2012) *Model Code 2010: Final Draft.* International Federation for Structural Concrete, Lausanne.

Giaccio, G., Zerbino, R., Ponce, J. M. & Batic, O. R. (2008) Mechanical behavior of concretes damaged by alkali-silica reaction. *Cement and Concrete Research*, 38, 993–1004.

Haddad, R. H., Shannag, M. J. & Aa-Hambouth, M. T. (2008) Repair of reinforced concrete beams damaged by alkali-silica reaction. *ACI Structural Journal*, 105, 145–153.

Koenders Instruments (2015) *Zijlweg Monitoring System.* Koenders Instruments, Almere, The Netherlands.

Lantsoght, E. O. L. (2017) Chapter 9: Field assessment of a concrete bridge. In: Pacheco-Torgal, F., Melchers, R., Shi, X., Saez, A., De Belie, N. & Van Tittelboom, K. (eds.) *Eco-Efficient Repair and Rehabilitation of Concrete Infrastructures.* Woodhead Publishing, Cambridge, UK.

Lantsoght, E. O. L., Van Der Veen, C., De Boer, A. & Walraven, J. C. (2013) Recommendations for the shear assessment of reinforced concrete slab bridges from experiments. *Structural Engineering International*, 23, 418–426.

Lantsoght, E. O. L., De Boer, A. & Van Der Veen, C. (2017a) Distribution of peak shear stress in finite element models of reinforced concrete slabs. *Engineering Structures*, 148, 571–583.

Lantsoght, E. O. L., De Boer, A. & Van Der Veen, C. (2017b) Levels of approximation for the shear assessment of reinforced concrete slab bridges. *Structural Concrete*, 18, 143–152.

Lantsoght, E. O. L., Koekkoek, R. T., Hordijk, D. A. & De Boer, A. (2017c) Towards standardization of proof load testing: Pilot test on viaduct Zijlweg. *Structure and Infrastructure Engineering*, 16.

Lantsoght, E. O. L., Van Der Veen, C., De Boer, A. & Alexander, S. D. B. (2017d) Extended strip model for slabs under concentrated loads. *ACI Structural Journal*, 114, 565–574.

Lantsoght, E. O. L., Van Der Veen, C., De Boer, A. & Hordijk, D. A. (2017e) Proof load testing of reinforced concrete slab bridges in the Netherlands. *Structural Concrete*, 18, 597–606.

Lantsoght, E. O. L., Van Der Veen, C., De Boer, A. & Hordijk, D. A. (2018a) Assessment of slab bridges through proof loading in the Netherlands. *ACI SP-323 Evaluation of Concrete Bridge Behavior through Load Testing: International Perspectives.* American Concrete Institute, Farmington Hills, MI.

Lantsoght, E. O. L., Van Der Veen, C. & Hordijk, D. A. (2018b) Proposed stop criteria for proof load testing of concrete bridges and verification. *IALCCE 2018*, Ghent, Belgium.

Li, Y., Vrouwenvelder, T., Wijnants, G. H. & Walraven, J. (2004) Spatial variability of concrete deterioration and repair strategies. *Structural Concrete*, 5, 121–129.

Nederlands Normalisatie Instituut (1963) *Voorschriften voor het ontwerpen van stalen bruggen.* Nederlands Normalisatie Instituut, Delft, The Netherlands.

Neville, A. (2012) *Properties of Concrete*, 5th edition. Pearson Education, London, UK.

Nijland, T. G. & Siemes, A. J. M. (2002) Alkali-silica reaction in the Netherlands: Experiences and current research. *Heron*, 47, 81–84.

Normcommissie 351001 (2011a) *Assessment of Structural Safety of an Existing Structure at Repair or Unfit for Use: Basic Requirements, NEN 8700:2011 (in Dutch)*. Civil Center for the Execution of Research and Standard, Dutch Normalisation Institute, Delft, The Netherlands.

Normcommissie 351001 (2011b) *Eurocode 1: Actions on Structures, Part 2: Traffic Loads on Bridges, EN 1991–2/NA:2011*. Civil Engineering Center for Research and Regulation, Dutch Normalization Institute, Delft, The Netherlands.

Projectteam RWS/TNO Bouw (1997) *Safety Evaluation Existing Structures: Reinforced Concrete Bridges (in Dutch)*. TNO, Delft, The Netherlands.

Rafla, K. (1971) Empirische Formeln zur Berechnung der Schubtragfahigkeit von Stahlbetonbalken. *Strasse Brucke Tunnel*, 23, 311–320.

Rijkswaterstaat (2002) *Beheer-& Onderhoudsplan Viaduct Zijlweg gelegen over Rijksweg A59*. Ministry of Infrastructure and the Environment, Utrecht, The Netherlands.

Rijkswaterstaat (2008) *Inspectierapport Programmernisinspectie Complex-Cide 44G-113-01*. Ministry of Infrastructure and the Environment, Utrecht, The Netherlands.

Rijkswaterstaat (2013) *Guidelines Assessment Bridges: Assessment of Structural Safety of an Existing Bridge at Reconstruction, Usage and Disapproval* (in Dutch), RTD 1006:2013 1.1.

Royal Dutch Meteorological Institute (2017) *Hourly Weather Data for the Netherlands*. [Online]. Available from: www.knmi.nl/klimatologie/uurgegevens/datafiles/350/uurgeg_350_2011-2020. zip [accessed 7th June 2017].

Schmidt, J. W., Hansen, S. G., Barbosa, R. A. & Henriksen, A. (2014) Novel shear capacity testing of ASR damaged full scale concrete bridge. *Engineering Structures*, 79, 365–374.

Siemes, T., Han, N. & Visser, J. (2002) Unexpectedly low tensile strength in concrete structures. *Heron*, 47, 111–124.

Steenbergen, R. D. J. M. & Vrouwenvelder, A. C. W. M. (2010) Safety philosophy for existing structures and partial factors for traffic loads on bridges. *Heron*, 55, 123–140.

Talley, K. G. (2009) *Assessment and Strengthening of ASR and DEF Affected Concrete Bridge Columns*. Ph.D. Thesis, The University of Texas at Austin, Austin, TX.

TNO Diana (2012) *Users Manual of DIANA, Release 9.4.4*. TNO Diana, Delft, The Netherlands.

Vergoossen, R., Naaktgeboren, M. T., Hart, M., De Boer, A. & Van Vugt, E. (2013) Quick scan on shear in existing slab type viaducts. *International IABSE Conference, Assessment, Upgrading and Refurbishment of Infrastructures*, Rotterdam, The Netherlands.

Walraven, J. C. (2010) Residual shear bearing capacity of existing bridges. *fib Bulletin 57, Shear and Punching Shear in RC and FRC Elements; Proceedings of a Workshop held on 15–16 October 2010*, Salò, Lake Garda, Italy.

Walraven, J. C. (2012) *Proof Loading of Concrete Bridges (in Dutch)*. Stevin Report 25.5–12–9. Delft University of Technology, Delft.

Walraven, J. C. (2013) Minimum Shear Capacity of Reinforced Concrete Slabs without Shear Reinforcement: The Value v_{min}. Stevin Report 25.5–12–4. Delft University of Technology, Delft.

Witteveen+Bos (2014) *Materiaalonderzoek Bruggen, zaaknummer 31084913: 44G-113–01 – Ongelijkvloerse kruising rijksweg – Zijlweg (Zijlweg)*. Deventer.

Yang, Y., Den Uijl, J. A., Dieteren, G. & De Boer, A. (2010) Shear capacity of 50 years old reinforced concrete bridge deck without shear reinforcement. *3rd fib International Congress*, Washington, DC, USA: fib.

Yang, Y. & Hordijk, D. A. (2015) *Acoustic Emission Measurement and Analysis on Zijlwegbrug*. Stevin Report 25.5–15–01. Delft University of Technology, Delft, The Netherlands.

Part II

Testing of Buildings

Chapter 4

Load Testing of Concrete Building Constructions

Gregor Schacht, Guido Bolle, and Steffen Marx

Abstract

In 2000 the Deutscher Ausschuss für Stahlbeton (DAfStb) published its guideline for load testing of existing concrete structures and therewith the experimental proof of the load-bearing capacity became a widely accepted method in Germany. Since then over 2000 proof load tests have been carried out, and so a lot of experience exists in the usage of the Guideline. Nevertheless, there are still reports about load tests carried out with mass loads or using mechanical measurement techniques. To prevent a misuse of the Guideline in future, to take into account the great experiences existing in the evaluation of the measuring results and the bearing condition, the DAfStb decided to revise the existing Guideline. In recent years there has also been an increased research activity focusing the development of evaluation criteria for possible brittle failures in shear and to extend the safety concept for the evaluation of elements not directly tested. These results will also be considered in the new Guideline. This chapter gives an overview of the rules of the existing Guideline and discusses these in the background of the gained experience during the last 17 years. The new criteria for evaluating the shear capacity are explained, and the thoughts of the new safety concept for load testing are introduced.

4.1 Historical development of load testing in Europe

4.1.1 Introduction

Since ancient times, humans have aimed to achieve the highest level of safety in all aspects of their lives. In civil engineering, this desire for safety is particularly strong, because the failure of structures is much less accepted by society than the failure of, say, motor vehicles or electronic devices. We feel a sense of safety particularly when we "know" or can "see" that structural elements can withstand the relevant loads.

The origin of civil engineering lies in the building trade, where experiences gained by trial and error used to be of particular importance. This way of proving that something works as intended is not only the etymological but also the historical origin of proof loading, and the first load tests performed by humans – for example, to determine the tensile resistance of a liana or the load-carrying capacity of a fallen tree – were carried out many millions of years ago. Load tests are as old as humanity itself. They serve to confirm that structures are capable of withstanding the relevant loads and to instill trust in the load-bearing capacity of important structures.

DOI: https://doi.org/10.1201/9780429265969

Supporting this trust in load-bearing capacity is the main goal of all proof load tests. It is achieved by carrying out a visual demonstration of the structural safety of the construction. New construction methods or novel structural elements had (and still have) to earn this trust by passing proof load tests.

4.1.2 The role of load testing in the development of reinforced concrete constructions in Europe

Owing to the large number of different construction types, in the early days of reinforced concrete construction (around the turn of the 20th century) ferroconcrete structures generally underwent proof load testing – their serviceability was confirmed by their ability to withstand the proof loads. For individual structures, proof load testing prior to operation was either mandated or strongly requested by the relevant department of building regulation. The tests were carried out according to the stipulations of the department representative. Construction firms were satisfied with obtaining an individual verification of the structural capacity of each particular construction and were rarely interested in deriving general knowledge about the structural behavior of ferroconcrete from the tests. To verify the adequate resistance of industrially manufactured structural elements against the expected loads, failure tests were also often carried out.

Figure 4.1 shows a failure test of so-called Visintini beams (von Emperger, 1908), which are one of the many different types of patented structural elements from the early days of reinforced concrete. Visintini beams are prefabricated reinforced concrete elements generally used in floor slabs and modeled on steel trusses. The loading machine developed by the engineer G. Hill for executing failure tests of structural elements (Fig. 4.2) represents a remarkable technical development in this area (von Emperger, 1908). Hill used a hydraulic press to apply the required compressive force. This not only allowed for the application of the correct load but also for a large number of tests to be carried out in a very small amount of time. Furthermore, it permitted the magnitudes of the applied load and the deformation of the test specimen to be read directly from the machine's gages. The special advantage of the hydraulic principle for preventing collapse, however, was not taken into consideration.

Figure 4.1 Loading test of Visintini beams
Source: G. (1904).

Figure 4.2 Hydraulic testing machine of Hill
Source: Von Emperger (1908).

The required load was generally applied by placing ballast mass – mostly sandbags or iron rails – on the structure. Even though the limited practical relevance of deflection measurements was known from the previous testing of iron bridges (Tetmajer, 1893; Zimmermann, 1884, 1881; Sarrazin, 1893), in reinforced concrete construction this parameter was also used as the sole criterion for assessing the structural safety.

While the proof load tests described above were used to prove that individual structural elements or entire structures exhibited sufficient structural capacity, systematic tests were carried out in research labs with the aim of developing theoretical methods for analyzing ferroconcrete. Emil Mörsch and Friedrich von Emperger were two of the main protagonists in Germany. These two engineers not only gained essential theoretical insights about ferroconcrete, but they also demonstrated the importance of load tests for the development of models to simulate the structural behavior of this composite material. Decades later, load tests were still being used to gain a better understanding of various aspects of the structural behavior of ferroconcrete structures. The so-called mushroom slabs, for example, were initially designed only with information gained from testing (Maillart, 1926).

There were, however, also problems and shortcomings in the execution of proof load tests. Due to patent protection, there was a large variety of construction types, and the sole objective of the developing engineers was to prove that their constructions were able to withstand the greatest load possible. Even laymen were allowed to conduct tests and advertise their products as being safe. In 1895, Bresztovszky (von Bresztovszky, 1895) therefore put on paper, for the first time, important requirements for the loading devices used in the tests. These included the requirement for the devices to be statically determinate to prevent them from contributing to the load transfer. This shows how early the requirements for the functional configuration of the loading elements were recorded. Unfortunately, however, they were rarely considered in in-situ tests.

Figure 4.4 shows an example of an incorrectly executed proof load test, in which the relevant cross-section area was increased by the tightly stacked sandbags, and therefore part of

Figure 4.3 Two loading tests on mushroom slabs
Source: Maillart (1926).

Figure 4.4 Faulty loading tests with insufficient movable loading
Source: Von Emperger (1902).

the load was not carried by the test specimen but transferred directly to the supports due to arching action (von Emperger, 1902).

Furthermore, the collapse of structures kept occurring despite the numerous proof load tests carried out (B., 1908). Von Emperger, for example, reports the collapse of a newly constructed factory in Berlin-Lichterberg, Germany, in which the top floor failed due to excessive load (von Emperger, 1914). The resulting subsequent failure of the lower floors caused the deaths of four factory workers. The sometimes-considerable consequences demonstrate the high risk associated with using ballast mass without anti-fall guards, and they show with how little care and knowledge some proof load tests were carried out at the time.

4.1.3   Development of standards and guidelines

As a consequence of the great number of load tests on concrete structures executed under various conditions and with greatly varying objectives, many experts questioned the sense of the tests and discussed their advantages and disadvantages (von Bresztovszky, 1895; Robertson, 1897).

There was a general consensus about the necessity for standardized rules for the design and execution of load tests. The initial attempts at developing such rules for the proof load testing of reinforced concrete structures are documented in the *Vorläufigen Leitsätzen zur Vorbereitung, Ausführung und Prüfung von Eisenbetonbauten* (Preliminary guidelines for the preparation, execution, and testing of ferroconcrete structures) released in 1904 (Deutscher Beton- und Bautechnik Verein, 1904). These guidelines, however, were of a very general nature. For instance, the rules were recommended to be used for structures with construction defects or structures that had been found to have insufficient structural capacity. The load was to be applied only 45 days after the pouring of the concrete, and an upper limit for the applicable load was given. The structure was deemed to be safe if no "appreciable permanent deformation had occurred" (*"nennenswerte bleibende Formänderungen nicht entstanden sind"*; translation from German by the authors) (Preußisches Ministerium der öffentlichen Arbeiten, 1907).

The *"Bestimmungen des Kgl. Preußischen Ministeriums der öffentlichen Arbeiten für die Ausführung von Konstruktionen aus Eisenbeton bei Hochbauten"* (Regulations of the Royal Prussian Ministry of Public Works for the execution of ferroconcrete constructions in buildings), issued in 1907 (Preußisches Ministerium der öffentlichen Arbeiten, 1907), contained little more specific information. Only in the *"Bestimmungen für Ausführung von Bauwerken aus Eisenbeton"* (Regulations for the execution of ferroconcrete structures), released in 1916 by the Deutscher Ausschuss für Eisenbeton (German Committee for Ferroconcrete) (DAfEB, 1916) can more detailed provisions be found. These provisions were adopted exactly or in a slightly modified form in many of the regulations issued at later dates. One of the recommendations was to carry out proof load testing only where absolutely necessary. The load was required to be "movable and capable of adapting to the deflection of the floor" (*"sie in sich beweglich ist und der Durchbiegung der Decke folgen kann"*; translation from German by the authors). The objective of this provision was to prevent the structural capacity from being overestimated due to the presence of arching action, as described in Section 4.1.2, or due to flexural action of elements of the loading device. The maximum load was to remain on the structure for a minimum of 12 hours. The most important measurement parameter was the deflection of the structure. The test was deemed to have been passed if the permanent deflection after removal of the load was less than 25% of the maximum measured deflection. These fundamental rules for the execution of load tests, for the requirements with respect to the load configuration, for the magnitude and duration of load application, and for the assessment of the maximum and permanent deflections were adopted by various other European regulations issued around the same time. The magnitude of the required applicable load, however, varied between 1.0 times the imposed load (Austria, the Netherlands, France, Belgium) and 1.5 times the imposed load (Germany, Denmark, Sweden, Norway) (von Emperger, 1928).

The execution of load tests under service load (1.0 times the imposed load) attracted some criticism. Von Emperger (von Emperger, 1921) draws attention to the particular structural behavior of non-ductile structures. He mentions prefabricated beams which were removed from an existing school building due to doubts regarding their static performance and which subsequently underwent failure tests. The beams failed at a load just above the service load, due to failure of the anchorage (Fig. 4.5).

The simple assessment criterion based on the ratio of the permanent deflection to the maximum occurring deflection was contained in all the regulations and attracted a particular amount of criticism right from the start. Mörsch suggests to "directly measure the magnitude

Figure 4.5 Tests of reinforced concrete beams with anchorage failure

Source: Von Emperger (1921).

of the stresses actually occurring in the iron and concrete during the proof load test, prefer-ably with suitable strain gages, so that one does not have to rely on the doubtful method of using deflection measurements if one is absolutely determined to prove the safety of the con-struction experimentally." ("*die Größe der tatsächlich bei der Probebelastung auftretenden Spannungen in Eisen und Beton am besten mit geeigneten Dehnungsmessern direkt zu messen, so dass man gar nicht auf den zweifelhaften Umweg über die Einsenkungen ange-wiesen ist, wenn man durchaus den Nachweis über die Sicherheit der Konstruktion experi-mentell erbringen will*"; translated from German by the authors) (Moersch, 1908). Despite the criticism, however, the ratio of the permanent and maximum measured deflections remained for many years the sole and almost unchanged assessment criterion in most of the regulations.

4.1.4 *Proof load testing overshadowed by structural analysis*

In the 1970s a great number of improved methods for analysis were developed, due mainly to the now widespread use of computers. This allowed the structural capacity of a newly built structure to be verified simply and safely by executing analytical calculations. Proof load testing for confirming the structural safety of new structures hence became obso-lete. This must be regarded as a positive development as it led to a decrease in effort and –

Figure 4.6 Foil tub for carrying the water loading
Source: Bader et al. (1982).

if the relevant rules were adhered to – an increase in safety. Additionally, the risk of collapse associated with load tests was eliminated.

The manner of dealing with existing structures, however, was and still is a critically important issue, particularly for structures with material or construction deficiencies, existing damage or similar issues that prevent the structural safety being determined with sufficient accuracy even with suitable analysis methods. Proof load testing was and still is considered by experts to represent a suitable approach to experimentally verify unclear analysis assumptions and to confirm that structures exhibit the required structural capacity.

A notable example from those years is described in Bader et al. (1982). The reinforced concrete floors of the regional psychiatric hospital in Wiesloch, Germany, exhibited construction deficiencies and deterioration due to corrosion. A load test using water ballast was carried out (Fig. 4.6) and the deflections were measured – the traditional approach was executed in all its glory.

4.1.5 *Further theoretical and practical developments of the recent past*

In Germany, the necessary steps for the development of the method of the experimental evaluation of the bearing safety of structural elements took place almost simultaneously in both the German Democratic Republic (GDR) and the Federal Republic of Germany (FRG), though the direction the developments were taking in the two countries was quite different. In the former GDR, Schmidt and Opitz developed the theoretical basis for the design and assessment of experimental evaluations (Opitz, 1988). This included the development of a corresponding safety concept based on the partial safety factors that are used in practically all of today's standards. This concept ensured a similar level of safety to that obtained with calculations of the structural capacity, yet it was tailored to the specific boundary conditions of load tests. Further to this concept, specific differentiated observation criteria were defined for each expected failure mode, which were to be used to assess the state of the structure during the test. All the rules were included in the regulation "TGL 33407/04 – *Nachweis der Trag-und Nutzungsfähigkeit aufgrund experimenteller Erprobung*" (Verification of the structural capacity and serviceability based on experimental tests) (TGL, 1986), which allowed the execution of load tests in the former GDR according to standardized rules from 1986 onwards.

One of the many practical examples of the time is described in Brandl and Quade (1988). In 1988, it was to be experimentally demonstrated that the floor above the basement of Großmarkthalle Leipzig, which had been reinforced with an additional concrete layer, was capable of withstanding an imposed load of 10 kN/m^2 (1.45 psi). The load was applied using product-filled pallets (Fig. 4.7) and the deflection was recorded with mechanical dial gages.

In the former FRG, a great amount of development in the area of testing technologies was carried out under the leadership of Steffens (Steffens, 2002). He designed a mobile loading device consisting of steel elements which, when combined with a hydraulic system, obviates the need for ballast mass to generate the load. The collapse of sufficiently ductile structures is prevented with this system. The rapid development of computers paired with electric measuring technology permitted in-situ evaluation of the recorded data and detailed assessment of the structural state of the tested object during the test.

German reunification gave rise to a flurry of activities in this area. In 2000, following further research, the German Committee for Structural Concrete (DAfStb) released the guideline "*Belastungsversuche an Betonbauwerken*" (Load testing of concrete structures)

Figure 4.7 Loading test of a basement floor at the market halls, Leipzig, 1988
Source: Brandl and Quade (1988).

Figure 4.8 Loading test of a basement floor at the market halls, Leipzig, 1992
Source: Steffens (2002).

(DAfStb, 2000), which provides a unified basis for the design, execution, and assessment of load tests.

In Steffens (2002), an interesting practical example of the proof load testing of a concrete floor is shown which demonstrates the developments mentioned above. Because of a change in use of the aforementioned Großmarkthalle Leipzig, the floor above its basement was required to undergo a proof load test with a superimposed load of 15 kN/m^2 (2.176 psi). In contrast to the tests carried out in 1988 (see Brandl and Quade, 1988), mobile hydraulic loading technology and online measurement technology were used instead of the considerable ballast mass and the mechanical measuring devices (see Fig. 4.8). The setup of the loading device shown in Figure 4.8 allowed for different spans to be loaded in quick succession.

4.2   Load testing of existing concrete building constructions

4.2.1   *Principal safety considerations*

The term "load testing" has been preserved due to its historical significance. Today, this term is also used for other types of tests using quasi-static loads. The following test types are distinguished:

- Systematic measurements at the service-load level (diagnostic load test)
- Application of load until the target test load or the limit test load limit is reached (proof load test)
- Application of load until the specimen fails (failure test).

It is not generally the aim of a load test to determine the actual structural capacity of a structural element; rather it is intended to verify the structural safety with respect to a defined action (Steffens, 2002). The goal is to apply the maximum imposed load F_k the structural element can sustain without suffering damage. The magnitude of the applicable load is the target test load F_{target}, which is determined by summing the relevant loads (including safety factors). The maximum allowable load of the test is the limit test load F_{lim}, that is, the load which if exceeded triggers a critical state that limits the structural capacity or serviceability of the test specimen. The challenge is hence to apply the highest possible test load in order to verify the ability of the structural element to resist the desired imposed load with a high factor of safety, while ensuring that the load does not cause damage in the structural element. The expressions and definitions mentioned above are commonly used in Germany and are used throughout this chapter to describe the standards in vigor in other countries.

Spaethe uses probability considerations to illuminate the influence of a successfully conducted load test on the structural capacity of the investigated structural element (Spaethe, 1994). He examines the level of safety, expressed by the safety index β, during and after proof load testing and concludes that, "from the safety-theoretical point of view, proof load testing can be useful, pointless or even harmful" ("*aus Sicht der Sicherheitstheorie eine Probebelastung nützlich, sinnlos oder gar schädlich sein kann*"; translation from German by the authors). The essential criterion for assessing the factor of safety achieved by a load test is the ratio of the maximum load applied during the test to the previously imposed load. Figure 4.9 shows that the application of the target test load leads to a decrease in the safety index during the test. Upon successful completion of the test, however, the level of safety increases because the understanding of the structural behavior of the construction has improved and the load has been sustained without any damage. When the tested construction has sustained a deterministic load s_p without showing any signs of

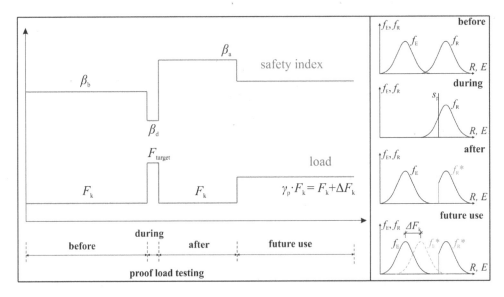

Figure 4.9 Explanation of the increase in safety and the identification of hidden bearing reserves by load testing, after Spaethe (1994)

damage, any structural capacity values smaller than the sustained loads are removed from the unknown probability density function. This causes the probability density function to be truncated (see Fig. 4.9) and marks a decrease in the uncertainty of the resistance. The gained safety can be used to increase the allowable imposed load ΔF_k. With his calculations, Spaethe also shows that a load test at the service-load level does not greatly increase the level of safety. The level of safety increases only when higher test loads are used, as weak spots can be identified and uncertainties anticipated in a controlled manner, making it easier to eliminate them (Spaethe, 1994).

4.2.2 Load testing in Germany

4.2.2.1 Introduction

Proof load testing still constitutes an important alternative for verifying the structural capacity of existing structures. In the early 1990s, systematic research was initiated to develop the scientific fundamentals of proof load testing by using modern experimental methods (Steffens, 2002). These research activities led to the creation of the guideline "*Belastungsversuche an Betonbauwerken*" (Load tests of concrete structures) (DAfStb, 2000), issued in 2000 by the German Committee for Structural Concrete. This guideline will henceforth be referred to as "the guideline for load tests." The use of this guideline resulted in numerous existing structures being preserved and their structural safety being proven experimentally (Gutermann and Guttenberg, 2011; Hampel et al., 2011; Czentner et al., 2010; Schacht et al., 2017; Hahn, 2013).

This guideline has not been endorsed by the building inspectorate. To base the structural design of a structure on experimental results only, the permissions of the owner and the competent authority are required (DIN EN 1990/NA:2010–12 [Eurocode 0, 2010], NCI to 5.2[1]).

According to (Eurocode 0, 2010), Section 5.2, the design of structures may be carried out in conjunction with tests if no suitable models are available, for instance, or if the design assumptions need to be verified. In Appendix D, the fundamentals of the test-based design are explained in more detail and information about the safety-theoretical evaluation of the test results is given. Additionally, an extensive classification of the various types of tests is proposed. The classification proposed in Table 4.1, however, is considered by the authors of this chapter to be more suitable for the experimental investigations of existing structures.

In this classification, the experimental investigations on existing structures can be divided into four groups (A to D), according to their character and the required effort. The scope of application of the guideline for load tests is restricted to investigation types B, C1, and C2.

The planning and execution of load tests generally needs to be coordinated beforehand with the responsible structural engineer and the competent building inspectorate, because a repetition of the tests in the case of disagreements is associated with very high costs. Furthermore, the tests may only be carried out by testing laboratories disposing of the necessary knowledge and experience, as well as of loading and measurement devices suited to the requirements.

If a structural element is tested according to test type B (global investigation; see Table 4.1), the entire structure is subjected to service loading. Various structural reactions of the loaded system, such as the deflections, support rotations, and strains, are recorded. This test is used for the calibration of calculation models and provides insight into

Table 4.1 Classification of experimental investigations on existing structures

Type of experimental investigation	Objective of the test	Comments
A Material tests and investigations of the geometry at selected locations	Determination of material properties, geometry, or the general state of the structure	Nondestructive tests or tests causing negligible damage to the specimens; basis for the analytical investigations of the limit states
B Global investigation at the service-load level	Assessment of the criteria of the serviceability limit state (SLS), determination of the global structural behavior, hybrid verification of the structural capacity (at the ULS) by computational extrapolation of the results	Verification of design assumptions, improvement of numerical models. Use extreme caution with extrapolations because of the nonlinear effects
C1 Proof load test executed to the target test load F_{target} (expected imposed load multiplied by partial safety factors)	Direct experimental verification of the structural capacity with respect to a stipulated live load, as well as verification of serviceability and sufficient durability	Standard application of proof load testing, assessment of the structural reserves for planned increases in the imposed load
C2 Proof load test executed to the limit test load F_{lim} (load which, if exceeded, causes damage to the test specimen)	Direct experimental verification of the structural capacity with respect to the maximum possible live load	Increased measurement and monitoring effort. High risk of structural damage, especially if the expected mode of failure is unknown
C3 Failure test	Direct experimental verification of the structural capacity	Destructive test, very useful for safety assessments, suitable if a great number of identical elements (such as prefabricated slabs) are to be assessed
D Monitoring	Verification of time-dependent analysis assumptions. Monitoring of durability or deterioration processes	Costly, durable measurement technology, large amounts of data, especially for dynamic measurements

system parameters that are difficult to quantify by other means (such as multiaxial load transfer and the contribution of additional cross-area sections). A thus improved calculation model is subsequently used to analytically reassess the serviceability and ultimate limit states. Caution is advised, however, if the insights gained from the measurements are to be extrapolated for assessment of the ultimate limit state, as the structural behavior exhibited by a system under service load and under ultimate load are not necessarily the same. Significant changes in the resulting nonlinear system behavior can occur if friction forces are overcome (in construction joints and composite joints, for example), there is local stability failure (lateral buckling of compression chords, for instance), local yielding occurs, or crack formation processes take place. All of the above can have a negative impact on the global structural capacity of the structural element, and the recalibrated calculation model might not be able to take these changes into account in a reliable manner.

Henceforth, the expression "load test" will be used only for test types C1 and C2. Tests for executing global measurements at the service-load level according to Table 4.1 will not be discussed further.

Where the exact structural behavior of existing structures is not known, analytical structural assessments are generally carried out by making conservative assumptions with regard to the acting loads, material properties, structural element geometry, and the analytical models used. Furthermore, structural reinforced concrete elements contain not only the reinforcement required for static reasons; some reinforcement is added purely for installation and construction purposes. This frequently leads to an accumulation of structural reserves. The actual structural capacity of a structural element is often greater than that determined by theoretical considerations.

It is often impossible to identify these structural reserves by analytical means, as not all the parameters required for the assessment of the structural capacity of an element can be determined with sufficient accuracy and reliability. These parameters include:

- Fixed supports, continuity effects, or biaxial load transfer;
- Mechanical and geometric properties of the reinforcing and prestressing steel or other types of structural steel inserts (in floor slabs supported by lightweight steel beams), unless these components can be extracted and tested;
- The effectiveness of strengthening measures or the interaction of various structural elements or various segments of structural elements (composite action);
- Wear, damage, and construction deficiencies, as well as design-relevant parameters for reinforced concrete elements that exhibit great scatter (concrete cover, concrete compressive strength, corrosion damage, etc.).

Frequently, it is exceedingly difficult just to determine the necessary information about the reinforcement type, amount, and layout if none or only part of the necessary building documentation is available. In such cases, load tests can be used to provide an in-situ stability analysis which allows for an integral assessment of the structural safety of the existing structure and for the activation of structural reserves for the transfer of the relevant permanent and variable loads.

4.2.2.2   Basics and range of application

Load tests may be carried out only in conjunction with analytical stability verifications. Typical cases in which load tests are carried out are:

- Conversion of buildings, resulting in additional dead load and imposed loads
- Where the building documentation is missing or incomplete
- To assess the effects of damage (for example resulting from fires or flooding) or age-related wear
- To assess structural elements for which insufficient design information is available or which cannot be analyzed using current design methods (such as reinforced block floors).

Load tests on existing structural members serve to simulate the ultimate limit state for the relevant imposed loads and the additional dead load, while making sure that neither failure of the structure occurs nor serviceability criteria are exceeded. To achieve this, the guideline for load tests (DAfStb, 2000) defines the limit test load $F_{\lim}$ as the load which, when exceeded, causes damage that will affect the structural capacity and serviceability of the structure during its service life. The limit test load must under no circumstances be exceeded during the test. It is hence, by definition, always smaller than the maximum load the structure can sustain.

To identify the limit test load, the guideline for load tests (DAfStb, 2000) lists a number of deformation criteria, which can be subdivided into the following categories:

- Concrete strain limits, i.e., the negative concrete strains must remain in the range of stable microcracking
- Steel strain limits, i.e., the reinforcing steel must remain in the elastic range
- Limiting values for the load-dependent crack widths and crack width changes
- Limiting values for the total deflection and the permanent deflection after unloading
- Limiting values for the deformations in the shear zone of the beams; this applies to the compressive strains in concrete compression struts and to the tensile strains in the shear reinforcement.

The limit test load $F_{\lim}$ is also considered to have been reached if

- Experimentally determined parameters, such as the load-deformation behavior or noise emissions, indicate critical changes which upon further load increase would cause damage to the structure;
- The stability of the structure is in danger;
- Critical support displacements have occurred.

Some of these criteria have been found not to be suitable or applicable to actual tests (Bolle, 1999). The criterion requiring the steel stress to be limited to 90% of the elastic limit, for example, cannot be ensured in a sensible manner during tests. Neither the elastic limit nor the existing pre-strain is known before the test; the location of the maximum strain also cannot be determined exactly.

The magnitude of the remaining structural reserve (the distance between the limit test load and the ultimate load) is strongly dependent on the properties of the actual system and the governing criteria for the limit test load. It cannot be quantified exactly, even upon completion of the test, and hence cannot be utilized.

The objective of load tests, however, is not generally the determination of the limit test load but the verification of the structural capacity with respect to the expected additional dead load and imposed loads. Therefore, the guideline for load tests defines the target test load F_{target}. The target test load is determined from the load the structure is expected to sustain (characteristic load) multiplied by the relevant safety factors. This load is generally significantly higher than the maximum load resulting from the past or future use of the structure and is hence a load that the structure has never experienced. The portion of the target test load that is externally applied to the structure (i.e. the actual test load) is called $\text{ext}F_{\text{target}}$.

To determine the target test load, the following equation can be used:

$$\text{ext}F_{target} = \sum_{j>1} \gamma_{Gj} \cdot G_{k,j} + \gamma_{Q,1} \cdot Q_{k,1} + \sum_{i>1} \gamma_{Q,i} \cdot \Psi_{0,i} \cdot Q_{k,i} \leq \text{ext}F_{ult} \tag{4.1}$$

$extF_{target}$ External part of the target test load (excluding existing self-weight)
$extF_{ult}$ External part of the ultimate test load (excluding existing self-weight)
$\gamma_{G,j}$ Partial safety factor for dead load (for existing dead loads: $\gamma_G = 1.0$, for superimposed dead load $\gamma_G = 1.35$)
$G_{k,j}$ Characteristic value of dead loads not present during the test
$\gamma_{Q,1}$ Partial safety factor for the dominant live load $\gamma_Q = 1.50$
$Q_{k,1}$ Characteristic value of the dominant live load
$\gamma_{Q,i}$ Partial safety factor for additional live loads $\gamma_Q = 1.50$
$\Psi_{0,i}$ Combination factor for additional live loads $\Psi_{0,i} = 0.4 - 0.8$
$Q_{k,i}$ Characteristic value of additional live loads

The self-weight of the structure is fully effective during a test and is therefore not considered in the calculation of $extF_{target}$. The parameter $G_{k,j}$ represents additional permanent loads that are not effective during the test (loads resulting from future floor constructions, for instance). These loads must be multiplied by the relevant partial safety factors and added to the test load. If the test results are extrapolated to assess other, not directly tested structural elements, additional safety elements further to the safety factors present in Equation 4.1 must be taken into account.

The minimum externally applied target test load on the right side of the equation governs for structural elements with very high self-weight and ensures that $extF_{target}$ does not, under any circumstances, exceed 1.35 to 1.00 times the permanent loads present during load tests.

The guideline for load tests (DAfStb, 2000) permits only the execution of nondestructive tests, which is why only structures exhibiting early signs of impending failure may undergo load testing. These signs are not just used to understand when failure is imminent but also to identify the limit test load so that damage and the resulting decrease in structural capacity and serviceability can be prevented.

Structural elements for which brittle failure is expected are hence ineligible for nondestructive load tests according to DAfStb (2000). In recent research projects (Marx et al., 2011, 2013; Schacht, 2014) it has been shown, however, that even structural elements with low ductility (such as those with insufficient or no shear reinforcement) exhibit measurable early signs of failure that can be recorded with special, combined measurement methods (acoustic emission measurements, close-range photogrammetry, section-wise deformation measurements).

In some cases it can be helpful to carry out failure tests or tests where the failure of the structural element is accepted (Schacht et al., 2017). These types of tests, however, are outside the scope of the guideline for load tests (DAfStb, 2000). For the execution of such tests, measures have to be taken to minimize the effects of the failure of the specimen (using auxiliary structures to catch falling pieces, for example).

Due to the relatively high costs, load tests should be carried out only if preliminary analysis indicates sufficient structural reserve, and successful verification of the structural capacity can reasonably be expected. It is particularly economical to carry out load tests on a small number of structural elements and use the results to assess a large number of similar elements of a structure (such as floor slabs or floor beams with identical dimensions and material properties). Particular attention must, however, be paid to the selection of the structural elements to undergo direct testing, as they must be sufficiently representative of the other, untested elements. In any case, additional safety factors must be applied to the target test load in order to account for the usually existing tolerances in the material and system properties across the entire system.

4.2.2.3   *Planning of loading tests*

Load tests should be carried out during the structural assessment phase of the structure, prior to rehabilitation, conversion, or extension. In some cases, the test results can contain information that is crucial for the type, scope, and economy of such measures. On the other hand, the planning must be at a stage at which the essential parameters of the intended future use are already known, as these influence the target test load and the validity of the test results. This applies to information such as

*   The type and extent of future use;
*   The type and magnitude of the expected additional dead load (self-weight of floor constructions and building services; dividing walls);
*   Decisions on whether to retain or replace parts of the building that contribute to the load transfer but cannot be considered in the structural analysis (floor constructions acting compositely with the floor slab, for example);
*   The definition of the necessary structural changes (such as additional openings or changes to load-bearing walls).

If, for example, a floor slab acting fully or partially compositely with the floor screed is subjected to a load test, the test results apply to the floor slab including the floor screed, and hence it is not allowed to remove the floor screed at a later time.

Before a load test can be planned in detail, the static and constructive characteristics of the structure, its state of structural health, and the static interactions between the individual structural elements must be determined in sufficient detail. This also includes the verification of the essential material parameters of the load-bearing structural elements.

The load to be applied to the tested elements must lead to the actions necessary for verification of the structural capacity to occur in a sufficiently large region of the construction. This means, for example, that the loaded area of floor slabs intended for uniaxial load transfer must be large enough to eliminate the possibility of load being transferred laterally in the central region of the slab. Where load can be applied only to some areas of the construction, transverse load transfer must be prevented in other ways (by executing cuts at the edges of the test span, for example). A similar approach is to be followed if, for example, beneficial effects of non-load-bearing, solid dividing walls located below the tested structural element are to be excluded. In such cases, at least one layer of brick or stone of the dividing wall must be removed.

The load must be generated and applied in such a way that the structure can be unloaded quickly and completely should it enter critical structural states. Therefore, it is imperative to use a special loading technology coupled with a hydraulic system (see Fig. 4.10). Using this technology, the hydraulically generated test loads are tied back in the vicinity of the tested structural element. Constraints are created, and this provides an intrinsic safety feature in the setup provided that the loading device is sufficiently stiff. If the structure deflects, the hydraulic pressure decreases and the entire system moves into a safe state of equilibrium. Rapid unloading is also possible by manually reducing the hydraulic pressure. Thus, collapse of the tested construction is rendered nearly impossible. A further

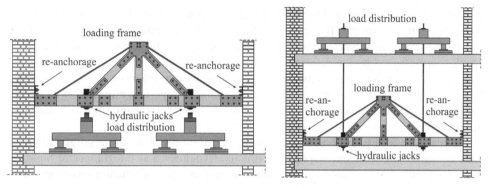

Figure 4.10 Different possibilities of self-protecting load generation
Source: Marx et al. (2015).

advantage is the very sensitive control of the load application, allowing the load-deformation behavior to be recorded over the entire load range, which is required for the assessment of the test results according to the guideline for load testing (DAfStb, 2000). Figure 4.10 show different options for implementing this closed, intrinsically safe force loop.

In load tests used to verify the structural capacity, the load is not allowed to be applied directly with ballast mass due to the unsafe handling required in the case of critical structural states and the latent risk of collapse.

During the test, the load must be applied to the construction in such a way that the verification-relevant load distribution is achieved with sufficient accuracy. For floor slabs of buildings, for example, for which uniformly distributed load can generally be assumed, the load is applied at 16 load introduction points across the testing area. A sufficiently stiff and statically determinate load distribution construction is required for all tests (see Fig. 4.10). Furthermore, it must be ensured that the loading device can adjust to the deformation of the structural element and no hidden alternative load paths exist.

Electronic load cells can be used to record the magnitude of the applied loads with high precision. An additional plausibility check can be carried out by a manometer integrated in the hydraulic loading system. It is generally not advisable, however, to use only manometers to measure the forces due to the low measurement accuracy of such devices.

The choice of measurement devices to be used for recording the response of the structural element always depends on the actual objective of the test and the boundary conditions, as well as the governing criteria for the limit test load (see Section 4.2.2.2).

In most of the cases (i.e. in tests on sufficiently ductile structural elements), traditional deformation measurement techniques may be used. Generally, the following types of deformations are distinguished:

- Global deformations (e.g. deflections)
- Local deformations (e.g. strains)
- Section-wise deformations (e.g. curvature or relative displacements).

Global deformations yield information about the global deformation behavior of the structure. Early structural changes, however, sometimes cannot be detected clearly. Local deformation measurements, on the other hand, offer an exact impression of the behavior of an albeit limited area of the structure. Therefore, it is generally best to use a combination of both types of measurements. It is important for deformation measurements to be carried out with respect to a deformation-neutral location, that is, with respect to a reference plane which is not affected by the deformation of the tested structure.

In addition, acoustic emission measurements can be used to monitor crack formation and friction processes. If the sensors are positioned suitably, important information about the magnitude, type, and region of microcrack and macrocrack formation due to a load can be garnered.

The planning of the test concludes with the development of a testing program, which must be agreed on with the responsible structural engineer and the competent building inspectorate prior to test execution and should at least contain information about the building owner, the tested structure, and the testing laboratory. Furthermore, it should contain the following information:

- Calculation of the limit test loads
- Test setup including loading device, load introduction, and load distribution
- Safety and retaining devices, if required
- Type and specifics of the measurement technology, and layout of the measurement locations
- Testing procedure
- Governing criteria for determining the limit test load and how to measure those criteria.

The person developing the testing program should ideally also be responsible for executing the test and assessing the test results (the main investigator).

4.2.2.4 Execution and evaluation

During the test, the structure undergoes a loading regime consisting of loading and unloading phases. The structure of the loading regime depends on the objectives of the test. According to DAfStb (2000), the load must be increased in at least three load steps to reach the target test load. After each step, the structure is to be unloaded to the minimum load level. At each loading and unloading step, the load must be kept constant for a certain (suitable) amount of time. It is recommended to execute several load steps up to the target test load F_{target}, with one representing the service load F_k. Furthermore, it is advisable to repeatedly load structures to important load levels, as this allows early nonlinear load-deformation behavior to be observed as early as possible (see Fig. 4.11).

The parameters measured during the test, that is, the applied force and the most significant structural responses, are recorded online, displayed graphically on a monitor as a load-structural response diagram or other type of graph which allows real-time assessment by the testing engineer. The measurement values should be recorded and displayed at an interval of one second or less to permit the timely capture and assessment of the present structural state.

The measurement results are evaluated using the limit test load criteria defined prior to the test. The experimental assessment of the structural state of the structure, however, also requires sufficient experience in carrying out experimental investigations, and specific knowledge of the structural and deformation behavior of the tested constructions.

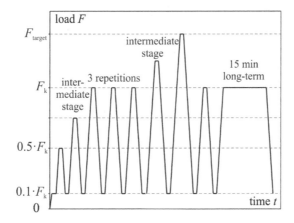

Figure 4.11 Typical loading sequence for an experimental proof load test

To evaluate the structural capacity of a structural element subjected predominantly to bending, load-deflection diagrams are generally used. The deformation measurements are used to identify beginning nonlinear deformation and changes in stiffness due to crack formation, and to determine elastic and plastic deformation components. Figure 4.12 shows the load versus midspan deflection of a very slender reinforced concrete building floor from the 1920s, recorded during a load test. By increasing the load step by step, the structural response under increasing load can be assessed in real time. For a complete assessment, however, the unloading curves are also required. These can be used to evaluate the permanent deformation induced by each increase in load and the energy consumed by irreversible structural changes (changes in the area of the hysteresis loop) (Bolle, 1999).

If the risk of a sudden failure of the tested structural element cannot be eliminated reliably despite careful preliminary investigations, an effective retaining structure underneath the construction is required for safety reasons and to limit damage consequences. This structure should be positioned as closely as possible to the tested structural element so that the impact loads occurring due to failure of the element are minimized. At the same time, the tested structural element must not be restricted from deflecting freely.

The structural capacity of the structural element with respect to the imposed loads and additional dead load used to determine the target test load is considered to have been verified if the target test load can be applied to the structure without any of the limit test load criteria being reached.

If one of the limit test load criteria is reached during the test, no further load is allowed. The reached limit test load can then be used to derive the allowable additional dead load and imposed loads, always taking into account the partial safety factors defined in the testing program.

The results of the load test are documented in a test report, which should include information about the testing laboratory, the responsible testing engineer, details of the test procedure and the measurement results, as well as the following information:

- Result of the structural health assessment from the preliminary investigations
- Essential information from the testing program
- Maximum test loads and the characteristic loads derived from this load

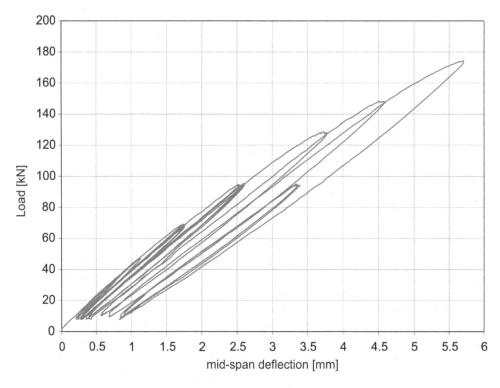

Figure 4.12 Load-deflection graph for a proof load test on a reinforced concrete floor slab. Conversion: 1 mm = 0.04 in., 1 kN = 0.225 kip

- Evaluation of the measurement results
- Information about special requirements and boundary conditions for future use and information about the validity of the test results.

The last point generally applies to issues such as:

- Restrictions with respect to changes to structural elements that have been identified in the test as load-bearing (such as floor constructions acting compositely with the floor slabs or edge beams that add restraints – elements that cannot readily be removed from the construction);
- Restrictions with respect to static-constructive changes (such as the adding of openings to existing structures);
- Limits to the validity of the test results if deterioration processes are taking place at the time of testing (progressive corrosion of the reinforcement, for example) or corresponding stipulations to ensure sufficient durability.

4.2.3   *Load testing in the United States*

In the United States, regulations for the execution of load tests have been part of the concrete standards since their inception. Today they are contained in standard ACI 318M-14 (ACI,

2014). The historical development of the regulations for load tests in the United States has been described in ACI (2007). With respect to importance and objectives, it shows strong similarities with the development of load tests in Germany (Bolle et al., 2010a, 2010b). Although load tests were still providing the ultimate verification of the structural capacity of new structural elements in the early 20th century, their importance has since decreased due to the increase in accuracy of analytical methods. Contrary to the situation in Germany, however, the rules regarding load tests were never removed from the US standards.

Starting in 2004, the regulations for load tests were revised by ACI Committee 437 (ACI, 2007), and an approach to the execution and assessment of load tests on existing structures was proposed as an alternative to the approach given in ACI 318, Section 20. In 2013, the revised regulations contained in ACI 437 (ACI, 2013a) were included in ACI 562–13 (ACI, 2013b). The regulations contained in the two standards are described in detail in Chapter 3 of Vol. 12.

4.2.4 Load testing in Great Britain

Before the Eurocodes were introduced, standard BS 8110-II (British Standard, 1989) contained the regulations for the execution of experimental investigations on reinforced and prestressed concrete structures. Even though the rules were generally intended for tests carried out during or shortly after construction, they were also used for the experimental evaluation of existing structures (Bungey and Millard, 2006; Menzies, 1978). Despite the current lack of explicit rules, which is due to the fact that BS 8110 was replaced by a Eurocode, it can be assumed that the recommendations contained in BS 8110-II (British Standard, 1989) are still being used today. According to these recommendations, failure tests or tests in which the structure is loaded to its design loads should only be executed in exceptional cases. The objective is the verification of the structural safety by carrying out tests with loads at or slightly above the service-load level. The measured deformations should be used for calibrating analytical models and the verification should be carried out using the calibrated model. The target test load should be greater than the characteristic loads from self-weight G_k and live load Q_k, and meet the following requirements:

$$G_k + Q_k \leq F_{target} \leq \min\left\{ \begin{array}{l} 1.0 \cdot G_k + 1.25 \cdot Q_k \\ 1.125 \cdot (G_k + Q_k) \end{array} \right\}$$ (4.2)

The test loads should be increased and decreased incrementally. The maximum test load should be applied at least twice, and the interval between the two load applications should be no less than one hour. The measurement values should be recorded 5 minutes after each load increase in order for them to reach stable values. Even though it is not compulsory to execute a test in which the structure is loaded to the target test load and this load is kept constant over 24 hours, such a test can be carried out if necessary. The measured deformations are compared with the corresponding calculated values. If significant deflections take place during a test, the elastic recovery after the second load application must be approximately equal to the elastic recovery after the first load application, and it must be at least 75% of the total deformation for reinforced concrete and at least 85% of the total deformation for prestressed concrete. Additionally, the deflections and crack widths observed under service load must fulfill the requirements stipulated in the standard. If the measured deflections are smaller than $l/1000$, with l the span length, the evaluation of the elastic recovery can be omitted.

In Bungey and Millard (2006), the authors describe several load tests executed in Great Britain and recommends higher target test loads, as the factor of safety achieved using a load determined with Equation 4.2 is too small. The Institute of Structural Engineers (Institution of Structural Engineers, 1976) and Lee (Lee, 1977) propose target test loads of $1.25 \cdot (G_k + Q_k)$ and $1.50 \cdot (G_k + Q_k)$, respectively. Menzies (1978) also suggests a higher target test load, namely 1.4 to $1.5 \cdot (G_k + Q_k)$, in order to achieve a sufficient safety distance to the service load. According to Bungey and Millard (2006), ballast mass is generally used to generate the test loads, since hydraulic systems are considered too expensive and tie-back systems too cost-intensive. Incrementally recorded load-deformation diagrams, from which various indicators of early nonlinear deformation behavior can be extracted, are used for the assessment. Specific parameters and their corresponding limiting values, however, are not given in Menzies (1978).

4.2.5   Load testing in other countries

In Canada, load tests on buildings are dealt with in Section 20 of Canadian Standard A23.3–04 (Canadian Standard, 2004). Tests on unbonded post-tensioned structures and structures with suspected corrosion damage are not allowed. The target test load is required to be at least 90% of the design loads, and if the test results are used to assess similar, untested structural elements, this needs to be 100% of the design loads. Further stipulations with regard to applying test results to untested structural elements are not given. The load must be applied to the structural element during 24 hours, and the elastic recovery must be evaluated after a further 24 hours. At that time, the deflection of prestressed and reinforced concrete elements must have decreased by at least 80% and 60%, respectively. If these conditions are not fulfilled, the test on the structural element can be repeated. This second test should be carried out approximately 72 hours after the first test. The decrease in deflection after this test must be at least 75%.

Australian standard AS 3600 (Australian Standard, 2009) contains only a few rules regarding the execution of load tests on structural reinforced and prestressed concrete elements. In tests carried out to determine the structural safety of existing structural elements, the target test load should be equal to the design load and should be applied to the structural element in constant increments. Deformation (deflection) and crack formation are to be observed carefully during load application. At the first signs of damage the test must be aborted. The structural element is considered to have sufficient structural capacity if the structural element has been subjected to the design load for 24 hours and no sign of damage or excessive cracking has been observed. Tests to verify the serviceability are carried out under service load and are evaluated using the serviceability criteria stipulated in the standard.

In Switzerland, SIA 269 (SIA, 2011) allows the experimental determination of the structural behavior of existing structures. In the tests it must be shown that only elastic deformations take place under a defined load and that no excessive cracking, vibration, or displacements occur. The determined structural response of the structural element must be interpreted by means of a suitable structural model. No information is given about the magnitude, duration or type of loading or the deformations to be measured. In Egger (1998), a practical example is described, in which hydraulic load generation is used, the test load is increased in a few steps to 1.8 times the service load, and the structural state is assessed using the load-deformation behavior. It is not disclosed how the test load was determined.

In Italy, the reinforced concrete standard (Norme, 2005) allows load tests on new structural elements, but no specific information about the planning, execution, or evaluation of such tests is given. Tests may be carried out only by engineers with at least 10 years of relevant experience. The engineers are hence responsible for how the tests are executed and for the interpretation of the test results. Some very general information on this is given in Casadei et al. 2006). For static tests, the test load needs to be incremented in small steps and kept constant while deformation measurements are taken. The target test load is calculated from the infrequent combination of the service loads. The deformations should be proportional to the load and no sign of impending failure should be visible (cracks, spalling). The permanent deformation, including initial plastic settlements, shall not exceed a certain percentage of the maximum deformation. If this latter condition is not met, the structure can be loaded again up to the maximum load, and the structure must exhibit largely elastic behavior. Furthermore, the measured maximum deformations shall not exceed the corresponding calculated deformations (Casadei et al., 2006).

In (federation internationale du béton, 2002) the International Federation for Structural Concrete (fib) provides fundamental information on how to execute load tests carried out to verify calculation models or to confirm the structural capacity of a structure. To verify the serviceability, the target test load must be equal to or greater than the service load and must not be greater than $0.25 \cdot G_k + 1.25 \cdot Q_k$. The load must be kept constant for 24 hours in order for the deformations to stabilize. Tests for determining the structural safety should demonstrate that the safety distance to the service load is sufficiently large. However, no minimum value is given. The load should be increased in five to ten increments and the deformations at each load increase should be recorded. There is no intermediate unloading of the structure. If the deformations enter nonlinear territory, the load should be kept constant so that the development of the deformations can be observed. If the deformations increase, the structural element must be unloaded immediately and the elastic recovery must be recorded. The load may be generated by ballast mass; water is considered to be a suitable medium (federation internationale du béton, 2002). The use of hydraulic loading systems is considered to be too cost-intensive due to the required tie-back system. The measured deformations are to be compared with the calculated limit values determined prior to the test, and the serviceability of the structure must not be compromised by the test. A significant increase in cracking and deformation during the test is not permitted to occur. As in the Italian standard, many years of experience are required for anyone executing such tests.

4.2.6 Comparison and assessment

Comparing the various regulations, it can be seen that the approach in most of the standards is quite traditional. The most striking attribute of these approaches is the recommendation for the use of ballast mass to generate the test load. Because the load is applied incrementally and in a temporally and spatially rather unregulated manner, it is not possible to obtain a continuous load-deformation curve. Therefore, the quality of the test result evaluation is rather modest. Only simple assessment criteria can be used, such as the ratio of the permanent to the maximum deformation. These criteria, however, are not real limiting criteria. Furthermore, the use of ballast mass poses a great risk of material damage and injuries to persons, due to the latent risk of collapse.

Due to these reasons, the described approach was already considered unsatisfactory as early as 60 years ago (Roebert, 1957), and this sentiment has not changed to this day. The complete lack of relevant publications describing executed load tests suggests that this sentiment might be the reason that tests on existing structures are rarely executed in countries with traditional or very "soft" regulations.

At the same time, only a very few publications on the further development of the experimental evaluation of the structural safety, carried out in the 1980s and the 1990s in Germany, are widely known in other countries. The guideline for load tests (DAfStb, 2000) is known only in a few of Germany's neighboring countries.

The opinion of some authors (such as those of federation internationale du béton [2002]) that load tests with hydraulic loading technology are more cost- and labor-intensive than those using ballast mass, is not shared by the authors of this chapter. They believe that, if such loading technology is available, the opposite is true. The load tests can be executed much more rapidly and safely. Especially in investigations on one sample of multiple identical structural elements, the hydraulic loading technology offers significant economic advantages due to its quick load application and removal and the small mass to be moved within the structure.

Furthermore, the quality of the test result evaluation is higher, because continuously recorded loading and unloading curves can be obtained by means of the hydraulic loading system combined with an electronic measurement data acquisition system. These curves can subsequently be used to perform checks against various extended assessment criteria, for example those given in the guideline for load tests (DAfStb, 2000) or in ACI 437 (ACI, 2013a).

The use of this technology, however, requires a comparatively high initial investment, and special training must be given to the persons working with it. The cost is therefore only justified if load tests are a part of engineering practice and not just used in highly exceptional cases.

A further important attribute of load tests given in the individual standards is the magnitude of the applicable load for tests serving to confirm the structural capacity of structural elements. Various approaches can be distinguished. In some cases, there is no clear correlation between the safety factors applied in the calculations and those of the tests. Some standards, such as BS 8110-II (British Standard, 1989), assume that the analytical ultimate limit state can be extrapolated from the structural state reached under the comparatively low test loads. For reinforced concrete structures, however, this assumption only holds in rare cases due to the generally small amount and low quality of information available about existing structures. The applied limit test loads are too small for the verification of the structural safety (i.e. without extrapolation of the observed structural state) (Spaethe, 1994).

Furthermore, it can be observed that the traditional assessment criteria and also most of the modern ones are based on detecting the onset of nonlinear deformations from the recorded load-deformation behavior exhibited by the tested structural element during the test. The criteria consider the onset of nonlinear deformation to be a structural change and use it to identify when the structure approaches its ultimate limit state. In applying these criteria, it is very important to clearly separate flexural crack development and plastic deformations so as to avoid misinterpretations when the structural element transitions into the cracked state. The traditional, simple criteria combined with the load generation by ballast mass are not or not very well suited to this task, as the correlation between the load and deformation is measured only in very few locations (Schacht et al., 2016a).

In DAfStb (2000), acoustic emission analysis is described as an effective complementary tool for monitoring crack development. This type of analysis, however, is not mentioned in any of the international standards and guidelines analyzed in this chapter.

The authors believe that it would be useful to take the criteria that are intended to be used for the deflection (the global deformation parameter) and apply them also to region-wise deformations (such as section-wise curvature or strains/relative displacements in the shear zone). This would allow not only the identification of the limit test load but also the localization of the structural change. However, confirmation would be needed that limit values similar to that for the deflection can be developed for these additional deformation parameters.

The admittedly very difficult question of how far the test results of a sample can be used to assess identical, untested structural elements is not addressed in many regulations. Notable exceptions are ACI 437 (ACI, 2013a) and the Canadian standard (CSA, 2004). This issue is dealt with in a very undifferentiated manner by the two standards: they stipulate a general load increase. The actually occurring scatter in the governing parameters is thus reflected in a limited manner.

In every guideline and standard analyzed, it is stressed that load tests may be carried out only by engineers with sufficient experience in the structural assessment of structural reinforced and prestressed concrete elements. This is an essential requirement, since the structural responses of existing structures can vary widely and each structural element has its own specific boundary conditions which require individual assessment (Schacht et al., 2016a).

4.3 New developments

4.3.1 Safety concept

Due to economic reasons, only a limited number of load tests are generally carried out on a structure, and the results are assumed to also apply to the structural elements in the structure or part of the structure that are identical to those tested. Choosing representative samples is difficult. In most cases, it is possible to test only some of the structural elements due to constructive or operational constraints. In order to apply the test results of a sample to other structural elements, the applicable target test load must be multiplied by additional safety factors. To carry out verification for all structural elements, even those not tested directly, a higher test load is applied to compensate for the scatter in the resistance of the individual structural elements.

In the simplest case a single structural element is tested to show that the element can withstand x times a certain load without exhibiting any signs of damage, and it is concluded that all the other structural elements can therefore safely withstand 1 times this load. It stands to reason that if an individual structural element withstands a very high test load (three times the service load, for example), there is a sufficient margin of safety to apply the assessment results to other, untested structural elements. It does not always make sense, however, to use this approach in practice, as in theory the structural elements do not have any structural reserves, and an unnecessarily high test load is more likely to induce failure than to prolong the life of the elements. Testing a larger number of structural elements with a lower test load is often the more sensible approach. Thus, it must be decided how many structural elements must be tested under what load in order to allow the reliable assessment of other, untested structural elements.

In a load test, only very few structural elements of a structure are tested. They are not tested to failure but loaded to a target test load defined prior to the test, under which no structural damage is expected to occur. The target test load must also represent the resistance of all the similar structural elements with a sufficient factor of safety. The question of where the tested structural element is positioned within the actual scatter band of the resistance is very important too. The authors of this chapter are currently developing a method for determining the safety factors that need to be applied if the test results of tested structural elements are to be extrapolated to assess untested structural elements (extrapolation factors). They are using the properties of the structural elements and the boundary conditions of the test to derive these factors.

The objective is to achieve, with the help of these extrapolation factors, the same safety level as that of new structures. The extrapolation factors take into account the ratio of the number of tested samples to the number of total structural elements, and the number of individual tests. A higher number of individual tests increases the reliability of the extrapolation, which allows a lower test load to be used.

The extrapolation factors can be reduced further for individual structures by taking a detailed survey of the structure's condition, choosing particularly "bad" samples, estimating the damage occurring during application of the target test load, and by taking into account additional information obtained from analytical investigations or the test measurements. The decision of whether to do this falls to the responsible engineer and must be justified for each case. The method for determining the extrapolation factors will be included in the revised version of the German guideline for load testing (Schacht et al., 2018).

4.3.2   Shear load testing

In order to achieve a sufficiently reliable assessment of the structural state during load tests, the structure must show some signs when failure is imminent. A warning of impending failure is traditionally considered to have been given if a structure exhibits clearly visible signs prior to its transition into a critical state. If the load is subsequently increased further, some structural reserve is still available initially; however, the load increase leads to large deformations and excessive cracking – that is, to irreversible damage (plastic deformation) of the structure. Not all existing structures can be assumed to exhibit early signs of failure, as many of them are under-reinforced and there is a risk of brittle shear failure. One mode of failure with particularly few signs of impending failure is the shear failure of structural reinforced concrete elements with no or very little shear reinforcement, which is why this type of failure and its measurable warning signs have been the subject of several research projects. The focus of these projects has been on the early detection of shear crack formation, as the crack formation process is of crucial importance for the structural behavior and cannot always be inferred from deflection measurements.

For structural elements without shear reinforcement, shear crack formation (local damage) occurs at the same time as the ultimate limit state (global damage) is reached. Whether the load can be increased further after the formation of shear cracks in structural elements without shear reinforcement depends on many influencing factors (load type, effective width, shear slenderness, etc.) and cannot be determined in a definite manner. Therefore, the initiation of shear crack formation must be detected as early in the test as possible. Hence to facilitate the detection of cracking, it is particularly important to predict the location and development of the shear cracks with the highest possible accuracy.

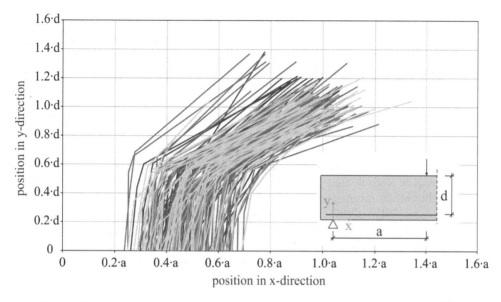

Figure 4.13 Position and development of shear cracks from based on assessment of 171 shear force tests

Source: Schacht (2014).

The origin and progression of the critical shear crack in beams without shear reinforcement can be predicted by referring to images of crack patterns published in the literature. The evaluation of 171 images of crack and failure patterns available in the literature showed that the critical flexural shear crack formed in the region between $0.25 \cdot a$ and $0.70 \cdot a$ (Fig. 4.13). In 157 beams the cracks formed in the region between $0.3 \cdot a$ and $0.65 \cdot a$, in nine beams it formed in the region beyond $0.65 \cdot a$, and in five beams it developed at a distance of less than $0.3 \cdot a$ from the support (Schacht, 2014). The region in which the orientation of the cracks reached 45° was between $0.25 \cdot a$ and $0.73 \cdot a$ in the x-direction and between $0.13 \cdot d$ and $0.70 \cdot a$ and $0.56 \cdot d$ in the y-direction.

The following crack states have been defined as critical structural states with respect to shear damage:

- Development of diagonal cracks with orientations between 45° and 60° with respect to the longitudinal axis of the beam;
- Development of diagonal cracks which intersect the line of action between the load and the support or which extend beyond mid-height of the beam and the flexural cracks;
- Development of dowel cracks.

The same definitions apply to structural elements with shear reinforcement. However, elements with shear reinforcement can sustain significantly more load after local damage has occurred, because the shear reinforcement can sustain the forces released by the formation of the crack. Whether shear failure occurs depends on the amount of shear reinforcement, but failure will always be ductile.

In order to examine whether the theoretical considerations for the assessment of the structural behavior are suitable for engineering practice and to derive fitting indicators and criteria for the detection of the limit test load, experimental investigations on reinforced concrete beams with differing amounts of shear stirrups (including beams with no stirrups) were carried out (Schacht, 2014). To allow the damage development in the beams to be recorded under increasing load, a combination of measurement techniques was used (Schacht, 2014): photogrammetry, acoustic emissions, traditional deformation measurements in the shear field, inclination sensors, and so forth. The local deformations in the web of the reinforced concrete elements were recorded with a photogrammetric measurement system.

Using photogrammetry to detect and visualize cracks is a very effective way to track crack development during the tests. The utilization of photogrammetric measurement systems, however, leads to more labor-intensive test preparation, execution, and evaluation, but the quality of the results justifies the increased effort (Schacht et al., 2017). The displacements measured at discrete points can be used to determine any desired displacement or strain. To demonstrate the possibilities of photogrammetric image analysis, the crack development in the right-hand shear zone of a tested beam tracked with this method is shown in Figure 4.14. Acoustic emission analysis is another accompanying measurement technique that can prove useful in experimental investigations for detecting developing cracks and for localizing regions with increased cracking activity.

The limit test load was detected correctly in all the tests. The following threshold load tests were used to prove that the local damage present at the time of the detection of the limit test load did not affect the structural behavior at the service-load level. The detection of the limit test load was carried out as soon as local damage in the shear zone occurred,

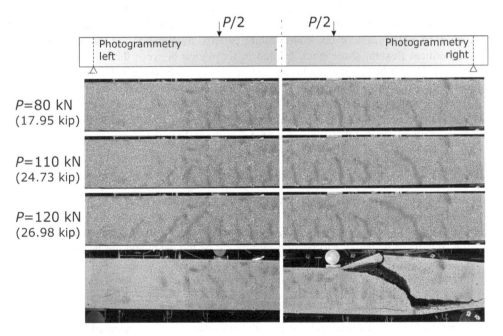

Figure 4.14 Observation of the crack development with photogrammetry
Source: Schacht et al. (2016b).

that is, at the state at which the diagonal cracks grew past the mid-height of the beam or extended beyond the flexural cracks, or the crack angle exceeded 45°. The crack state, that is, the development or inclination of the diagonal crack or the vertical position of the crack tip, was the governing criterion in all the tests.

The beams without stirrups did not exhibit very ductile behavior after the limit test load had been reached: the load was increased only between 10% and 50% of the determined limit test load before failure occurred. The scatter of the load increase–shear slenderness relation is quite large and shows that a distinctly conservative limit test load criterion is required for structural elements without shear reinforcement. The limit test loads identified in the tests on the beams with stirrups represent very conservative limit values. The stirrups permitted a significant amount of additional load to be applied beyond the limit test load (between 59% and 82%). A suitable criterion for determining the limit test load for this type of beam is a limit to the deformation in the diagonal cracks which signals plastic deformation of the shear reinforcement (Schacht et al., 2016b).

For the revised version of the guideline for load tests, a criterion of when to abort the test was defined: the limit test load is assumed to have been reached as soon as diagonal cracks with crack angles <60° with respect to the longitudinal axis of the structural element develop in the shear zone. To monitor the structural state during the test, the use of photogrammetry and acoustic emission analysis is recommended.

4.4 Practical recommendations

The possible range of application of load testing for experimental evaluation of the bearing safety is wide and covers almost all fields of concrete constructions. Load testing is always a good solution if the calculation doesn't allow the exploration of all bearing reserves, for example because:

- An adequate mathematical model is missing
- Construction information is missing
- Material parameters are missing
- The influence of damage or faulty construction on the bearing behavior is unknown.

The experimental evaluation of the bearing safety can lead to considerable reduction of costs for the reconstruction of existing building structures by fulfilling objectives of preservation and guaranteeing a future use. A combined use of calculation and experimental methods can lead to further benefits and uses the advantages of both approaches.

For the planning of load tests a detailed investigation and assessment of the existing structures to be tested has to be carried out. The following should be checked in detail:

- Geometry, bearing condition, existing deflection or deformation
- Existing reinforcement (type, amount, condition, position, and anchorage)
- Existing cracks (crack widths, development, load history)
- Material investigations (concrete, reinforcement)
- Damage, corrosion, moisture penetration.

Based on this information, the calculation has to be updated and the decisive failure modes have to be identified.

The loading equipment has to fulfill the requirements of a possible rapid removal of the load in case of a beginning failure. This imperatively requires hydraulic loading devices. The load has to be applied stepwise in loading and unloading cycles. For planning, execution, and evaluation of load tests, an experienced team is required.

In general, the recommendations and rules of the German guideline for testing of concrete structures can be also applied for steel, timber, or masonry buildings. The special modes of failure for these specific construction materials have to be accounted for to be able to early detect minimal structural changes with appropriate measuring equipment to be able to avoid any structural failures. Specific criteria have to be determined for every individual case.

4.5 Summary and conclusions

This chapter presents the current practice of load testing of building constructions in Germany. The historic origin and the meaning of load testing as proof of ultimate bearing safety is discussed and it is shown which theoretical and practical developments during the last 20 years have led to a comeback of load testing as an accepted alternative proof of bearing safety of existing concrete structures. Load testing allows the further use of existing structures and a minimization of necessary strengthening. With the help of the German Guideline of load testing, the bearing safety of an enormous amount of existing structures could be proven successfully and therewith many existing structures could be preserved.

References

ACI (2007) *Load Tests of Concrete Structures: Methods, Magnitude, Protocols and Acceptance Criteria* (ACI 437.1R-07). American Concrete Institute Committee 437.

ACI (2013a) *Code Requirements for Load Testing of Existing Concrete Structures* (ACI 437–13). ACI Standard and Commentary. ACI Committee 437.

ACI (2013b) *Code Requirements for Evaluation, Repair, and Rehabilitation of Concrete Buildings* (ACI 562–13). ACI Standard and Commentary. ACI Committee 562.

ACI (2014) *Building Code Requirements for Structural Concrete (ACI 318M-11)*. ACI Standard and Commentary. American Concrete Institute Committee 318.

Australian Standard (2009) *Concrete Structures: AS 3600*. Standards Australia, NSW, Australia.

B., G. (1908) Einsturz eines Eisenbetongebäudes in Mailand. *Schweizerische Bauzeitung*, 51(18), 235–236.

Bader, U., Feuchter, G. & Stiglat, K. (1982) Versuche an Stahlbetondecken aus dem Jahre 1905 im Psychiatrischen Landeskrankenhaus in Wiesloch. *Beton- und Stahlbetonbau*, 77(3), 74–79.

Bolle, G. (1999) *Zur Bewertung des Belastungsgrades biegebeanspruchter Stahlbetonbauteile anhand von Last-Verformungs-Informationen*. Dissertation, Bauhaus Universität Weimar, Weimar, Germany.

Bolle, G., Schacht, G. & Marx, S. (2010a) Geschichtliche Entwicklung und aktuelle Praxis der Probebelastung – Teil 1: Geschichtliche Entwicklung im 19. und Anfang des 20. Jahrhunderts. *Bautechnik*, 87, 700–707.

Bolle, G., Schacht, G. & Marx, S. (2010b) Geschichtliche Entwicklung und aktuelle Praxis der Probebelastung – Teil 2: Entwicklung von Normen und heutige Anwendung. *Bautechnik*, 87, 784–789.

Brandl, H. & Quade, J. (1988) Erprobung der Trag- und Nutzungsfähigkeit einer Kellerdecke, Experimentelle Erprobung nach TGL 33407/04 an der Großmarkthalle Leipzig. *Bauplanung – Bautechnik*, 42(10), 443–446.

British Standard (1989) *Structural Use of Concrete, Part 2: Code of Practice for Special Circumstances*. British Standard Institution, London, UK.

Bungey, J. H. & Millard, S. G. (2006) *Testing of Concrete in Structures*, 3rd edition. Blackie Acadamie, Glasgow, Scotland.

Canadian Standard (2004) *Design of Concrete Structures (A23.3–04)*. Canadian Standards Association.

Casadei, P., DeLuca, A., Serafini, R. & Agneloni, E. (2006) Assessment of concrete structures through in-situ load testing: An Italian prospective. *The Second FIB International Congress*, Naples, Italy.

Czentner, G., Fiedler, L., Kapphahn, G. & Steffens, K. (2010) Deutsches Archäologisches Institut DAI in Rom – Hybrider Tragsicherheitsnachweis für Bibliothekslasten und Erdbeben. *Bautechnik*, 87, 127–132.

DAfEb (1916) *Bestimmungen für Ausführung von Bauwerken aus Eisenbeton, Bestimmungen des DAfEb*. DAfEb, German Standard for Reinforced Concrete Constructions, Berlin.

DAfStb (2000) *Belastungsversuche an Betonbauwerken*. Beuth Verlag GmbH, German Guideline for Load Testing, Berlin, Deutschland.

Deutscher Beton- und Bautechnik Verein (1904) *Vorläufige Leitsätze zur Vorbereitung, Ausführung und Prüfung von Eisenbetonbauten*. German Standard for Reinforced Concrete Constructions, Berlin, Germany.

Egger, G. (1998) Statische Belastungsversuche an Gebäudedecken. *Schweizer Ingenieur und Architekt*, 116(36), 644–647.

Eurocode 0 (2010) *DIN EN 1990: Grundlagen der Tragwerksplanung inkl.* Nationaler Anhang DIN EN 1990. Beuth Verlag.

federation internationale du béton (2002) *Management, Maintenance and Strengthening of Concrete Structures*. Schriftenreihe der fédération internationale du béton (fib) bulletin 17, Lausanne.

G., O. (1904) Weitere Versuche mit Gitterträger System Visintini. *Beton und Eisen*, 3(1), 42–44.

Gutermann, M. & Guttenberg, U. (2011) Groß, Größer, Großmarkt. Wie man den experimentellen Tragsicherheitsnachweis für 40.000 m^2 Geschossdecke führen kann. *6. Symposium Experimentelle Untersuchungen von Baukonstruktionen, Schriftenreihe Konstruktiver Ingenieurbau Dresden*, (24), 95–106.

Hahn, O., Meichsner, E., Krämer, W. D. & Lorenz, P. (2013) Experimentelle Untersuchungen zum Tragverhalten des Fußbodens des Marmorsaals im Neuen Palais Potsdam. *7. Symposium Experimentelle Untersuchungen von Baukonstruktionen, Schriftenreihe Konstruktiver Ingenieurbau Dresden*, (32), 143–154.

Hampel, T., Opitz, H., Michler, H., Popp, T. & Scheerer, S. (2011) Experimentelle Überprüfung der Tragsicherheit von Stahlbetonrippendecken. *Bautechnik*, 88, 42–46.

Institution of Structural Engineers (1976) Report of working party on high alumina cement. *Structural Engineer*, 54, 352–361.

Lee, C. R. (1977) *Load Testing of Concrete Structures, with Particular Reference to CP110 and Experience in the HAC Investigations*. Build. Research, Dok. B507.

Maillart, R. (1926) Zur Entwicklung der unterzugslosen Decke in der Schweiz und in Amerika. *Schweizerische Bauzeitung*, 87, 263–265.

Marx, S., Maas, H.-G., Schacht, G., Koschitzki, R. & Bolle, G. (2011) Versuchsgrenzlastindikatoren bei Belastungsversuchen. *Abschlussbericht zum Forschungsvorhaben "Zukunft Bau"*. IRB Verlag Fraunhofer. Available from: http://nbn-resolving.de/urn:nbn:de:bsz:14-qucosa-76111.

Marx, S., Schacht, G., Maas, H.-G., Liebold, F. & Bolle, G. (2013) Versuchsgrenzlastindikatoren bei Belastungsversuchen II. *Abschlussbericht zum Forschungsvorhaben "Zukunft Bau"*, IRB Verlag Fraunhofer. Available from: http://nbn-resolving.de/urn:nbn:de:bsz:14-qucosa-130176.

Marx, S., Bolle, G. & Schacht, G. (2015) Kapitel 3.2 – Bestandsaufnahme und Bestandsbewertung und Kapitel 7 – Bewertung der Tragfähigkeit auf Grundlage von Belastungsversuchen. In: Fingerloos, F., Schnell, J. & Marx, S. (Hrsg.) *Tragwerksplanung im Bestand*. Betonkalender, Ernst & Sohn, Berlin.

Menzies, J. B. (1978) Loading testing of concrete building structures. *Structural Engineer*, 56, 347–353.

Moersch, E. (1908) *Der Eisenbetonbau – seine Theorie und Anwendung*. Konrad Wittwer Verlag, Stuttgart.

Norme (2005) Norme tecniche per le costruzioni. *Decreto del Ministerio dei trasporti e delle Infrastrutture*, Gazzetta Ufficiale.

Preußisches Ministerium der öffentlichen Arbeiten (1907) Bestimmungen des Kgl. Preußischen Ministeriums der öffentlichen Arbeiten für die Ausführung von Konstruktionen aus Eisenbeton bei Hochbauten. *Prussian Standard for Reinforced Concrete Constructions*, Berlin.

Robertson, J. R. (1897) Die Nutzlosigkeit der Probebelastungen eiserner Brücken. *Bulletin de la Commission internationale du Congrès des chemins de fer*.

Roebert, S. (1957) *Kritische Einschätzung der Probebelastungen an Stahlbetonbiegeträgern nach DIN 1045 §7 unter besonderer Berücksichtigung der Verformungsberechnung*. Dissertation, TU Dresden, Dresden, Germany.

Sarrazin, O. (1893) Brücken-Einsturz in Frankreich. *Centralblatt der Bauverwaltung*, 308.

Schacht, G. (2014) *Experimentelle Bewertung der Schubtragsicherheit von Stahlbetonbauteilen*. Dissertation, Technische Universität Dresden, Fakultät Bauingenieurwesen, Dresden, Germany.

Schacht, G., Bolle, G. & Marx, S. (2016a) Belastungsversuche – Internationaler Stand des Wissens. *Bautechnik*, 93(2), S.85–97. doi: 10.1002/bate.201500097.

Schacht, G., Bolle, G., Curbach, M. & Marx, S. (2016b) Experimentelle Bewertung der Schubtragsicherheit von Stahlbetonbauteilen. *Beton-und Stahlbetonbau*, 111(6), 343–354. doi: 10.1002/best.201600006.

Schacht, G., Bolle, G. & Marx, S. (2017) Shear load testing of damaged hollow core slabs. *Structural Concrete*, 18, 607–617. doi: 10.1002/suco.201600082.

Schacht, G., Bolle, G. & Marx, S. (2018) Load testing of concrete structures in Germany: General practice and recent developments. In: Hordijk, D. A. & Lukovic, M. (eds.) *Proc. of the 2017 fib Symposium. High Tech Concrete: Where Technology and Engineering Meet*, Maastricht, Netherlands, June 12–14, 2017.

SIA (2011) *Grundlagen der Erhaltung von Tragwerken (SIA 269)*. Ingenieur-und Architektenverein, Schweiz.

Schmidt, H. & Opitz, H. (1988) Experimentelle Erprobung von Stahlbetonbauwerken in situ. *13. Kongress des IVBH*, Helsinki.

Spaethe, G. (1994) Die Beeinflussung der Sicherheit eines Tragwerks durch Probebelastung. *Bauingenieur*, 69, 459–468.

Steffens, K. (2002) *Experimentelle Tragsicherheitsbewertung von Bauwerken*. Ernst & Sohn, Berlin, Germany.

Tetmajer, L. (1893) Ueber die Ursachen des Einsturzes der Morawa-Brücke bei Ljubitschewo. *Schweizerische Bauzeitung*, 21/22, 55–58.

TGL (1986) *TGL 33407/04 – Nachweis der Trag-und Nutzungsfähigkeit aufgrund experimenteller Erprobung*. German Standard for Reinforced Concrete Constructions, Berlin, Germany.

von Bresztovszky (1895) Probebelastung von Decken und Gewölben. *Centralblatt der Bauverwaltung*, 433–434.

von Emperger, F. (1902) Die Durchbiegung und Einspannung von armierten Betonbalken und Platten. *Beton & Eisen*, 4 Theil, 21–35, [10].

von Emperger, F. (1908) *Handbuch für Eisenbeton, Band 1, Entwicklungsgeschichte und Theorie des Eisenbetons*. Verlag von Wilhelm Ernst & Sohn, Berlin.

von Emperger, F. (1914) Bestimmungen des Polizei-Präsidenten von Berlin vom 21. Oktober 1914, betr.: Unzulässige Deckenbelastungen. *Beton & Eisen*, 19(12), 365–366.

von Emperger, F. (1921) *Handbuch für Eisenbetonbau, Band 8 – Feuersicherheit, Bauunfälle*. Verlag von Wilhelm Ernst & Sohn, Berlin.

von Emperger, F. (1928) *Handbuch für Eisenbetonbau, Band 9 – Die in-und ausländischen Eisenbetonbestimmungen*. Verlag von Wilhelm Ernst & Sohn, Berlin.

Zimmermann, H. (1881) Ueber Unterhaltung und Dauer von Drahtseilhängebrücken. *Centralblatt der Bauverwaltung*, 346–347.

Zimmermann, H. (1884) Einsturz einer Straßenbrücke bei Salez in der Schweiz. *Centralblatt der Bauverwaltung*, 548–549.

Part III

Advances in Measurement Techniques for Load Testing

Chapter 5

Digital Image and Video-Based Measurements

Mohamad Alipour, Ali Shariati, Thomas Schumacher, Devin K. Harris, and Charles J. Riley

Abstract

In order to capture the deformation response of a bridge member during load testing, the structure has to be properly instrumented. Image- and video-based measurement techniques have significant advantages over traditional physical sensors in that they are applied remotely without physical contact with the structure, they require no cabling, and they allow for measuring displacement where no ground reference is available. This chapter introduces two techniques used to analyze digital images: digital image correlation (DIC) and Eulerian virtual visual sensors (VVS). The former is used to measure the static displacement response and the latter to capture the natural vibration frequencies of structural members. Each technique is introduced separately, providing a description of the fundamental theory, presenting the equipment needed to perform the measurements, and discussing the strengths and limitations. Case studies provide examples of real-world applications during load testing to give the reader a sense of the advances and opportunities of these measurement techniques.

5.1 Introduction

Image- and video-based measurements represent an emerging tool in the area of structural health monitoring (SHM) and assessment, providing an approach for remotely monitoring the behavior of bridges. These tools are particularly well suited for in-service load testing applications as they provide a form of non-contact measurement, meaning that the measurement tool (camera or video) need not be in contact with the bridge during testing. These methods of measurement have been successfully deployed on an increasing number of bridge load tests in recent years and are becoming more of a common tool for load testing (Alipour et al., 2019; Harris et al., 2015; Hebdon et al., 2017; Jauregui et al., 2003; Jiang and Jauregui, 2010; Ojio et al., 2016; Schumacher and Harris, 2016; Yoneyama and Murasawa, 2009). This chapter describes two specific approaches within these classes of measurement techniques, namely digital image correlation (DIC) to measure deformations and Eulerian virtual visual sensors (VVS) to capture natural vibration frequencies. The main difference between the two approaches lies in how the data, which are digital images in both cases, are analyzed. DIC tracks the displacement of a patch of pixels between two images, therefore, using Lagrangian coordinates; Eulerian VVS simply monitor the changes in intensity for a pixel or a patch of pixels with a fixed coordinate.

DOI: https://doi.org/10.1201/9780429265969

· The following sections describe the fundamental theory, equipment needed for deployment, and strengths and limitations for each method. Additionally, each approach is also illustrated through field case studies.

5.2   Digital image correlation (DIC) for deformation measurements

5.2.1   *Theory*

In recent years, the application of photogrammetry and photogrammetric assessment methods has shown a great deal of promise for describing the behavior (Citto et al., 2011; Ghrib et al., 2014; Murray et al., 2015; Lawler et al., 2001; Peddle et al., 2011; Xiao et al., 2012; Yoneyama et al., 2007) and condition state (Lattanzi and Miller, 2012; Vaghefi et al., 2015, 2012) of civil infrastructure systems due to their non-contact nature, relative ease of deployment, and recent improvement of imaging technologies. Digital image correlation (DIC) is a technique that extends from the principles of photogrammetry and can be described as a full-field non-contact surface measurement technique. DIC utilizes image correlation and tracking algorithms on a series of sequential images to match and track displacements/deformations of a specimen subjected to loading (Sutton et al., 2009). Figure 5.1 provides a schematic of the DIC concept, but a more comprehensive description and illustration of the method can be found in the literature (Chu et al., 1985; Peters et al., 1983; Sutton et al., 1983; Yoneyama and Murasawa, 2009). DIC remains an area of extensive study in the field of experimental mechanics, but only a summary of the key principles of DIC are described here, with an emphasis on features relevant to large-scale behavior measurement such as bridge live load testing.DIC leverages the rich dataset of pixel intensity inherent to a digital image and their corresponding spatial locations within dataset to find correspondence and essentially track the movements and deformations of pixel patterns. Structural components being measured using DIC typically include a unique and stochastic contrasting pattern on the surface to facilitate the correlation of pixel intensity, but to mitigate correspondence challenges associated with the identification of pixel uniqueness within the pattern, the images are analyzed over series of smaller regions, typically designated as subsets or facets. This discretization ensures the

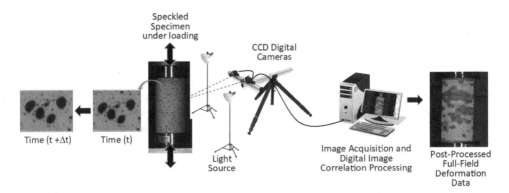

Figure 5.1 Illustration of digital image correlation (DIC) methodology to estimate deformation

uniqueness of the correspondence or "correlation" within an image and across an image sequence. The correspondence between the undeformed image (subsets within image) and the deformed image is determined using a cross-correlation criterion or a sum-squared difference correlation criterion, which ultimately attempt to match and align pixel intensity within the region of interest to describe pattern movement (Pan et al., 2009). Between the sequence of images, the DIC technique tracks the spatial position of the pattern movements within these subsets to describe displacement and the pattern deformation to describe strain.

DIC was originally developed using a single camera for measurement, which allows for in-plane displacements and deformations to be measured (2D-DIC); however, more recent advances have included the extension of the general approach to three dimensions (3D-DIC) using stereo-paired cameras to enable out-of-plane measurements. A physical scale is required in 2D-DIC, which can be established within the image or specific details of the experimental setup (i.e. sensor size, focal length, distance from point of measurement), to translate pixel displacements into real-world dimensions. The extension to 3D-DIC utilizes a calibration process through which the intrinsic properties of the camera pair are determined and ultimately used to generate real-world scaling through the calibration process. Regardless of the approach (2D or 3D) used, both have potential applications for bridge load testing; however, this chapter will primarily emphasize 2D-DIC applications for load testing with some references to 3D-DIC where appropriate.

5.2.2 Equipment

The physical equipment required for DIC-based measurement is primarily limited to an appropriate camera or imaging system for acquisition of images. For DIC, the fundamental principles behind image-based measurement stem from the relationships between internal imaging system parameters (i.e. focal length, sensor size, resolution) and the field of view (FOV). Available literature provided by commercial vendors suggests that the measurement resolution that can be achieved using DIC is on the order of multiple hundred thousands of the FOV for in-plane measurements of displacement and deformation and roughly half of the in-plane resolution for out-of-plane measurements of deformation (Correlated Solutions Inc., 2017, 2013; Dantec Dynamics A/S, 2018). This resolution is a lower bound on the correlation process and assumes an ideal setup with minimal noise in the measurement; however, this noise is ultimately inherent to the measurement scenario with contributions from features such as the camera, lighting, wind, and so forth. Using the pinhole camera relationship, Equation 5.1 coupled with known camera parameters Figure 5.2, the relationship between lower bound of resolution becomes a direct relationship with camera distance and field of view Equation 5.2. The specifics of the camera capabilities are ultimately a function of the experimental setup but can include most modern digital camera systems with sufficient spatial resolution. For DIC, images can be acquired from a number of sources including charge-coupled device (CCD) cameras, digital single-lens reflex (SLR) cameras, point and shoot, or even high-resolution video; however, applications for load testing are likely to place additional constraints on the camera due to environmental constraints (distance, reference point, etc.).

$$\frac{w}{W} = \frac{h}{H} = \frac{f}{d} \rightarrow W = FOV = w\frac{d}{f} \tag{5.1}$$

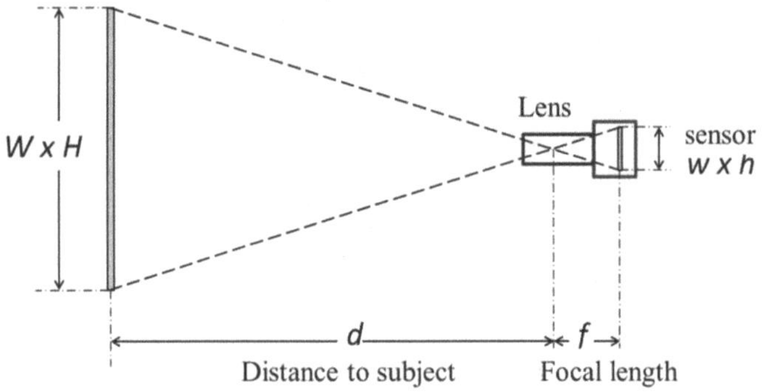

Figure 5.2 Relationship between camera system configuration and field of view (FOV)

$$\varepsilon = \frac{1}{50,000} \; W = \frac{1}{50,000} \, w \frac{d}{f} \qquad (5.2)$$

Beyond the physical equipment, three additional components are necessary for DIC-based measurements: (1) suitable pattern for image correlation; (2) lighting; and (3) digital image correlation software or algorithms for post-processing of the images. The pattern used for most DIC applications is typically a high-contrast speckle pattern (e.g. a white base with random black dots) applied to the surface of the area of interest. The applied pattern creates the necessary intensity contrast needed for the correlation algorithm to distinguish, map, and track the unique patterns of pixels within a subset region. The use of natural texture inherent to the surface in lieu of an applied speckle pattern is also feasible (Reu, 2015; Yang et al., 2010), which may be of high value for bridge load testing. As DIC has historically been used in controlled laboratory settings, artificial lighting is often part of the experimental setup to ensure uniform illumination of the region of interest. However, in outdoor applications, lighting may not be necessary and in fact shielding may be required to ensure that images are not washed out or over-saturated. In scenarios where illumination is required (e.g. night time testing, or heavily shaded area), additional lighting may be required to illuminate the area of interest. Commercial DIC software is available from a number of vendors including Correlated Solutions, LaVision, GOM, Dantec, MatchID, and Trilion, to name a few. This list is not exhaustive, but provides an illustration of the availability of off-the-shelf tools, with many vendors offering both 2D and 3D packages. There also exist a number of user-developed programs available as open source codes that are readily available for use, albeit typically without integrated image acquisition and control capabilities native to the commercial applications.

5.2.3   Strengths and limitations

The strengths and limitations of DIC-based displacement measurements from a load testing perspective are summarized in the following subsections.

5.2.3.1   *Strengths*

- *Non-contact measurement*: The technique does not require sensors to be physically affixed to the structure being tested. Measurements can be taken from a remote stand-off distance, so long as a suitable pattern for tracking is within the field of view.
- *No fixed ground reference required*: Unlike traditional deflection measurement techniques, the application of DIC can be deployed without a fixed ground reference. For this, cameras need to be mounted at a location which does not deflect or deform during live loading.
- *High resolution measurement with sub-pixel accuracy*: Deformations on the order of 1/100,000 of the field of view can be achieved in theory. Additionally, DIC provides the ability to resolve full-field deformations.
- *Flexibility and ease of deployment*: The camera systems can be simply mounted to the structure or supported by a tripod for measurement. Speckle patterns can be applied to the structure directly or provided via a mounted target for correlation and tracking.

5.2.3.2   *Limitations*

- *Cost of software*: Commercial DIC software has a relatively steep initial capital cost. However, open source codes are also freely available.
- *Potential noise in measurement*: Measurements derived from DIC are potentially impacted by both experimental and environmental noise. Experimental noise can be attributed to factors such as variability in the equipment setup, proper lighting, focusing, and scaling/calibration. Environmental noise can be attributed to factors such as fluctuations in lighting, wind, and ambient vibration.
- *Durability of equipment*: The primary equipment used are cameras, which are generally not well suited for environmental exposure. Cameras are potentially vulnerable to the environment from factors such as moisture and temperature. Additionally, the cameras also utilize fixed or variable focus lenses, which can accumulate dust and debris, potentially affecting quality of the images collected.
- *Lack of real-time results*: Commercial DIC systems are capable of capturing real-time results, but in this scenario cameras must be connected to a computer system during acquisition. For the scenarios described in this chapter, real-time results are not easily collected as most of the camera systems do not have onboard processing.
- *Data storage requirements*: With the system collecting images as the primary data source, the potential for large datasets is likely and is dependent on the image file size, speed of image acquisition, and duration of testing.

5.2.4   *Case study*

5.2.4.1   *Structural system details and instrumentation*

The case study selected for illustration of the DIC technique for load testing is the Hampton Roads Bridge-Tunnel (HRBT) which carries I-64 and US-60 over Hampton Roads from Norfolk, to Hampton, Virginia, as shown in Figure 5.3. The bridges that make up the approaches to the tunnel are low level trestle bridges with a concrete deck on top of pre-stressed concrete girders. The bridges were constructed as side-by-side structures with the first crossing being built in 1956 and the second being built in 1976. The westbound

Figure 5.3 Hampton Roads Bridge-Tunnel (HRBT): (a) Google map and (b) photo courtesy of Tom Saunders, VDOT

Table 5.1 Camera system specifications

Item	Specifications
Camera model	GoPro Hero 3+ Silver modified with Ribcage mount to accommodate C-mount SLR lens
Image sensor	CMOS 1/2.7"
Imaging capabilities	Single image (up to 10 MP) Time lapsed images (max 2 image/second) Video (60 frames/second) at 1080p resolution
Lens	Tamron Auto Focus 70–300 mm f/4.0–5.6 Di LD Macro Zoom Lens (Model A17NII)
Focal length	70 to 300 mm
Aperture	f/4 to 5.6 (aperture manually modified to f/8 to accommodate low light scenarios)

crossing structure was widened by 10 ft (3.05 m) in 1995. The bridges were tested in 2016 as part of a study to assess their load rating, as the structures have shown significant deterioration including concrete cracking and spalling as well as exposed prestressing strands with corrosion and section loss. The testing included spans in each direction of HRBT system and was performed with the intent of minimizing traffic disruption along the busy corridor. This objective resulted in conducting instrumentation from below using a spud barge and conducting the live load testing at night with staged lane closures. Span 57 was tested for the eastbound structure and Span 34 was tested for the westbound structure. Testing included traditional affixed instrumentation mounted on the superstructure members for measuring strain and acceleration as well as a multi-camera system mounted to the pier cap to measure deflections at midspan. Details of the traditional instrumentation are not included herein as this summary emphasizes the DIC-based measurement, but results will be included in a future publication. Due to the complex testing environment, the project team used commercial GoPro 3+ Silver cameras for image acquisition and fixed size target cards to establish the scaling factor. These cameras are low cost and compact with a camera resolution up to 10 MP and a video resolution up to 1080p. To accommodate the distance between camera and measurement location, the cameras were modified to accommodate telephoto lenses (Tamron AF 75–300 mm f/4.0–5.6). Details of the camera and lens system are presented in Table 5.1 and Figure 5.4. Figure 5.5 provides a basic

Figure 5.4 GoPro DIC system components: (a) GoPro camera, (b) GoPro camera with lens mount, and (c) variable telephoto lens used in conjunction with (b)

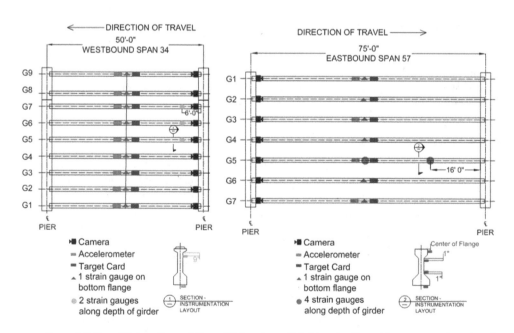

Figure 5.5 Span 34 (westbound)/Span 57 (eastbound) bridge details and instrumentation plan. 1 ft = 0.305 m

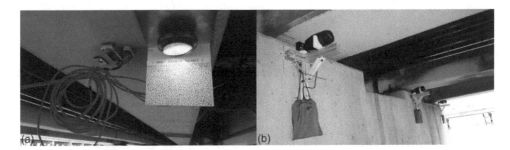

Figure 5.6 Field-deployed 2D-DIC displacement measurement system: (a) DIC target with local lighting, (b) GoPro camera with bracket mount

illustration of the measurement schematic and basic bridge details. The eastbound structure has 75 ft (22.9 m) spans and girders are labeled from left to right when looking in the direction of traffic (east to west). The westbound structure has 50 ft (15.2 m) spans and girders are labeled right to left when looking in the direction of traffic (east to west).

5.2.4.2   *Testing*

During the load tests, cameras were mounted at the pier cap directly beneath each girder and aligned to acquire images of targets placed at midspan (Fig. 5.6).The cameras were mounted on a rigid aluminum frame and speckled targets were adhesively bonded to the girders at midspan. The targets were made in a predetermined size (6 in. × 6 in. [152 × 152 mm]) to provide scale and speckled with a random dot pattern for the correlation analyses. Because the testing was performed at night, each target was outfitted with a battery-powered light to sufficiently illuminate the target for image acquisition. During the live load testing, two image acquisition modes were used including video (30 fps) and time lapsed pictures (two images per second) which were both sufficient for the quasi-static testing under trucks moving at crawl speed (approximately 5 mph [8 km/h]). For higher speed applications, specialized cameras with higher acquisition frame rates may be required.

5.2.4.3   *Load testing sequence*

For the live load testing, a series of VDOT tandem axle trucks filled with gravel were used. The weights of the test trucks were measured prior to testing and ranged from 23.1 to 28.0 tons (21.0 to 25.4 metric tons) with a distribution between front and rear axles of approximately 33% to 67%, respectively. Configurations 1 through 4 utilized a single truck either driving in a lane, or as close to the shoulder barrier as was physically possible. Configuration 5 used two trucks, one in each driving lane at the same time, at the same pace. Figures 5.7 and 5.8 illustrate each of the truck configurations.

5.2.4.4   *Results*

Each set of images collected in a truck event in the load test were analyzed to produce the deflections of the midspan of the girders. In this study, the commercial 2D-DIC package (VIC 2D) from Correlated Solutions was used for the analyses. It should be noted that

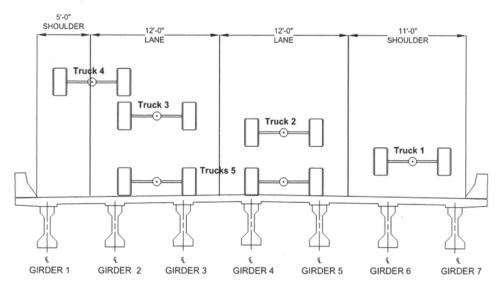

Figure 5.7 Truck configurations 1 to 5 on the eastbound span. 1 ft = 0.305 m

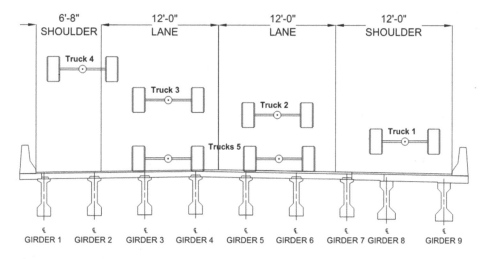

Figure 5.8 Truck configurations 1 to 5 on the westbound span. 1 ft = 0.305 m

prior to testing, the resolution of the measurement technique was evaluated in the laboratory using a variety of lighting methods. The results of the laboratory investigation provided confidence that the technique could consistently resolve deflections on the order of
0.005 in (0.13 mm) under ideal conditions. It was expected that this resolution would
decrease in the field due to environmental and lighting conditions, but the 2D-DIC technique is known to have in-plane resolution on the order of 1/100,000 of the field of
view for in-plane measurements, which is sufficient to resolve strains on the order of 50
$\mu\varepsilon$. A sample of the collected images used for correlation and deflection measurement
together with the correlation parameters is shown in Figure 5.9. Figure 5.10 depicts a

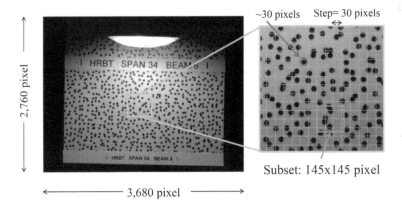

Figure 5.9 Sample of captured images and the correlation parameters employed

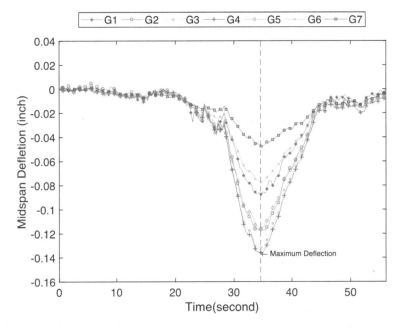

Figure 5.10 Midspan deflection of the eastbound span girders under truck configuration 5. I inch = 25.4 mm

time history of the midspan deflections of the girders of the eastbound span as the two trucks of configuration 5 cross the bridge. The different deflection levels of the curves high-light the load distribution in the cross section whereby girders experience deflections cor-responding to their distance from the truck load application (e.g. the closest girders G3 and G4 experience the largest deflections and G7 has the smallest). The deflection profiles are extracted for all the five different truck configurations, each of which was repeated three times, and the maximum deflection was obtained from the profiles as shown in Figure 5.10.

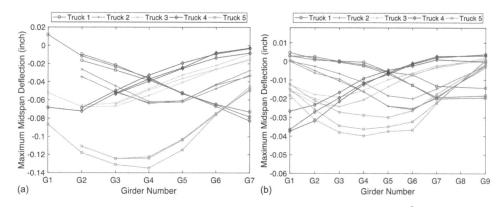

Figure 5.11 Maximum midspan deflection of the girders (a) eastbound span (b) westbound span (each truck configuration includes three runs). 1 inch = 25.4 mm

These maximum deflections are then plotted across the width of the cross section and shown for each one of the spans in Figure 5.11. This figure clearly shows the load distribution behavior among the girders with maximum deflections of up to 0.14 in (3.6 mm) in the longer eastbound span. Note that points without markers on the plots of Figure 5.10 denote missing data due to camera malfunction, as a result of a low-strength wireless signal between a few of the cameras and the remote control. The case study presented here demonstrates a successful application of DIC for load testing a high-profile structure in a complicated environment. Two spans of a bridge structure over water were instrumented with cameras in four to six hours during the day and tested with loaded trucks at night in about 1.5 hours per span. The use of this optical technique enabled the measurement of deflections without a fixed ground reference at the point of interest as would be required with traditional mechanical deflection sensors.

5.3   Eulerian virtual visual sensors (VVS) for natural frequency measurements

5.3.1   Theory

The methodology used here is based on the work introduced in (Schumacher and Shariati, 2013; Shariati et al., 2015). The concept is that every pixel in a digital video represents a candidate virtual visual sensor (VVS) that can be monitored. In contrast to the technique presented in Section 2, this methodology uses an Eulerian specification whereby a specific pixel (or patch of pixels) is selected and its intensity value (or the average intensity value across the patch) are tracked over time, as illustrated in Figure 5.12.

By using the discrete Fourier transform (DFT), natural frequencies of vibration can be extracted. When a black and white target is used, the displacement amplitudes can be estimated as well (Shariati and Schumacher, 2016). Such targets do increase the signal-to-noise ratio and can thus be helpful even for the case where only vibration frequencies are to be captured (Shariati et al., 2015). Pixel intensity is computed by converting the red-

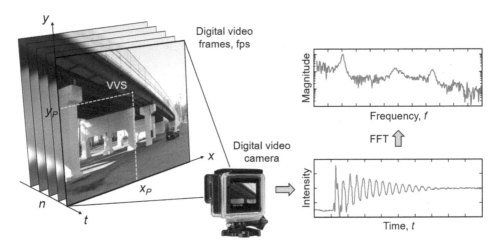

Figure 5.12 Illustration of Eulerian-based methodology to estimate frequencies of vibration. VVS = Virtual visual sensor, which can be one pixel or a patch of pixels

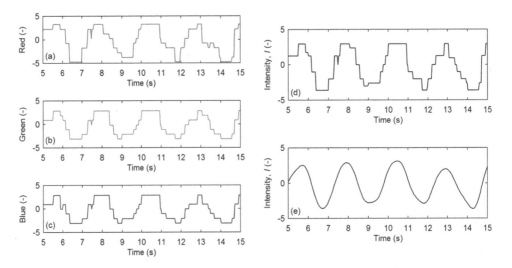

Figure 5.13 Example of experimental data extracted from a VVS: brightness of (a) red; (b) green; and (c) blue; (d) computed intensity (used for subsequent analyses), *I*; and (e) smoothed intensity (for illustrative purposes)

Source: Schumacher and Shariati (2013).

green-blue content of a pixel using the following equation:

$$I = 0.2989R + 0.587G + 0.114B \tag{5.3}$$

Figure 5.13 shows an example of collected and processed data from a laboratory experiment discussed in (Schumacher and Shariati, 2013).

5.3.2 *Equipment*

For the results presented in this section, a commercially available GoPro Hero3 Black Edition camera was used. This camera is widely used within different communities for being small, lightweight, and comparatively inexpensive. An image of a standard GoPro camera is shown in Figure 5.4a. The camera comes with a built-in Wi-Fi connection and can shoot videos with a maximum frame rate of 120 fps at 848×480 pixels and a maximum resolution of 1920×1080 pixels at 30 fps. Although the fisheye effect was present in the recorded videos, this did not affect the measurements. If necessary, it could be corrected during post-processing. To improve the signal-to-noise ratio, simple black-and-white targets can be employed, which can be printed on paper on a laser printer. For video processing and vibration analysis, a MATLAB code was developed by the authors. Details of how the data is analyzed is provided in Section 5.3.1.

5.3.3 *Strengths and limitations*

5.3.3.1 *Strengths*

As a non-contact sensor, digital video measurements provide an opportunity to inexpensively monitor the vibrations of an object without the hardships of attachment, wiring, and maintenance. The proposed Eulerian-based analysis approach (Schumacher and Shariati, 2013; Shariati et al., 2015) is particularly well suited to capture natural frequencies of vibration without directly having to measure a physical quantity such as displacement, which can be computationally expensive. The approach turns each pixel in the video into a candidate sensor; in other words, this makes the spatially continuous natural frequency measurement possible. This is especially useful when there is no known location for optimal monitoring. For example, in the case of a bridge cable, if the sensor is located at a point close to a vibration node, the respective natural frequency will not be detectable. Using this approach, however, several points along the cable can be selected and analyzed, which helps identifying several peak frequencies. Through laboratory and real-world experiments, the authors have demonstrated that the accuracy is comparable to what would be achieved with standard accelerometers, with the added benefit that multiple objects can be monitored simultaneously. This omits the need for tedious wiring, attachment, and data acquisitioning especially when multiple objects are considered. Another advantage is that this approach works without any targets added to the structure. The addition, however, can help increase the signal-to-noise ratio, if needed.

5.3.3.2 *Limitations*

Like with any remote non-contact sensor, the motion of the camera can interfere with the vibrations of the structure. Although the peak frequencies of both camera and the structure should exist in the frequency domain, since the motion of the camera is often dominant especially when it is far from the vibrating object, distinguishing the small structural vibration can be difficult. This can be resolved using an independent sensor on the camera or having a stationary point in its view, so that one can measure the natural frequencies of the camera. Also, identification of peak frequencies depends on a number of parameters, the spatial gradient of the target point, the resolution of the CCD sensor, and the amount of displacement as compared to the projected pixel size to name the most important ones. According to a recent study, in the lab setting, if the amount of displacement is bigger than 20% of the projected pixel size, the frequency can be captured using the proposed approach given a certain amount of contrast in the target. It is shown that nonlinearity in the intensity of pixels can form spurious higher harmonics and without a target the signal to quantization noise ratio (SQNR) can be low and make the detection of peak frequencies difficult. As mentioned in Section 5.3.3.1, some of these issues can be addressed by working with a target that is attached at the location of interest on the structure.

5.3.4 *Case studies*

5.3.4.1 *Estimation of cable forces on a lift bridge using natural vibration frequencies*

The Hawthorne Bridge in Portland, Oregon, spans across the Willamette River and is, as it was built in 1910, the oldest vertical lift bridge in the United States. Figure 5.14 shows the bridge, which is owned and maintained by Multnomah County. The bridge has a movable 244 ft (74 m) center span that can be lifted vertically for ship passages. Each side of the span is attached to $2 \times 12 = 24$ steel cables connected to an 880 kip (3914 kN) concrete counterweight, which is suspended from a 165 ft (50 m) steel truss tower. Cable length and mass of 121 ft (36.9 m) and 0.117 slug/ft (5.6 kg/m) were provided by the Multnomah County Bridge Services Section.

On November 4, 2016, the authors collected digital video data using a GoPro Hero 3 camera from the vertical lift cables of the east tower on the Hawthorne Bridge (Fig. 5.14b) during regular in-service conditions. These cables experienced sufficient ambient vibrations not requiring any additional external stimuli. The objective of this investigation was to determine the distribution of the cable forces within a cable group. The focus of this study were the 12 cables marked by an ellipsoid in Figure 5.14b. The video was taken from the location on the east tower marked with an "X" (Fig. 5.14a and 5.14b) acquired at 60 fps.

Figure 5.14 Photo showing: (a) lift span and (b) east tower with the 24 vertical cables of the Hawthorne Bridge in Portland, Oregon

Figure 5.15 Sample snapshot of full frame from digital video. The enlarged window shows the selected pixel (or Eulerian VVS) on cable 3 for frequency analysis

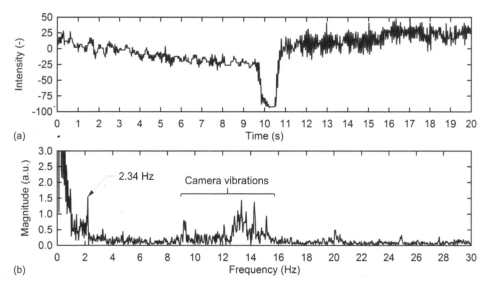

Figure 5.16 Sample time history (a) and frequency spectrum (b) for vertical lift cable 3 extracted from the digital video. Labeled are the first natural frequency of vibration and camera vibrations caused by traffic and wind

In Figure 5.15, a sample snapshot of the recorded digital video is shown. As has been demonstrated by the authors, although the vibrations of the cables are barely observable in the video visually, and despite the large pixel size compared to the vibration amplitudes, it was possible to extract the frequency of vibration.

Figure 5.16 presents the time history and frequency spectrum for the sample VVS introduced in Figure 5.15. The frequency spectrum was obtained via discrete Fourier transform (DFT). Several features of interest are revealed in the frequency spectrum (Fig. 5.16b). A distinct peak at 2.34 Hz (labeled), which is associated with the fundamental frequency of vibration. Additional peaks are visible between 9 Hz and 15 Hz and at 20 Hz, which are

Table 5.2 Estimated cable tension forces. Cable 3, which was introduced in Figures 5.14 and 5.15, is shaded

Cable # (−)	Measured cable frequency (Hz)	Error (Hz)	Estimated tension force (kip (kN))	Error (kip (kN))
1	2.23	0.06	34.1 (151.8)	±0.9 (4.0)
2	2.23	0.06	34.1 (151.8)	±0.9 (4.0)
3	2.34	0.06	37.8 (168.2)	±0.9 (4.2)
4	2.34	0.06	37.8 (168.2)	±0.9 (4.2)
5	2.17	0.06	32.3 (143.9)	±0.9 (3.9)
6	2.34	0.06	37.8 (168.2)	±0.9 (4.2)
7	2.17	0.06	32.3 (143.9)	±0.9 (3.9)
8	2.34	0.06	37.8 (168.2)	±0.9 (4.2)
9	2.34	0.06	37.8 (168.2)	±0.9 (4.2)
10	2.17	0.06	32.3 (143.9)	±0.9 (3.9)
11	2.23	0.06	34.1 (151.8)	±0.9 (4.0)
12	n.a.	0.06	n.a.	n.a.
Avg. estimated cable force for cables 1 through 11			35.2 (156.6)	n.a.
Avg. cable force computed from counterweight			36.7 (163.3)	n.a.

Note: n.a. means "not available."

associated with structural vibrations of the platform to which the camera was attached. Both of these observations could be confirmed by the researchers visually during the test. Furthermore, an independent estimation of the fundamental frequency of vibration of 2.33 Hz was provided by the Multnomah County Bridge Services Section and further confirmed the validity of the peak observed in Figure 5.16b.

Table 5.2 presents the measured cable frequencies obtained for 11 of the 12 cables on the east tower and the associated estimated tension forces. Note that one cable was located behind another cable and its frequency could therefore not be determined. The equation used to compute the cable tension force, T_{f_k} is as follows:

$$T_{f_k} = \frac{4\bar{m}L^2 f_k^2}{k^2} \tag{5.4}$$

where $\bar{m}$ is the mass per unit length of the cable, f_k is the natural frequency of vibration of mode k, and L is the total unsupported length of the cable. Equation 5.4 assumes pinned boundary conditions. Table 5.2 reveals that the counterweight is not equally shared by all the cables but the average force estimated by our video measurements (35.2 kip/cable [156.6 kN/cable]) is close to the computed average cable force using the counterweight, (880 kip/24 cables = 36.7 kip/cable [3914 kN/24 cables = 163.3 kN/cable]). The small difference can be explained by the assumed pinned end conditions in the cable. The minor difference, however, confirms that this assumption is reasonable for this application. The error was computed based on the frequency resolution in the FFT. Moreover, the differences between the estimated forces for each cable are relatively minor and thus likely not significant enough to affect the functionality or safety of the lift mechanism. Continued monitoring will allow the agency to make decisions should the force distribution change over time.

Figure 5.17 Photo showing bridge, shaker, black and white targets, and location of camera. The insert shows a snapshot from Google Maps with camera and target locations marked by "x" and "o," respectively

5.3.4.2   Identifying bridge natural vibration frequencies with forced vibration test

A 30° skewed single-span composite steel girder bridge carrying Eberlein Avenue over the A Canal in Klamath Falls, Oregon, is representative of a large fraction of short-span highway and road bridges in the United States (see Fig. 5.17). Constructed in 2003, the bridge spans 94.2 ft (28.7 m) with an out-to-out width of 44.6 ft (13.6 m) and includes two lanes of traffic, pedestrian sidewalks and steel rail. Five 40.6 in. (1,033 mm) deep girders, spaced at 9.19 ft (2.80 m) support an 8.25 in. (210 mm) thick concrete deck. The National Bridge Inventory indicates that the deck, superstructure, and substructure are in very good condition and the bridge has a sufficiency rating of 91.9% (FHWA, 2016).

On 4 November 2017, forced vibration testing was conducted with a Sony RX10-III (3840 × 2160, 30 fps, zoom lens with 600 mm focal length) placed off of the structure shooting longitudinally along the north sidewalk (Fig. 5.17). The distance between camera and target was approximately 50 ft (15.2 m). An APS Electroseis 113 shaker with a 15 lb (6.8 kg) oscillating mass was placed at midspan and forced at 4.02 Hz and 1.25g. Traffic was light at the time of testing and no traffic crossed the structure during harmonic forcing. Ambient traffic sufficient to excite the structure was, at most, a half-ton pickup truck. The structure has a design dead load deflection of 5.31 in. (135 mm) and vibrations were detectable by both Eulerian VVS as well as an iPhone accelerometer (Fig. 5.18). The resonant frequency for the fundamental flexural mode was determined by forcing at frequencies around a modeled frequency of 4.07 Hz. Peak response values measured with an iPhone accelerometer occurred at a forcing frequency of 4.02 Hz. An a priori 3D finite element model had been previously adjusted based on field measurements by iPhone; including stiffness contributions from the haunch and sidewalk, the first flexural mode was modeled to be 4.07 Hz.

Figure 5.19 shows a close-up photo of shaker and target. The structural response frequency was identified using a patch of pixels around the horizontal portion of the black

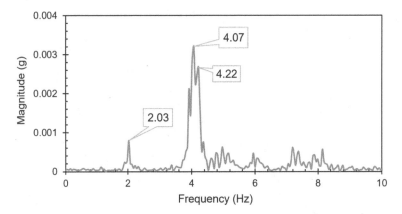

Figure 5.18 Frequency spectrum from iPhone accelerometer placed on the sidewalk at center span; excitation by ambient traffic

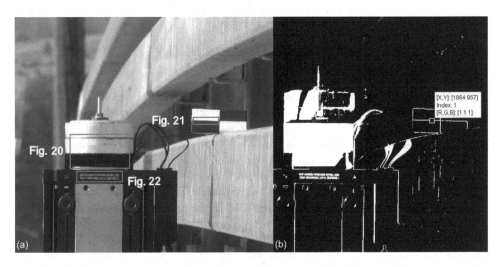

Figure 5.19 Sample photo snapshot of (a) partial frame from digital video and (b) black-and-white version of (a) to highlight significant intensity change. The numbers shown in (a) represent the figure numbers of the corresponding data

and white target (Fig. 5.21). The shaker frequency was confirmed by evaluating a patch of pixels around the oscillating shaker armature and mass (Fig. 5.20). Possibly valuable in isolating global response characteristics, the high-contrast white font on the shaker body (Fig. 5.22) can also be used to determine structural response as the shaker body moves with the structure; the advantage is that the signal is not as contaminated with other component vibrations like the target placed on the railing. A second torsional mode at 11.3 Hz was forced at the quarter point of the structure and structural response was measured similarly (not discussed in this study). Finite element (FE) model calibration can be an effective means of improving the accuracy of bridge load ratings, especially for deteriorated structures. Modal properties (frequencies, mode shapes, and modal damping values) measured

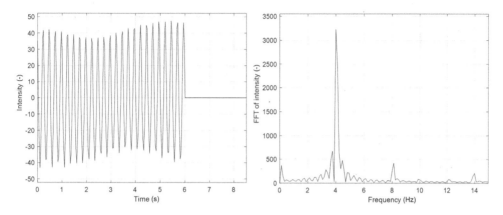

Figure 5.20 Time history (left) and frequency spectrum (right) for a patch of 700 × 300 pixels at oscillating shaker head; local peak at 4.00 Hz

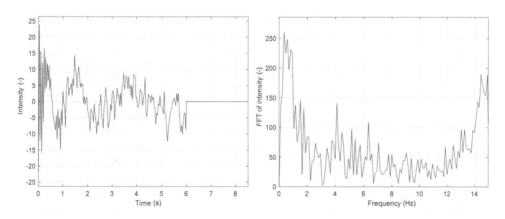

Figure 5.21 Time history (left) and frequency spectrum (right) for a patch of 240 × 20 pixels around horizontal junction on black-and-white target; local peak at 4.11 Hz

in-situ are one way of calibrating such models with many methods suggested in the literature on the subject. Forced vibration resonance testing and VVS monitoring can be effectively used as a method of confirming or identifying modal frequencies to support FE model updating. These results also suggest that a frequency sweep and the half-power bandwidth method could be used in combination with the vision sensing data to determine modal damping parameters.

5.4   Recommendations for practice

In order to successfully use video-based techniques, it is critical that their strengths and limitations as discussed in the respective subsections are well understood. For example, the accuracy of the results are not only a function of camera resolution and frame rate, distance from the bridge, and displacement amplitude, but also of atmospheric and lighting

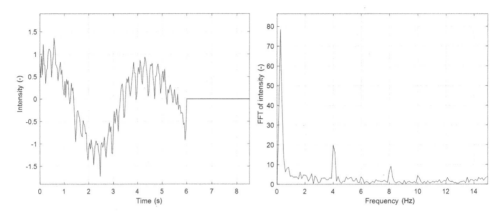

Figure 5.22 Time history (left) and frequency spectrum (right) for a patch of 400 × 80 pixels around white font on shaker body; local peak at 4.00 Hz

conditions. Subsequently, recommendations specific to each presented technique are presented separately.

5.4.1 *Digital image correlation (DIC) for deformation measurements*

An important factor for the successful use of DIC for the assessment of bridge structures is the consideration of environmental effects. As DIC has traditionally been regarded as a laboratory measurement technique, its performance can be affected by external factors in the field. As such, its field deployment requires attention to the factors that can affect its performance, which include external sources of noise (ambient vibrations, wind, etc.), the presence and variability of natural sources of light, as well as conditions that are unfavorable for the camera use (humidity, dust, heat, etc.). The effect of these factors can be alleviated and minimized by careful planning of the field deployment. For example, appropriate environmental protection and isolation combined with the use of durable materials and equipment should be considered. Additionally, in the study described in Section 5.2.4, the analysis of the results was not performed in real time. Rather than that, deflections based on the images collected during testing were obtained through post-processing at a later time. This feature could be improved in the future with the integration of real-time tracking and onboard data processing/analysis.

5.4.2 *Eulerian virtual visual sensors (VVS) for natural frequency measurements*

To reduce the effect of camera motion in the measurements and to use the camera's resolution efficiently, its placement should be as close as possible to the region of interest and proper optical zooming should be incorporated. Since this technique is employed to measure natural vibration frequencies, it is also important to note that the duration of the recording corresponds directly to the resolution of measurement in the frequency domain. The laboratory measurements presented in Shariati et al. (2015) and Shariati and Schumacher (2016) demonstrate that by using this technique, subpixel displacement magnitudes are detectable given sufficient contrast on the target. The use of a black-and-white target is recommended in the absence of a natural high spatial gradient in the medium. Finally, while camera movements and vibrations do not necessarily pose a problem, as long as those are different from the vibrations of the target, it is still recommended to minimize any movement. One possibility is to mount an accelerometer to the camera to help discriminate vibrations of interest from noise. If a smartphone is used for the measurement, built-in accelerometers can be used for this purpose.

5.5 Summary and conclusions

In this chapter, two techniques were presented that analyze digital images. The first technique is based on digital image correlation (DIC) to measure the static displacement response of a difficult-to-access bridge during in-service load testing. A series of cameras were mounted at the pier cap directly beneath each girder and aligned to acquire images of targets placed at midspan during the passing of loaded trucks. The use of DIC enabled the measurement of deflections without a fixed ground reference at the point of interest as would be required with traditional mechanical displacement sensors. The second technique uses an Eulerian-based technique to measure the natural vibration frequencies of structural members using digital videos. Referred to as Eulerian virtual visual sensors (VVS), two case studies are presented to highlight the potential of this technique. The first is the measurement of vibrating steel cables on a lift bridge and the second one the natural vibrations of a simple steel-concrete composite bridge during a forced harmonic vibration test. In both cases, the natural vibration frequencies were captured accurately. The first technique usually requires a target at the measurement location; the latter does not. Additionally, both techniques presented are still within their early stages of development for field applications and provide novel approaches to collecting data that would otherwise require physical sensors and in the case of DIC a ground reference. With the continuous advancement of camera and lens hardware, it can be expected that the availability and accuracy of the measurements using these techniques will improve.

5.6 Outlook and future trends

It can be expected that video-based techniques will continue to be advanced and developed to the point where they become a common measurement technique. There are several reasons for this. First, camera technology is going to be improved continuously. While image and video quality will increase, the costs for camera hardware will decrease. For example, ultra-high-resolution video cameras could alleviate the practical difficulties of implementing an array of physical sensors, which could potentially lead to cheaper and more sensitive measurements allowing for the use of damage location algorithms. Also, considering the rate of innovations in this field, a future when traffic cameras can simultaneously monitor traffic flows as well as the health of bridges is not difficult to imagine. Second, new algorithms are being developed constantly that improve the accuracy and reliability of video-based sensing techniques, for example to deal with changing lighting conditions (Lee et al., 2017). While the existence of high-contrast patterns on a structural component provides the best measurements, further research on using natural surface textures could be helpful for facilitating an easier use of the presented techniques.

Furthermore, integration of suitable patterns in new or existing structures could be considered. Third, new methodologies will continue to emerge. One example is a recently proposed solution that combines lasers and video cameras to produce a sensor that can accurately measure displacements and work under any lighting, including changing conditions and night (Vicente et al., 2018). Another example is drone-based systems, which open the possibility to provide close-up access to difficult-to-access locations on bridges (Yoon et al., 2017). Finally, current video-based techniques require post-processing of images and videos to extract the desired data. Systems that incorporate real-time processing are likely to follow soon, given increasingly powerful computing platforms. To conclude, the ubiquity of cameras in the urban environment and infrastructure, and further introduction of environmentally resistant imaging tools can translate into image-based and video-based infrastructure assessment systems becoming a standard tool.

Acknowledgments

The authors thank the following agencies for supporting their work presented in this chapter (in alphabetical order): City of Klamath Falls, OR; Hampton Roads District of the Virginia Department of Transportation, Suffolk, VA; and Multnomah County Bridge Section, Portland, OR.

References

Alipour, M., Washlesky, S. J. & Harris, D. K. (2019) Field deployment and laboratory evaluation of 2D digital image correlation for deflection sensing in complex environments. *Journal of Bridge Engineering*, 24(4), 04019010.

Chu, T. C., Ranson, W. F. & Sutton, M. A. (1985) Applications of digital-image-correlation techniques to experimental mechanics. *Experimental Mechanics*, 25(3), 232–244.

Citto, C., Wo, S. I., Willam, K. J. & Schuller, M. P. (2011) In-place evaluation of masonry shear behavior using digital image analysis. *ACI Materials Journal*, 108(4), 413–422.

Correlated Solutions Inc (2013) *VIC-2D v6 Full-Field Deformation Measurement System*. [Online]. Available from: www.correlatedsolutions.com/wp-content/uploads/2013/10/VIC-2D-Datasheet. pdf [accessed 4th April 2018].

Correlated Solutions Inc (2017) *VIC-3D 8 Measurement System*. [Online]. Available from: www.corre latedsolutions.com/wp-content/uploads/2017/12/VIC-3D%208%20System%20Specs.pdf [accessed 4th April 2018].

Dantec Dynamics A/S (2018) *Measurement Principles of (DIC)*. [Online]. Available: www.dantecdy namics.com/measurement-principles-of-dic [accessed April 4th 2018].

FHWA (2016). *Oregon National Bridge Inventory Data*. Available from: https://www.fhwa.dot.gov/ bridge/nbi/2016/OR16.txt. Accessed on December 30, 2017.

Ghrib, F., El Ragaby, A., Boufama, B., Li, L. & Memar, S. (2014) A novel technique for displacement measurements in RC beams using digital image correlation. *ACI Special Publication*, 298.

Harris, D. K., Kassner, B. L., Gheitasi, A. & Hansen, M. E. (2015) *Performance Measurement of Voided Slab Concrete Bridge Synergy of Current and Emerging SHM Technologies*. [Presentation] American Concrete Institute (ACI) Convention, Philadelphia, PA.

Hebdon, M., Harris, D. K., Roberts-Wollmann, C., Alipour, M., Riley, J., Edwin, J. E. B. A., et al. (2017) *Execution of a Collaborative Live Load Test on Concrete Bridges in Complex Environments*. [Presentation] American Concrete Institute (ACI) Convention, Anaheim, CA.

Jauregui, D. V., White, K. R., Woodward, C. B. & Leitch, K. R. (2003) Noncontact photogrammetric measurement of vertical bridge deflection. *Journal of Bridge Engineering*, 8(4), 212–222.

Jiang, R. & Jauregui, D. V. (2010) Development of a digital close-range photogrammetric bridge deflection measurement system. *Measurement*, 43(10), 1431–1438.

Lattanzi, D. & Miller, G. R. (2012) Robust automated concrete damage detection algorithms for field applications. *Journal of Computing in Civil Engineering*, 28(2), 253–262.

Lawler, J. S., Keane, D. T. & Shah, S. P. (2001) Measuring three-dimensional damage in concrete under compression. *ACI Materials Journal*, 98(6), 465–475.

Lee, L., Lee, K.-C., Cho, S. & Sim, S.-H. (2017) Computer vision-based structural displacement measurement robust to light-induced image degradation for in-service bridges. *Sensors*, 17(10).

Murray, C., Hoag, A., Hoult, N. A. & Take, W. A. (2015) Field monitoring of a bridge using digital image correlation. *Proceedings of the ICE: Bridge Engineering*, 168(1), 3–12. Available from: www.icevirtuallibrary.com/content/article/10.1680/bren.13.00024 [accessed 4th April 2018].

Ojio, T., Carey, C. H., Obrien, E. J., Doherty, C. & Taylor, S. E. (2016) Contactless bridge weigh-in-motion. *Journal of Bridge Engineering*, 21(7).

Pan, B., Qian, K., Xie, H. & Asundi, A. (2009) Two-dimensional digital image correlation for in-plane displacement and strain measurement: A review. *Measurement Science and Technology*, 20(6).

Peddle, J., Goudreau, A., Carson, E. & Santini-Bell, E. (2011) Bridge displacement measurement through digital image correlation. *Bridge Structures: Assessment, Design & Construction*, 7(4), 165–173.

Peters, W. H., Ranson, W. F., Sutton, M. A., Chu, T. C. & Anderson, J. (1983) Application of digital correlation methods to rigid body mechanics. *Optical Engineering*, 22(6).

Reu, P. (2015) All about speckles: Contrast. *Experimental Techniques*, 39(1).

Schumacher, T. & Harris, D. K. (2016) *Monitoring of Concrete Bridges Using Digital Videos and Imagery*. [Presentation] American Concrete Institute (ACI) Convention, Philadelphia, PA.

Schumacher, T. & Shariati, A. (2013) Monitoring of structures and mechanical systems using virtual visual sensors for video analysis: Fundamental concept and proof of feasibility. *Sensors*, 13(12), 16551–16564.

Shariati, A. & Schumacher, T. (2016) Eulerian-based virtual visual sensors to measure dynamic displacements of structures. *Structural Control and Health Monitoring*, 24(10).

Shariati, A., Schumacher, T. & Ramanna, N. (2015) Eulerian-based virtual visual sensors to detect natural frequencies of structures. *Journal of Civil Structural Health Monitoring*, 5(4), 457–468.

Sutton, M. A., Wolters, W. J., Peters, W. H., Ranson, W. F. & McNeill, S. R. (1983) Determination of displacements using an improved digital correlation method. *Image and Vision Computing*, 1(3), 133–139.

Sutton, M. A., Orteu, J. J. & Schreier, H. (2009) *Image Correlation for Shape, Motion and Deformation Measurements: Basic Concepts, Theory and Applications*. Springer Science & Business Media, New York, NY.

Vaghefi, K., Oats, R., Harris, D. K., Ahlborn, T., Brooks, C. N., Endsley, K., et al. (2012) Evaluation of commercially available remote sensors for highway bridge condition assessment. *Journal of Bridge Engineering*, 17(6), 886–895.

Vaghefi, K., Ahlborn, T., Harris, D. K. & Brooks, C. N. (2015) Combined imaging technologies for concrete bridge deck condition assessment. *Journal of Performance of Constructed Facilities*, 29(4).

Vicente, M. A., Gonzalez, D. C., Minguez, J. & Schumacher, T. (2018) A novel laser and video-based displacement transducer to monitor bridge deflections. *Sensors*, 18(4).

Xiao, J., Li, W., Sun, Z. & Shah, S. P. (2012) Crack propagation in recycled aggregate concrete under uniaxial compressive loading. *ACI Materials Journal*, 109(4), 451–462.

Yang, D., Bornert, M., Gharbi, H., Valli, P. & Wang, P. (2010) Optimized optical setup for DIC in rock mechanics. *The European Physical Journal Conferences*, 6.

Yoneyama, S. & Murasawa, G. (2009) *Digital Image Correlation*. [Online]. Available from: www. eolss.net/Sample-Chapters/C05/E6-194-04.pdf [accessed 4th & 10th April 2018].

Yoneyama, S., Kitagawa, A., Iwata, S., Tani, K. & Kikuta, H. (2007) Bridge deflection measurement using digital image correlation. *Experimental Techniques*, 31(1), 34–40.

Yoon, H., Hoskere, V., Park, J.-W. & Spencer, Jr., B. F. (2017) Cross-correlation-based structural system identification using unmanned aerial vehicles. *Sensors*, 17(9).

Chapter 6

Acoustic Emission Measurements for Load Testing

Mohamed K. ElBatanouny, Rafal Anay, Marwa A. Abdelrahman, and Paul Ziehl

Abstract

Several methods for analyzing acoustic emission (AE) data to classify damage in reinforced and prestressed concrete structures during load tests were developed in the past two decades. The majority of the methods offer relative assessment of damage, for example, classifying cracked versus uncracked conditions in prestressed members during or following a load test. In addition, significant developments were made in various AE source location techniques including one-, two-, and three-dimensional source location as well as moment tensor analysis that allow for accurate location of damage through advanced data filtering techniques. This chapter provides an overview of the acoustic emission technique for detecting and classifying damage during load tests. Recent efforts to apply the method in the field are also presented along with recommended field applications based on the current state of practice.

6.1 Introduction

Acoustic emission (AE) technique is used to detect transient stress waves emitted from deformations or fractures within a material. The method uses sensors with high sensitivity, in the kHz range, and high-speed data acquisition cards that are capable of processing the data in millions of samples per second. This makes AE one of the most sensitive damage detection methods that are currently available.

AE is extremely sensitive to damage formation and growth which enables detecting cracks in the micro-range long before they are visible. In concrete, the method has been traditionally used to detect and locate wire breaks in prestressed and post-tensioned concrete structures. Due to its sensitivity and suitability for long-term monitoring applications, significant research has been conducted to use AE to detect various concrete degradation mechanisms including corrosion and alkali-silica reaction (Abdelrahman, 2013, 2016; Ziehl and ElBatanouny, 2016). While the majority of this research was conducted in laboratory settings (Zdunek et al., 1995; Li et al., 1998; Mangual et al., 2013a, 2013b; ElBatanouny et al., 2014a; Abdelrahman et al., 2015; Appalla et al., 2016), a recent study showed the successful implementation of long-term AE monitoring to detect and evaluate corrosion damage in a decommissioned nuclear facility (Abdelrahman et al., 2018a).

Condition evaluation using load tests of aging in-service structures and sometimes of new structures are widely used (Ohtsu, 2015). Over the past two decades several studies were conducted to investigate the use of AE during load tests of reinforced and prestressed concrete structures (Schumacher, 2008; Nair and Cai, 2010). This chapter discusses the

DOI: https://doi.org/10.1201/9780429265969

basics of the technique and its potential use for detecting, locating, and evaluating damage during load tests.

6.2   Acoustic emission–based damage identification

6.2.1   *Definitions*

Acoustic emission (AE) is defined by the American Society of Testing and Materials (ASTM E1316–16) as "the class of phenomena whereby transient elastic waves are generated by the rapid release of energy from localized sources within a material." AE sensors consist of piezoelectric crystal mounted to the surface of the specimen. The sensors detect mechanical shock waves and convert them into an electrical signal that is amplified and processed by the associated data acquisition systems. The piezoelectric principle states that some materials produce a voltage when mechanically strained (direct effect) (ElBatanouny, 2012).

   As AE is passive material, if cracks or damage do not occur, AE signals will not be generated. When cracks or plastic deformations occur in a material, AE waves are emitted which can be captured by the AE sensors as shown in Figure 6.1. Each AE waveform is called a hit and can be processed to obtain several parameters. Figure 6.2 shows a typical AE waveform with some parameters defined as follows (Abdelrahman, 2013):

- *Hit*: The process of detection and measurement of an AE signal on an individual sensor channel (ASTM E1316).
- *Event*: The rise of AE activity that will cause multiple hits on different sensors (ASTM E1316).
- *Signal amplitude*: The magnitude of the peak voltage of the largest excursion attained by the signal waveform from a single emission event, usually reported in decibels (dB).
- *Duration*: The time between AE signal start and the signal end (the time between the first threshold crossing and the last threshold crossing of the signal).

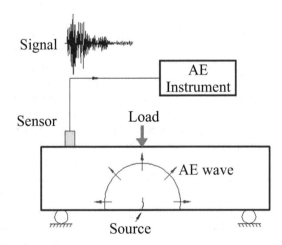

Figure 6.1 Schematic of AE monitoring

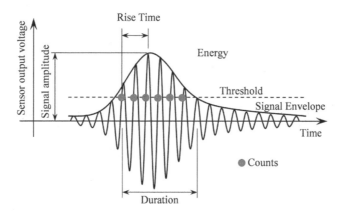

Figure 6.2 AE signal features

- *Rise-time*: The time between AE signal start and the peak amplitude of that AE signal (measured in microseconds).
- *Signal strength*: The measured area of the rectified AE signal, with units proportional to volt-sec.
- *Signal energy*: The energy contained in a detected acoustic emission burst signal with units usually reported in joules or values that can be expressed in logarithmic form (dB).
- *Count*: The number of times the acoustic emission signal exceeds a preset threshold during any selected portion of a test, and the count rate is the number of counts during a fixed period of time.
- *Frequency*: The number of cycles per second of the pressure variation in a wave.

6.2.2   AE parameters for damage detection

Earlier studies focused on the use of AE waveform parameters, such as AE energy or signal strength, to detect damage in reinforced concrete structures. This simple analysis was successful in identifying if damage started to occur or is progressing in the test specimen. One of the first phenomena observed in AE is the Kaiser effect, which is defined as the absence of detectable acoustic emission until previously applied stress levels are exceeded (Abdelrahman, 2013). An example of the effect is shown graphically in Figure 6.3. In this figure, when a specimen is loaded (from A to B), unloaded (from B to C), and reloaded (from C to D), AE activity is generated on the first loading (AB) but there are no emissions on the unloading phase (BC) or in the reloading (CB) until the previous load is exceeded. This irreversibility of AE activity indicates no or minor damage. The absence of Kaiser effect is called the Felicity effect, in which some emissions are detected during the reloading stage (Fowler, 1986). The Felicity effect is clarified in Figure 6.3 as when the specimen is loaded to point D, unloaded to point E, and then reloaded to point G; AE activity is detected at point F during the reloading stage, which is at lower load than the previous reached load D. The presence of the Felicity effect may indicate structural damage in the loaded specimen (Abdelrahman, 2013).

A simple application for the use of AE parameters for damage detection is the use of number of AE hits recorded to detect damage as shown in Figure 6.4, where the higher

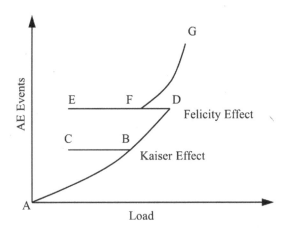

Figure 6.3 AE activity during repeated loading (Abdelrahman, 2013)

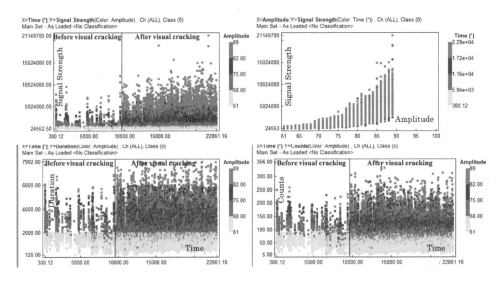

Figure 6.4 AE activity before and after visual cracking (time in seconds and signal strength in pVs [picovolt seconds])

amplitude and larger volume of AE hits can be related to initiation of visual cracking in a prestressed concrete beam (Ziehl and ElBatanouny, 2015). Other parameters that have been successfully used in detecting crack initiation or growth are cumulative signal strength and cumulative AE energy.

6.2.3   *Damage indicators*

During the last two decades, many studies were conducted to develop analysis methods capable of quantifying damage in structures during load tests using AE (Ono, 2010). The majority of these methods were developed from tests conducted on fiber-reinforced polymer (FRP) or reinforced concrete, while limited research has included prestressed specimens (Xu, 2008; ElBatanouny et al., 2014b). It is noted that all the methods are not yet standardized with a specific loading protocol. Some of the available AE evaluation methods are highlighted in the following sections.

6.2.3.1 Intensity analysis

Intensity analysis is a method developed to characterize damage in structural elements. The intensity analysis has been previously used to detect the degree of damage in FRP vessels (Fowler et al., 1989). Intensity analysis is performed by calculating the historic index and severity from signal strength. Historic index ($H(t)$), shown in Equation 6.1, is a form of trend analysis that estimates the changes of slope in the AE signal against time by comparing the signal strength of the most recent hits to a value of cumulative hits. The parameter attempts to provide a historical approach to the level of damage present in the element. The severity (S_r), shown in Equation 6.2, is defined as the average signal strength for the 50 events having the largest numerical value of signal strength (Golaski et al., 2002). By plotting the maximum severity–historic index values the intensity of acoustic emission data may be obtained. The chart can then be divided into sections corresponding to different damage levels.

$$H(t) = \frac{N}{N-K} \frac{\sum_{i=K+1}^{N} S_{oi}}{\sum_{i=1}^{N} S_{oi}} \tag{6.1}$$

$$S_r = \frac{1}{50} \sum_{i=1}^{i=50} S_{oi} \tag{6.2}$$

This method was used in previous studies to provide a qualitative index of damage in reinforced and prestressed concrete test specimens (Lovejoy, 2008; Nair and Cai, 2010). ElBatanouny et al. (2014b) used this method to classify damage in scaled prestressed concrete beams. The study included eight specimens with varying levels of corrosion damage. Figure 6.5 shows an example figure for classifying damage in pristine specimens. As shown in the figure, the method enables classifying the damage into before and after cracking regions which can be helpful in the case of the Class U prestressed design, which should remain uncracked under service loads. The same technique was also applied by Anay et al. (2015) to classify damage during a load test of a bridge prestressed concrete girder.

6.2.3.2 CR-LR plots

This method was proposed recently to quantify the degree of damage in structures using AE (Ohtsu et al., 2002). The assessment criterion uses plots of calm ratio (*CR*) versus load ratio (*LR*) divided into four sections to classify damage. These sections correspond to three levels of damage: minor, intermediate, and heavy. The literature provides different methods to calculate both *CR* and *LR*. The appropriateness of these methods varies based on the results provided in the literature. One such method has chosen to define the two parameters where load ratio (*LR*) is equal to load at the onset of AE activity in the subsequent loading divided by the previous maximum load, and calm ratio (*CR*) is total AE hits during the unloading phase divided by total AE hits during both loading and unloading

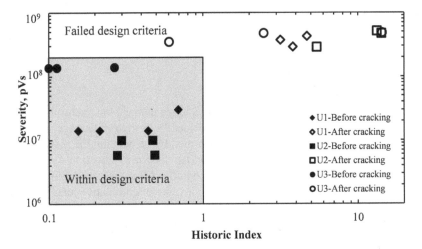

Figure 6.5 Intensity analysis condition assessment charts: (a) control (pristine) specimens, and (b) cracked specimens (ElBatanouny et al., 2014b)

Source: Reprinted with permission.

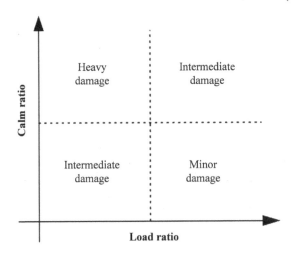

Figure 6.6 Example of NDIS-2421 damage assessment criteria

phases (Ohtsu et al., 2002). An example plot of the method is shown in Figure 6.6. The method is also recommended by the Japanese Society for Non-destructive Inspection (NDIS 2421, 2000).

6.2.3.3 Peak cumulative signal strength ratio

This method is proposed as a qualitative damage criterion for structural elements tested with the cyclic load test (CLT) method adopted by ACI Committee 437. In cases where AE is used, the standard CLT method is slightly modified to accommodate AE data. This modification is made by inserting additional cycles into the test. These AE cycles are performed by ramping load on a structure or element to a predefined load and holding the load for a small period of time, typically on the order of 4 minutes. This process is repeated in a second cycle at a lower load level. Peak cumulative signal strength (PCSS) is calculated as the summation of signal strength during a peak of the reload hold period divided by the summation of signal strength during the peak of the initial hold period. The specimen is assumed to fail the test if PCSS is greater than 40% (Ridge and Ziehl, 2006).

6.2.3.4 Relaxation ratio

The relaxation ratio parameter was proposed to classify a specimen's behavior based on activity during loading and unloading phases. The average energy during unloading is compared to the average energy during loading. If the ratio is calculated to be less than one (emitting more energy during loading than unloading), the load cycle is said to be loading dominant. The element becomes unloading dominant once more energy is emitted during an unloading phase than a loading phase (Colombo et al., 2005). During an unloading phase, AE activity rises primarily as a result of crack closure and friction. Higher AE activity during unloading phases as compared to the loading phase will indicate higher damage in the specimen. It was reported that the method can be used to determine the level of load as compared to the ultimate capacity in reinforced concrete beams, yet no clear correlation was found between the magnitude of the load and the results of relaxation ratio (Colombo et al., 2005). The method was also used to classify damage in prestressed concrete beams but was only successful in beams that were not previously loaded (ElBatanouny et al., 2014b).

6.2.3.5 b-value analysis

The b-value was proposed to detect damage using the magnitude-frequency relation of the recorded hits. The method was developed in seismology by Gutenberg and Richter (1954) to understand the relation between the magnitude and frequency of earthquakes. They found that earthquakes of larger magnitude have less frequency than earthquakes with smaller magnitude, which occur more often. The method was used in acoustic emission in the area of rock mechanics (Cox and Meredith, 1993; Sammonds et al., 1994) and concrete structures (Shiotani et al., 2001; Aggelis et al., 2011; ElBatanouny et al., 2014b, 2014c). Equation 6.3 shows the b-value relation, where N is the number of hits with amplitudes greater than A, A is the amplitude in dB, a is an empirical constant, and b is the b-value.

$$\log N = a - b \log A \qquad (3)$$

The b-value is used as an indication of damage associated with cracking. When cracks are forming, the number of hits with high magnitude increases causing decrease in the magnitude of the b-value. Literature indicates that b-values below 1 correspond to damage in the specimen (Ono, 2010).

6.2.3.6 Modified index of damage

This method was proposed to detect yielding of prestressed concrete specimens during cyclic load testing (Abdelrahman et al., 2014). The suggested index is a modification to

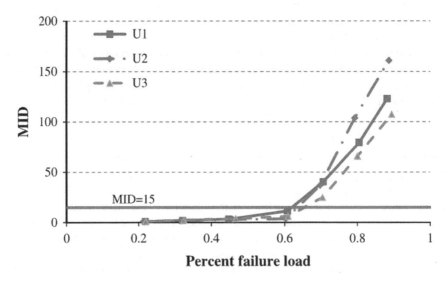

Figure 6.7 Modified index of damage results for uncracked specimens (Abdelrahman et al., 2014)
Source: Reprinted with permission.

the Benavent-Climent et al. (2011) index of damage parameter. The modified index of damage (MID) is computed by normalizing the cumulative AE energy at any point of the test by the cumulative AE energy at the end of the initial load cycle. The method was applied to eight prestressed concrete specimens and two different limits were developed to detect yielding in the specimens based on their initial condition, cracked and uncracked. An example of the MID results is shown in Figure 6.7.

6.3 Source location during load tests

Several AE source location techniques were developed to enable localization of damage during load tests including zonal, 2D (planar), and 3D methods. Source localization techniques are essential in AE analysis to extract the source coordinates of AE events (Grosse and Ohtsu, 2008). It has become an important tool for structural health monitoring (SHM) in research and field applications. AE localization can be applied to all kinds of construction materials (Grosse and Ohtsu, 2008). AE sources are located depending on the difference in the time that AE-generated signals reach the sensors with known wave velocity (Goszczyńska et al., 2013). In order for a source location to be justified, signals must be detected in a minimum number of sensors: one for zonal, two for linear, three for planar, and four for volumetric. AE source localization methods, which will be presented in the following sections, were developed based on earthquake seismology with some modifications (Grosse and Ohtsu, 2008; Bormann, 2002; Aki and Richards, 1980; Shearer, 2009). Knowing the exact origin of AE wave can help in determining (1) the source type such as friction between different parts of a structure, impact damage, and so forth, (2) evaluating the damage, and (3) understanding the damage mechanism and propagation (Aljets et al., 2010).

6.3.1   Types of source location

The most important difference in the location modes is the geometry they are designed for, which affects the number of hits per event used to calculate the location. Different source location methods have been developed with a variety of applications and accuracies. Crack maps can be produced after post-processing of AE data using different methods to obtain the required resolution in two or three dimensions. Selecting one of them depends on the objective of the experiment, the required solution, and the structural element type. Zonal localization is used where the exact source coordinates are not required. When more accuracy is required, 2D source localization is used to determine the coordinates of the source, while 3D source localization is used when depth information is important for understanding the problem.

6.3.2   Zonal and one-dimensional source location

Zonal location is the simplest way to locate the source of AE, where the sensor with the first arrival or the highest amplitude of the wave is said to be the closest to the source (Aljets et al., 2010), or it just retunes the coordinates of the first sensor in the list of hits that make up the event (AEwin Manual, 2014). This method is adequate when the inspected area or the sensor spacing is small or the damage initiation point is known (Aljets et al., 2010). However, an exact source location determination is not possible. Therefore, it is often used in inspecting large-scale structures such as buildings, suspension bridges, and pipes. If AE is recorded by a particular sensor, the technician should inspect the area nearby this sensor and look for damage (Behnia et al., 2014; Lovejoy, 2006). The inspected structural element is divided into zones which can be lengths, areas, or volumes depending on the dimensions of the array, and the source can be assumed to be within the region and less than halfway between sensors (one sensor covers each zone).

Another method uses two sensors to locate an AE source on a line which is called one-dimensional (1D) or linear source location. It is a step up from zonal location that performs linear interpolation between two sensors' coordinates based on the differences in the arrival times of the first two hits in the event (AEwin Manual, 2014). Its disadvantage is that in 2D area it limits the source to points on the line segments that connect the sensors when real sources do not face those restrictions. However, it is still good enough for source location of long structures such as pipelines (AEwin Manual, 2014). The 1D source can be determined between two sensors as follows (Fig. 6.8).

The distance between the two sensors is known $(x_1 + x_2)$ as well as the time at each sensor $(t_1$ and $t_2)$ and the sound velocity (v) of the material. The source time (t_0) and the 1D source location (x_0) are unknown. The distances x_1 and x_2 from the source to the sensors can be expressed as shown in Equation 6.4. By solving these equations, the unknown source can be determined.

$$x_1 = (t_1 - t_o) * v, \; x_2 = (t_2 - t_o) * v \text{ and } d = x_1 + x_2 \tag{6.4}$$

Goszczyńska et al. (2012) conducted an experimental validation of reinforced concrete beams loaded until failure based on the measurements of AE. AE data was recorded with four sensors spaced in the way shown in Figure 6.9. AE sources related to crack initiation and progression were identified and located using linear location method between sensors 2 and 3, as shown in Figure 6.10.

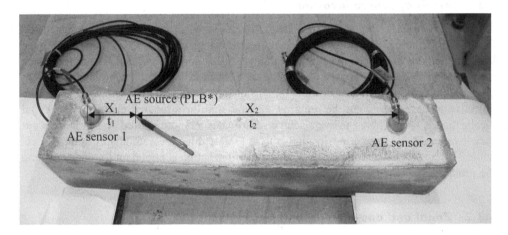

Figure 6.8 Principle of a one-dimensional localization on a structure

Note: *PLB: pencil lead break.

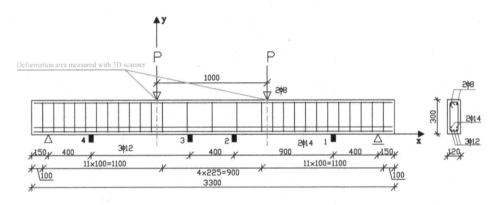

Figure 6.9 Reinforced concrete beam loading with marked sensor spacing (Goszczyńska et al., 2012). Units: mm. Conversion: 1 mm = 0.04 in.

Source: Reprinted with permission.

On-site AE monitoring for damage localization and assessment have been conducted on different existing structures such as steel and concrete bridges. Prestressed, reinforced concrete, and composite bridges having different conditions were load tested using AE (Golaski et al., 2002). It was difficult to assign each recorded signal to a particular crack because of the large scale of the tested element. Therefore, a zonal location was applied by dividing the tested element into 10 measuring areas (one AE sensor/area), and each area was evaluated separately. Figure 6.11 shows the monitored bridge built of post-tensioned beams.

Another study was performed on the Anthony Wayne suspension bridge that carries State Route 2 over the Maumee River in Toledo, Ohio, using the AE zonal localization technique (Gostautas et al., 2012). A major rehabilitation was scheduled, as the main cables have never been inspected since the bridge went into service over 80 years ago. Therefore, and to support this step, AE monitoring of the main cables for wire breaks was done prior to

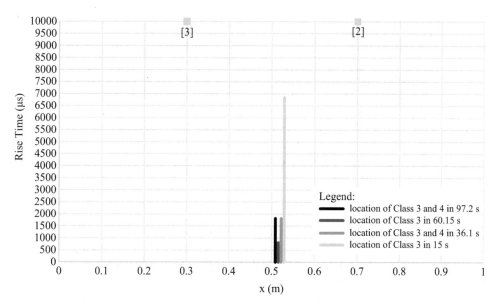

Figure 6.10 Linear source location of crack initiation signals (class 3) and crack progression signals (class 4) (Goszczyńska et al., 2012). Conversion: 1 m = 3.3 ft

Source: Reprinted with permission.

Figure 6.11 Severely damaged bridge built of post-tensioned beam (Golaski et al., 2002)

this invasive inspection. The primary goals of the AE monitoring were to (1) obtain an overview of the health of the entire cable volume by ascertaining the frequency and location of wire breaks, (2) help identifying areas with active corrosion, and (3) aid in locating the areas to be considered for invasive internal inspection. AE sensors were mounted at 100 ft (30.5 m) intervals on each on the main cables (see Fig. 6.12). The AE signature showed

Figure 6.12 Overview of Anthony Wayne Bridge, Toledo, Ohio (Gostautas et al., 2012)

that there were several located sections on both the north and south cables that are potential locations for future inspection.

6.3.3   2D source location

The next improvement in localization technique is to perform a 2D source location to determine the x and y coordinates of an AE event. It is applied when the accuracy of zonal or linear location is not sufficient. It is also referred to as a planner localization since no information about the depth of the source is provided (where the thickness of a structure is small compared to the extent of the object) (Grosse and Ohtsu, 2008). Three sensors at least are needed, assuming constant wave velocity at all of them. With three measured arrival times of the propagated wave from a source to AE sensors, the source can be determined. Sensor layout is an important factor in 2D source location and it affects the accuracy of the solution. For example, a better solution is to use a layout that minimizes the chances of linear events being observed. Figure 6.13 shows two possible 2D planner layouts. In Figure 6.13a, the horizontal sensor spacing is much smaller than the vertical spacing, while in Figure 6.13b all distances in between the sensors are approximately equal but exact equality is not required. Therefore, in this case, a better source location is expected from the second layout in Figure 6.13b. Note that triangles with one very long side or one with very short side should be avoided as their approaches become a straight line of sensors. Moreover, a random sensor distribution is preferred to locate a source more accurately.

If an event (crack) occurs in a structure (Fig. 6.14), stress waves propagating in all directions at the same velocity can be detected by AE sensors. Two-dimensional source location is achieved by using the time-distance relationship, and a brief overview of the equations behind the source location is shown below:

$$d = v.t \tag{6.5}$$

where d is the distance, v is the acoustic wave velocity, and t is time.

Also, in a 2D (x, y) plane, the distance between two locations of coordinates (x_1, y_1) and (x_2, y_2) is:

$$d = \sqrt{\{(x_2 - x_1)^2 + (y_2 - y_1)^2\}} \tag{6.6}$$

For the first AE hit at time t_1, and the second at time t_2, the arrival delay between the second AE hit sensor relative to the first one is:

$$t_2 - t_1 = (d_2 - d_1)/v \tag{6.7}$$

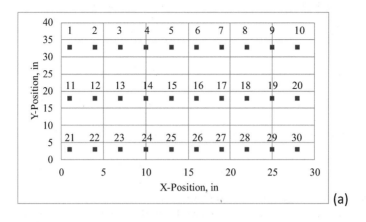

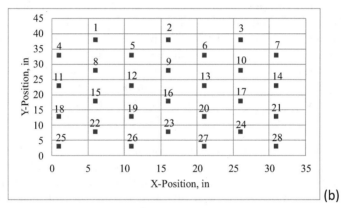

Figure 6.13 2D sensors layout, (a) layout one, (b) layout two Metric (SI) conversion factors: 1 in. = 25.4 mm

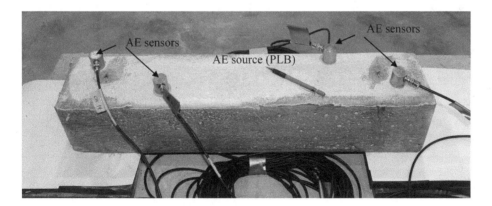

Figure 6.14 Principal of 2D localization approach

Coordinates (x_s, y_s) for the unknown source are calculated from:

$$t_2 - t_1 = \sqrt{[\{(x_2 - x_s)^2 + (y_2 - y_s)^2\} - \{(x_1 - x_s)^2 + (y_1 - y_s)^2\}]/v} \qquad (6.8)$$

and

$$t_3 - t_1 = \sqrt{[\{(x_3 - x_s)^2 + (y_3 - y_s)^2\} - \{(x_1 - x_s)^2 + (y_1 - y_s)^2\}]/v} \qquad (6.9)$$

Each additional hit adds an extra equation. In general:

$$t_i - t_1 = \sqrt{[\{(x_i - x_s)^2 + (y_i - y_s)^2\} - \{(x_1 - x_s)^2 + (y_1 - y_s)^2\}]/v} \qquad (6.10)$$

Solving these equations gives the unknown location of (x_s, y_s).

Source triangulation techniques were used to produce crack maps during load testing of prestressed concrete beams (ElBatanouny et al., 2014b). AE data was recorded using 16 AE resonant sensors (R6i) mounted on each specimen, as shown in Figure 6.15. Prior to data analysis, it is important to filter AE data to eliminate data not related to the response of the structure such as AE wave reflection, environmental noise and friction between the loading apparatus and the specimens. Figure 6.15a shows source location of the tested specimen using unfiltered data. The extensive scattering can be seen in the recorded data due to wave reflections. After filtering, crack maps were developed after 10 cycles (Fig. 6.15b) and at the end of the test (Fig. 6.15c), with a reasonable agreement between AE source location and visually observed cracks.

AE source location technique was used to monitor the cracking mechanisms and generate crack maps of scaled concrete beams having glass fiber reinforced polymer (GFRP) longitudinal reinforcement and no shear reinforcement (ElBatanouny et al., 2014c). Resonant (R6i) and broadband (WDi) sensors were attached on the surface of each specimen in an array of 10 AE sensors for small-scale specimens and 16 sensors for medium-scale specimens. The test setup and sensors layout for a specimen are shown in Figure 6.16. After applying the appropriate filters to eliminate non-genuine data, two-dimensional AE

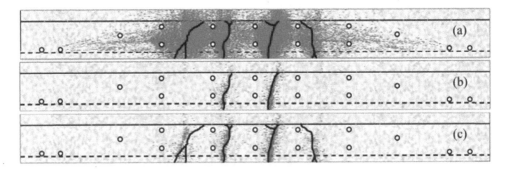

Figure 6.15 AE source location (dots) versus visual cracks (solid lines) for one specimen; (a) unfiltered data, (b) cracks developed after cycle 10, and (c) cracks at the end of the test (ElBatanouny et al., 2014b)

Source: Reprinted with permission.

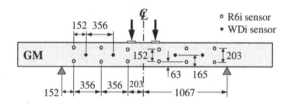

Figure 6.16 AE sensor layout in one of the specimens (dimensions in mm) (ElBatanouny et al., 2014c). Units: mm. Conversion: 1 mm = 0.04 in.

Source: Reprinted with permission.

source location was performed since all the sensors were mounted on one side of the specimen, as shown in Figure 6.17.

A similar technique was used to perform 2D source location in the field. A proof load test was performed on a 40-year-old simple-span prestressed concrete bridge with no available design plans located in southern New Mexico using the acoustic emission technique (Anay et al., 2015). AE data was collected under several loading conditions from two groups of sensors placed near the support and midspan of an interior double-tee beam. 2D source localization technique using AEwin software was applied to develop crack maps for the instrumented girder (Fig. 6.18) at both shear and moment regions. The AE data indicated that damage in the form of crack growth was more prevalent in the region near the supports than the midspan (Fig. 6.19).

Another study was conducted to evaluate the condition of a three-span, prestressed concrete girder bridge located in Guadalupe County, New Mexico, during a load test. The 15-year-old bridge has shear cracks in all four girders of the superstructure. Some shear cracks were injected with epoxy, however most of the cracks extend beyond the epoxy regions and some girders have developed new shear cracks. AE data was collected from sensors attached on three girders toward the obtuse corner of an exterior span under different levels of load. AE data analysis and source location algorithm were applied to assess the response of the structure under load increases and during load holds, and to develop crack maps. The results showed signs of crack propagation beyond the existing cracks, as shown in Figures 6.20 and 6.21.

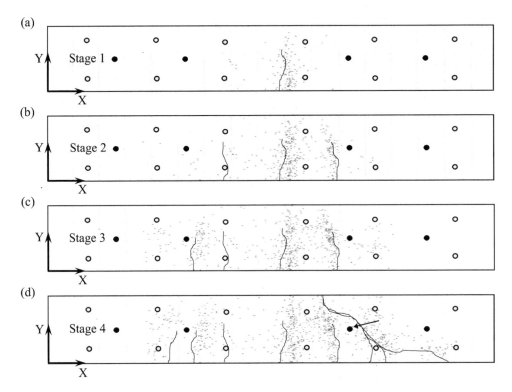

Figure 6.17 AE source location and crack development during load stages: (a) Stage 1, (b) Stage 2, (c) Stage 3, and (d) Stage 4 (ElBatanouny et al., 2014c)

Source: Reprinted with permission.

6.3.4   *3D source location and moment tensor analysis*

6.3.4.1   *3D source location*

Numerous researchers have utilized 3D source localization techniques to AE in civil engineering applications (Kurz et al., 2005; Labuz et al., 1988; Köppel and Grosse, 2000) especially when the area of interest lies beneath the surface. Three-dimensional source location is extension of 2D planner, assuming that sound travels directly through space rather than via an external shell (AEwin Manual, 2014). To determine 3D source location of an AE event accurately, at least four sensors are required, therefore four travel times are available to calculate the three coordinates and the source time of an event. The arrival times of the first wave (compressional wave [P-wave]) and second wave (shear wave [S-wave], if it can be detected) arrivals are used in 3D localization (Behnia et al., 2014). To get better results, sensors should be set up in a way to minimize the number of sensors that line up across the body of the test structure. It means that using a rectangular grid layout that lines up sensors on opposing faces of the test structure should be avoided. Instead sensors have to be staggered so that no two sensors are directly across from one another.

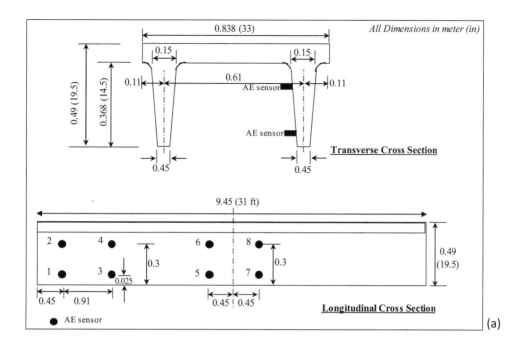

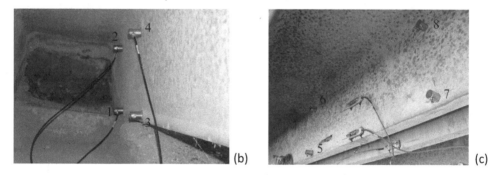

Figure 6.18 AE sensor positions: (a) sensor layout, not to scale (Anay et al., 2015), units in m (in.), unless otherwise specified; conversion: 1 m = 3.3 ft; (b) view at support; (c) view at midspan

A study was conducted by Parmar and Sharp (2012) on a bridge concrete wall on a highway crossing over a freight rail line that serves a cargo terminal to detect and locate the presence of microcracks and newly generated cracks using 2D and 3D AE source location. The techniques were employed to determine crack activity under several loading conditions detected by AE sensors attached along the width of the test wall. AE located activity was found to be related to expansion of the preexisting cracks and formation of new cracks, in addition to friction.

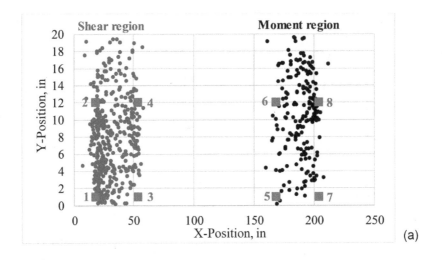

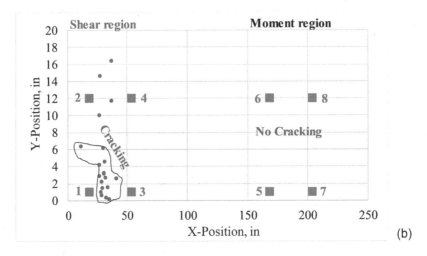

Figure 6.19 AE source location under four trucks back-to-back loading: (a) unfiltered data; (b) filtered data. Metric (SI) conversion factors: 1 in. = 25.4 mm

Another study by Alam et al. (2014) focused on fracture examination in concrete specimens having small, medium, and large sizes using AE technique to identify the location of fracture growth in 3D due to microcracks and macrocracks. During the load test, AE data was recorded by eight (R15i) resonant sensors, four of which were attached on the front surface and four of which were attached on the back surface (opposite surface), as shown in Figure 6.22. The 3D location of the AE sources for a medium-sized specimen is shown in Figure 6.23 at the end of the test, and a greater number of events is present above the notch, which shows the localization of microcracking.

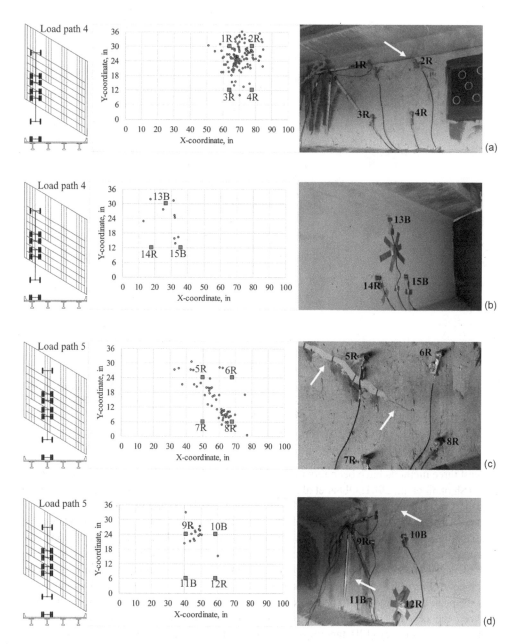

Figure 6.20 AE source location under maximum load (two trucks back-to-back on each girder): (a) sensors 1R–4R (girder 1: north face); (b) sensors 5R–8R (girder 2: south face); (c) sensors 9R–12R (girder 2: north face); (d) sensors 13B–15B (girder 1: south face). Metric (SI) conversion factors: 1 in. = 25.4 mm

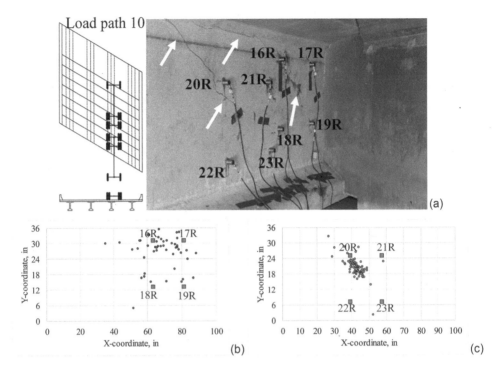

Figure 6.21 AE source location under maximum load (two trucks back-to-back of interior girder 3): (a) view of sensors 16R–23R (all on south face); (b) sensors 16R–19R; (c) sensors 20R–23R. Metric (SI) conversion factors: 1 in. = 25.4 mm

6.3.4.2   Crack classification and moment tensor analysis

At least three methods have been reported to identify cracking mechanisms in concrete structures (Shiotani et al., 2001; Ohtsu et al., 1998; Ohno and Ohtsu, 2010; Ohtsu and Ono, 1984; Aggelis, 2011; Aggelis et al., 2011; ElBatanouny et al., 2014c). The first method, which is a parameter-based method, uses the average frequency versus RA value. The two indices are calculated using AE parameters of rise time, maximum amplitude, counts, and duration. The literature indicates that tensile cracks tend to have low RA value and high average frequency, with the shear cracks being the opposite. Yet the limit for this empirical relation is not fully developed (Ohtsu et al., 1998). The second method is also a parameter-based method and uses the peak frequency data to calculate the P-value such that P is calculated as the distance between the centroid of data points located above the average peak frequency and the average peak frequency (ElBatanouny et al., 2014c). The method is based on the observation that the signature of peak frequency of AE hits during formation of shear cracks and flexural cracks is different.

The third approach for identification of crack mechanism is preformed using moment tensor analysis that uses 3D source location. The simplified Green's functions for Moment Tensor Analysis (SiGMA) procedure has been successfully applied to identify crack location, crack type, and crack orientation in concrete (Ohno et al., 2008; Tomoda et al., 2010). The theory of AE wave motions was generalized based on the elastodynamics

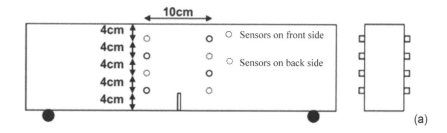

Figure 6.22 (a) AE sensor layout and (b) test setup (Alam et al., 2014). Conversion: 1 cm = 0.4 in.
Source: Reprinted with permission.

and the dislocation theory, hence a crack motion vector and a unit vector normal to the crack plane were used to define crack kinematics at the AE source (Ohtsu et al., 1999), as shown in Figure 6.24. The product of the two vectors lead to the moment tensor. Based on the eigenvalue analysis of the moment tensor, the classification of crack type and the determination of crack orientation can be performed by the SiGMA procedure.

In the SiGMA procedure, the amplitude $A(x)$ of the first motion is represented by (Ohtsu et al., 1999):

$$A(x) = \frac{Cs}{R} \; Ref(t,r)(r_1, r_2, r_3) \begin{pmatrix} m_{11}, & m_{12}, & m_{13} \\ m_{12}, & m_{22}, & m_{23} \\ m_{13}, & m_{23}, & m_{33} \end{pmatrix} \begin{pmatrix} r_1 \\ r_2 \\ r_3 \end{pmatrix} \tag{6.11}$$

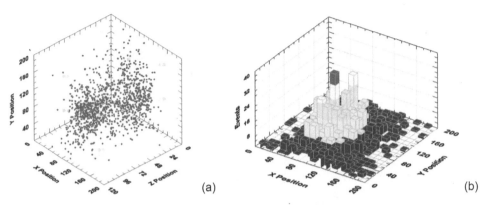

Figure 6.23 (a) 3D source location of AE events in medium size beam; (b) distribution of AE events on the rectangular grid at final load step (all dimensions in mm) (Alam et al., 2014). Metric (SI) conversion factor: 1 mm = 0.04 in.

Source: Reprinted with permission.

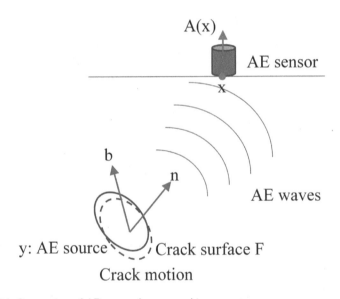

Figure 6.24 Generation of AE waves due to cracking

where *Cs* is the calibration coefficient, *R* is the distance between the source *y* and the observation point *x*, and *r* (r_1, r_2, r_3) is its direction vector. *t* is the direction of the sensor sensitivity, and *Ref*(*t*, *r*) is the reflection coefficient.

As shown in Equation 6.11, the moment tensor is symmetric and of the second order, therefore the number of independent unknowns m_{pq} is six. To solve Equation 6.11, two parameters of the arrival time (P_1) and the amplitude of the first motion (P_2) are calculated from recorded AE waveforms (Fig. 6.25), and the location of the source is determined from the arrival time difference. A multi-channel AE system (with at least six channels) is required to provide sufficient information and to determine all moment tensor components.

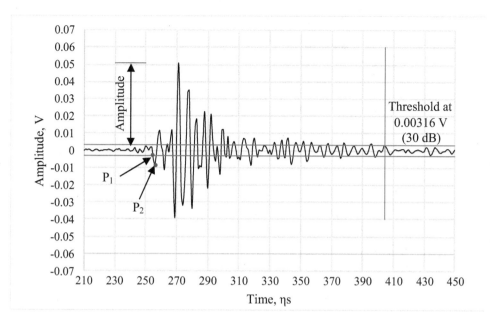

Figure 6.25 Sample of detected AE waveform

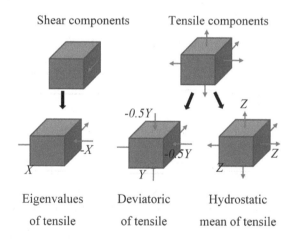

Figure 6.26 Unified decomposition of eigenvalues of the moment tensor

The eigenvalue analysis of the moment tensor was developed to classify crack type (Ohtsu, 1991). The eigenvalues of the moment tensor are represented by the combination of the shear crack and the tensile crack (Fig. 6.26), and the relative ratios X, Y, and Z are obtained as:

$$1.0 = X + Y + Z \tag{6.12}$$

$$\frac{The \ intermediate \ eigen \ value}{The \ maximum \ eigen \ value} = 0 - \frac{Y}{2} + Z \tag{6.13}$$

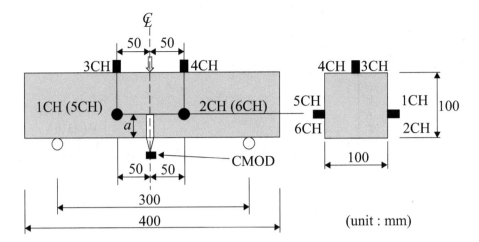

Figure 6.27 Geometry of specimen and sensors layout (Ohno et al., 2014). Conversion: 1 mm = 0.04 in.

Source: Reprinted with permission.

$$\frac{The\ minimum\ eigen\ value}{The\ maximum\ eigen\ value} = -X - \frac{Y}{2} + Z \tag{6.14}$$

When the shear ratios X are smaller than 40%, AE sources are classified as tensile cracks; when the ratios X are greater than 60%, they are referred to as shear cracks. In the case of ratios between 40% and 60%, they are classified as mixed-mode cracks. From the eigenvalue analysis, three eigenvectors e_1, e_2, and e_3 are determined as:

$$e_1 = l + n;\ e_2 = l \times n;\ e_3 = l - n \tag{6.15}$$

where l is the crack motion vector and n is the unit normal vector on the crack surface.

The fracture process zone in a notched concrete beam under three-point bending was investigated by employing the SiGMA procedure (Ohno et al., 2014). Six AE resonant sensors were attached to the surface of the specimen, as shown in Figure 6.27. To investigate the AE source mechanisms in the fracture process zone, the detected AE signals in the test were applied by the SiGMA procedure, and the AE source was classified into three types of tensile mode, mixed mode, and shear mode. Figure 6.28 shows results of the SiGMA analysis of $a/W = 0.3$ and 0.8 (a/W is the ratio of notch depth a to specimen height W). It was found that the width of the AE cluster increases with the increase in the maximum size of aggregate and the decrease in a/W.

6.4   Discussion and recommendations for field applications

It is widely known that a significant challenge with AE data is the noise that is typically present in all AE datasets (Abdelrahman et al., 2018b). The sources of noise in AE data include external sources such as electrical noises from power sources (if not filtered), friction between the specimen and loading apparatus, and rain and hail in field settings. In

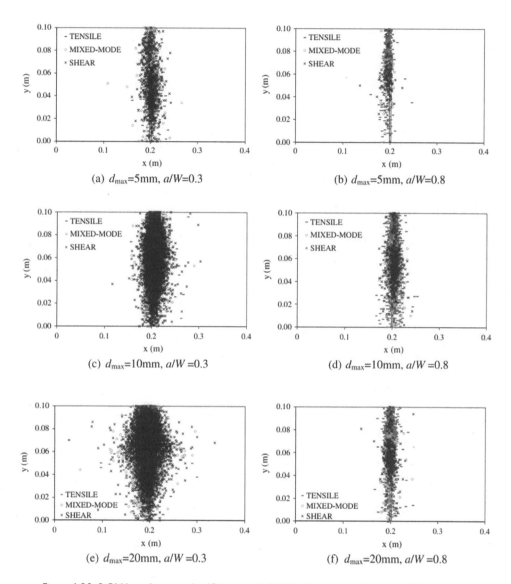

Figure 6.28 SiGMA analysis results (Ohno et al., 2014). Conversion: 1 mm = 0.04 in., 1 m = 3.3 ft
Source: Reprinted with permission.

addition, AE wave reflections represent a major source of noise as it occurs in both laboratory and field settings.

Several methods were proposed for AE damage identification including intensity analysis, calm and load ratio, peak cumulative signal strength ratio, relaxation ratio, b-value analysis, index of damage, and modified index of damage. The majority of the methods do not require significant AE data filtering as indicated in the literature; however, some simple data filters are recommended.

For AE source location, especially in 2D and 3D, significant data filtering is typically required. ElBatanouny et al. (2014b, 2014c) proposed and used parameter-based filters to clean AE data and plot the AE sources in two dimensions with good agreement with visually observed cracks. The same filters where applied by Anay et al. (2015) to locate damage in the field. This shows that these empirically developed filters have the potential to be applied on specimens with different sizes and tested in different setting. Abdelrahman et al. (2018b) proposed an improvement for AE data filters by using wavelet analysis to classify AE wave-forms into genuine and noise-related waves. The method uses the fact that the energy in the AE waveforms from noise sources are more scattered than those from genuine AE wave-forms This technique requires more advanced filtering but it also provides the opportunity for the development of standardized approach for AE data filtering.

The current state of practice of AE monitoring allows for several applications of AE in field settings. The use of AE can be implemented as a stop criterion during load tests where formation or growth of cracks is not desirable, such as the case for prestressed concrete that is designed to be uncracked. In such cases AE can be used to ensure that cracks are not forming while the loads are applied. Another case is steel fracture-critical bridges, where growth of fatigue related cracks at critical connections may lead to catastrophic failures.

References

Abdelrahman, M. (2013) *Assessment of Damage in Concrete Structures Using Acoustic Emission.* M.Sc. Thesis, University of South Carolina, Columbia, SC.

Abdelrahman, M. (2016) *Evaluation of Concrete Degradation using Acoustic Emission: Data Filtering and Damage Detection.* Ph.D. Dissertation, University of South Carolina, Columbia, SC.

Abdelrahman, M., ElBatanouny, M. & Ziehl, P. (2014) Acoustic emission based damage assessment method for prestressed concrete structures: Modified index of damage. *Engineering Structures*, 60, 258–264.

Abdelrahman, M., ElBatanouny, M., Ziehl, P., Fasl, J., Larosche, C. & Fraczek, J. (2015) Classification of alkali-silica reaction damage using acoustic emission: A proof-of-concept study. *Construction and Building Materials*, 95, October, 406–413.

Abdelrahman, M., ElBatanouny, M., Dixon, K., Serrato, M. & Ziehl, P. (2018a) Remote monitoring and evaluation of damage at a decommissioned nuclear facility using acoustic emission. *Applied Sciences*, 8(9), 1663.

Abdelrahman, M., ElBatanouny, M., Rose, J. & Ziehl, P. (2018b) Signal processing techniques for filtering acoustic emission data in prestressed concrete. *Research in Nondestructive Evaluation*, 0934–9847.

AEwin Software User Manual Version E5.5, Mistras Group Inc. (2014) Available from: www.physicalacoustics.com/by-product/aewin/.

Aggelis, D. G. (2011) Classification of crack mode in concrete by using acoustic emission parameters. *Mechanics Research Communications*, 38, 153–157.

Aggelis, D. G., Soulioti, D. V., Sapouridis, N., Barkoula, N. M., Paipetis, A. S. & Matikas, T. E. (2011) Acoustic emission characterization of the fracture process of fiber reinforced concrete. *Construction and Building Materials*, 25, 4126–4131.

Aki, K. & Richards, P. G. (1980) *Quantitative Seismology: Theory and Methods.* Freeman W. H., San Francisco, CA.

Alam, S. Y., Saliba, J. & Loukili, A. (2014) Fracture examination in concrete through combined digital image correlation and acoustic emission techniques. *Construction and Building Materials*, 69, 232–242.

Aljets, D., Chong, A., Wilcox, S. & Holford, K. (2010) Acoustic emission source location in plate-like structures using a closely arranged triangular sensor array. *Journal of Acoustic Emission*, 28.

Anay, R., Cortez, T. M., Jáuregui, D. V., ElBatanouny, M. K. & Ziehl, P. (2015) On-site acoustic-emission monitoring for assessment of a prestressed concrete double-tee-beam bridge without plans. *Journal of Performance of Constructed Facilities*, 30(4), 04015062.

Appalla, A., ElBatanouny, M., Velez, W. & Ziehl, P. (2016) Assessing corrosion damage in post-tensioned concrete structures using acoustic emission. *Journal of Materials in Civil Engineering*, 28 (2), 04015128, 10.

Behnia, A., Chai, H. K. & Shiotani, T. (2014) Advanced structural health monitoring of concrete structures with the aid of acoustic emission. *Construction and Building Materials*, 65, 282–302.

Benavent-Climent, A., Gallego, A. & Vico, J. M. (2011) An acoustic emission energy index for damage evaluation of reinforced concrete slabs under seismic loads. *Structural Health Monitoring*, 11(1), 69–81.

Bormann, P. (2002) *New Manual of Seismological Observatory Practice (NMSOP)*. Annexes. GeoForschungsZentrum.

Colombo, S., Forde, M., Main, I. & Shigeishi, M. (2005) Predicting the ultimate bending capacity of concrete beams from the "relaxation ratio" analysis of AE signals. *Construction and Building Materials*, 19, 746–754.

Cox, S. J. D. & Meredith, P. G. (1993) Microcrack formation and material softening in rock measured by monitoring acoustic emission. *International Journal of Rock Mechanics and Mining Sciences & Geomechanics Abstracts*, 30(1), 11–24, Pergamon.

ElBatanouny, M. K. (2012) *Implementation of Acoustic Emission as a Nondestructive Evaluation Method for Concrete Structures*. Ph.D. Dissertation, University of South Carolina, Columbia, SC.

ElBatanouny, M. K., Mangual, J., Ziehl, P. & Matta, F. (2014a) Early corrosion detection in prestressed concrete girders using acoustic emission. *Journal of Materials in Civil Engineering*, 26 (3), 504–511.

ElBatanouny, M. K., Ziehl, P. H., Larosche, A., Mangual, J., Matta, F. & Nanni, A. (2014b) Acoustic emission monitoring for assessment of prestressed concrete beams. *Construction and Building Materials*, 58, 46–53.

ElBatanouny, M. K., Larosche, A., Mazzoleni, P., Ziehl, P. H., Matta, F. & Zappa, E. (2014c) Identification of cracking mechanisms in scaled FRP reinforced concrete beams using acoustic emission. *Experimental Mechanics*, 54(1), 69–82.

Fowler, T. J. (1986) Experience with acoustic emission monitoring of chemical process industry vessels. *Progress in Acoustic Emission III, JSNDI*, 150–162.

Fowler, T. J., Blessing, J., Conlisk, P. & Swanson, T. L. (1989) The MONPAC system. *Journal of Acoustic Emission*, 8(3), 1–8.

Golaski, L., Gebski, P. & Ono, K. (2002) Diagnostics of reinforced concrete bridges by acoustic emission. *Journal of Acoustic Emission*, 20(2002), 83–89.

Gostautas, R., Nims, D., Tamutus, T. & Seyedianchoobi, R. (2012) Acoustic monitoring of the main cables of the Anthony Wayne suspension bridge. *Structural Materials Technology*, August, 111–119.

Goszczyńska, B., Świt, G., Trąmpczyński, W., Krampikowska, A., Tworzewska, J. & Tworzewski, P. (2012) Experimental validation of concrete crack identification and location with acoustic emission method. *Archives of Civil and Mechanical Engineering*, 12(1), 23–28.

Goszczyńska, B., Świt, G., Trampczyńsf, W., Krampikowska, A., Tworzewska, J. & Tworzewski, P. (2013) Experimental validation of the acoustic emission (AE) method for the cracking process determination and location in concrete element. *IABSE Symposium Report*, 99(27), 365–372. International Association for Bridge and Structural Engineering.

Grosse, C. U. & Ohtsu, M. (eds.) (2008) *Acoustic Emission Testing*. Springer Science & Business Media, Springer-Verlag, Berlin Heidelberg, Germany.

Gutenberg, B. & Richter, C. F. (1954) *Seismicity of the Earth and Associated Phenomena*. 2nd edition. Princeton University Press, Princeton, NJ.

Köppel, S. & Grosse, C. (2000) Advanced acoustic emission techniques for failure analysis in concrete. *15th World Conference on Nondestructive Testing, WCNDT Proceedings*, Roma, Italy, October 15–21.

Kurz, J. H., Grosse, C. U. & Reinhardt, H. W. (2005) Strategies for reliable automatic onset time picking of acoustic emissions and of ultrasound signals in concrete. *Ultrasonics*, 43(7), 538–546.

Labuz, J. F., Chang, H. S., Dowding, C. H. & Shah, S. P. (1988) Parametric study of acoustic emission location using only four sensors. *Rock Mechanics and Rock Engineering*, 21(2), 139–148.

Li, Z., Zudnek, A., Landis, E. & Shah, S. (1998) Application of acoustic emission technique to detection of reinforcing steel corrosion in concrete. *ACI Materials Journal*, 95(1), 68–76.

Lovejoy, S. C. (2006) *Development of Acoustic Emissions Testing Procedures Applicable to Conventionally Reinforced Concrete Deck Girder Bridges Subjected to Diagonal Tension Cracking*. Doctoral Dissertation, Oregon State University, Corvallis, OR.

Lovejoy, S. C. (2008) Acoustic emission testing of beams to simulate SHM of vintage reinforced concrete deck girder highway bridges. *Structural Health Monitoring*, 7, 327–346.

Mangual, J., ElBatanouny, M., Ziehl, P. & Matta, F. (2013a) Acoustic-emission-based characterization of corrosion damage in cracked concrete with prestressing strand. *ACI Materials Journal*, 110(1), 89–98.

Mangual, J., ElBatanouny, M., Ziehl, P. & Matta, F. (2013b) Corrosion damage quantification of prestressing strands using acoustic emission. *Journal of Materials in Civil Engineering*, 25(9), 1326–1334.

Nair, A. & Cai, C. S. (2010) Acoustic emission monitoring of bridges: Review and case studies. *Engineering Structures*, 32(6), 1704–1714.

NDIS 2421. (2000) Recommended practice for in situ monitoring of concrete structures by acoustic emission. *Japanese Society for Non-Destructive Inspection*, Tokyo, pp. 263–268.

Ohno, K. & Ohtsu, M. (2010) Crack classification in concrete based on acoustic emission. *Construction and Building Materials*, 24(12), 2339–2346.

Ohno, K., Shimozono, S., Sawada, Y. & Ohtsu, M. (2008) Mechanisms of diagonal-shear failure in reinforced concrete beams analyzed by AE-SiGMA. *Journal of Solid Mechanics and Materials Engineering*, 2(4), 462–472.

Ohno, K., Uji, K., Ueno, A. & Ohtsu, M. (2014) Fracture process zone in notched concrete beam under three-point bending by acoustic emission. *Construction and Building Materials*, 67, 139–145.

Ohtsu, M. (1991) Simplified moment tensor analysis and unified decomposition of acoustic emission source: Application to in situ hydrofracturing test. *Journal of Geophysical Research: Solid Earth*, 96(B4), 6211–6221.

Ohtsu, M. (2015) *Acoustic Emission and Related Non-Destructive Evaluation Techniques in the Fracture Mechanics of Concrete: Fundamentals and Applications*. Woodhead Publishing, Cambridge, UK.

Ohtsu, M. & Ono, K. (1984) A generalized theory of acoustic emission and green's functions in a half space. *Journal of Acoustic Emission*, 3(1), 124–133.

Ohtsu, M., Okamoto, T. & Yuyama, S. (1998) Moment tensor analysis of acoustic emission for cracking mechanisms in concrete. *ACI Structural Journal*, 95(2), 87–95.

Ohtsu, M., Kaminaga, Y. & Munwam, M. C. (1999) Experimental and numerical crack analysis of mixed-mode failure in concrete by acoustic emission and boundary element method. *Construction and Building Materials*, 13(1–2), 57–64.

Ohtsu, M., Uchida, M., Okamoto, T. & Yuyama, S. (2002) Damage assessment of reinforced concrete beams qualified by acoustic emission. *ACI Structural Journal*, 99(4), 411–417.

Ono, K. (2010) Application of acoustic emission for structure diagnosis. *Konferencja Naukowa*, Warsaw, Poland. pp. 317–341.

Parmar, D. S. & Sharp, S. R. (2012) Acoustic emission for nondestructive evaluation of active crack dynamics in concrete wall of a highway bridge due to interaction with freight traffic on railroad that the bridge crosses. *ASNT 21st Annual Research Symposium & Spring Conference*, Dallas, TX. pp. 73–77.

Ridge, A. & Ziehl, P. (2006) Evaluation of strengthened reinforced concrete beams: Cyclic load test and acoustic emission methods. *ACI Structural Journal*, 103(6), 832–841.

Sammonds, P. R., Meredith, P. G., Murrel, S. A. F. & Main, I. G. (1994) *Modelling the Damage Evolution in Rock Containing Pore Fluid by Acoustic Emission*. Eurock '94, Balkerna, Rotterdam, The Netherlands.

Schumacher, T. (2008) *AE Techniques Applied to Conventionally Reinforced Concrete Bridge Girder*, No. FHWA-OR-RD-09-04, Oregon DOT Report SPR633. Salem, OR. p. 199.

Shearer, P. M. (2009) *Introduction to Seismology*. Cambridge University Press, Cambridge.

Shiotani, T., Ohtsu, M. & Ikeda, K. (2001) Detection and evaluation of AE waves due to rock deformation. *Construction and Building Materials*, 15(5–6), 235–246.

Tomoda, Y., Mori, K., Kawasaki, Y. & Ohtsu, M. (2010) Monitoring corrosion-induced cracks in concrete by acoustic emission. *Proceedings of FraMCoS-7*, 23–28, May.

Xu, J. (2008) *Nondestructive Evaluation of Prestressed Concrete Structures by Means of Acoustic Emission Monitoring*. Dissertation, Dept. of Civil Engineering, Univ. of Auburn, Auburn, Alabama.

Zdunek, A. D., Prine, D. W., Li, Z., Landis, E. & Shah, S. (1995) Early detection of steel rebar corrosion by acoustic emission monitoring. *No.16, Corrosion 95, the NACE International Annual Conference and Corrosion Show*. Northwestern University Infrastructure Technology Institute, IL.

Ziehl, P. & ElBatanouny, M. (2015) Chapter 11: Low-level acoustic emission (AE) in the long term monitoring of concrete. In: Ohtsu, M. (ed.) *Acoustic Emission and Related Non-Destructive Evaluation Techniques in the Fracture Mechanics of Concrete: Fundamentals and Applications*. Woodhead Publishing, Cambridge, UK. pp. 216–236.

Ziehl, P. & ElBatanouny, M. (2016.) Chapter 10: Acoustic emission monitoring for corrosion damage detection and classification. In: Poursaee, A. (ed.) *Corrosion of Steel in Concrete Structures*. Woodhead Publishing, Cambridge, UK. pp. 193–209.

Chapter 7

Fiber Optics for Load Testing

Joan R. Casas, António Barrias, Gerardo Rodriguez Gutiérrez, and Sergi Villalba

Abstract

This chapter presents the application of fiber optic sensor technology in the monitoring of a load test. First, the description of the fiber optic technology is described with main emphasis in the case of distributed optical fiber sensors (DOFS), which have the potential of measuring strain and temperature along the fiber with different length and accuracy ranges. After that, two laboratory tests in reinforced and prestressed concrete specimens show the feasibility of using this technique for the detection, localization, and quantification of bending and shear cracking. Finally, the technique is applied to two real prestressed concrete bridges: in the first case, during the execution of a diagnostic load test; and in the second case, for the continuous time and space monitoring of a bridge subjected to a rehabilitation work. These experiences show the potential of this advanced monitoring technique when deployed in a load test.

7.1 Introduction

The most regularly practiced structural health monitoring (SHM) applications, where load testing is included, are based on electric strain sensors, accelerometers, inclinometers, global navigation satellite system-based sensors, acoustic emission, wave propagation, and so forth. Nevertheless, all of them present genuine challenges when deployed in real-world applications. In order to improve the accuracy and efficiency of the measurements acquired during those practices, optical fiber sensors (OFS) have been used for the past two decades and are one of the fastest growing and most promising monitoring areas. This is a result of some of their intrinsic features of durability, stability, small size, and insensitivity to external electromagnetic perturbations, which makes them ideal for the long-term health assessment of built environment (Bao and Chen, 2012).

Fiber optic sensor based monitoring methods are highly welcome for nondestructive assessment of all types of engineering structures and in particular for load testing, mainly because: they cannot be destroyed by lightning strikes; they can survive in chemically aggressive environments; they can be integrated into very tight areas of structural components; and they are able to deliver sensor chains using one single fiber (Habel and Krebber, 2011).

7.1.1 Background of fiber optics operation

An optical fiber is essentially a cylindrical symmetric structure composed by a central "core" with a uniform refractive index and with a diameter between 4 and 600 μm (Gupta, 2006).

DOI: https://doi.org/10.1201/9780429265969

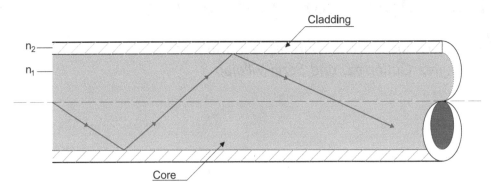

Figure 7.1 Light guiding and reflection in an optical fiber (Barrias et al., 2016)

This core is then enclosed by a "cladding" which has a relatively lower refractive index, enabling the reflection of the emitted light waves at the interface between core and cladding (Fig. 7.1). Most times, the cladding is then covered with an external plastic coating providing the optical fiber sensing with an enhanced environmental and mechanical protection.

Optical fiber cables and sensors are constantly exposed to external perturbations which induce geometrical and optical changes. For communication applications, it is easily understandable that the goal is to minimize these effects in order to provide a reliable signal transmission and reception. On the other hand, in fiber optic sensing the response to these external induced effects is intentionally enhanced (Gholamzadeh and Nabovati, 2008).

The change of some of the properties of the guided light can be produced inside or outside (in another medium) of the optical fiber. Therefore, two different types of sensors can be differentiated: extrinsic and intrinsic (Casas and Cruz, 2003). In their turn, each of these classes of fibers has various subclasses, and even sub-subclasses in some cases, resulting in a large offering of fiber sensors.

There are different ways to classify OFS depending on which property is being considered, such as modulation and demodulation process, application, measurement points, and so forth (Gholamzadeh and Nabovati, 2008). OFS can be categorized into three different classes: interferometric sensors, grating-based sensors, and distributed sensors (Guo et al., 2011) (Fig. 7.2).

The interferometric sensors are point sensors conceived by an intrinsic or extrinsic interferometric cavity along the optical path. When external physical changes occur to the monitored structure, they are transmitted and emulated in the sensor measurements by the optical phase difference between two interference light waves. A grating-based sensor, such as the Fiber Bragg Grating (FBG) sensor, which is the most applied and mature grating-based sensor, is also a point sensor. These sensors use the Bragg wavelength shift of the incoming light, which is affected by external environmental changes such as strain, temperature, and vibration. Finally, distributed sensors are based on the interaction between the emitted light and the properties of the optical physical medium, defined as scattering. For further reading on the difference between these types of sensors, the following references are suggested (Guo et al., 2011; Udd and Spillman, 2011).

The first two types (interferometric and grating) have been greatly investigated and used in civil engineering monitoring applications (Del Grosso et al., 2001; Tennyson et al., 2001;

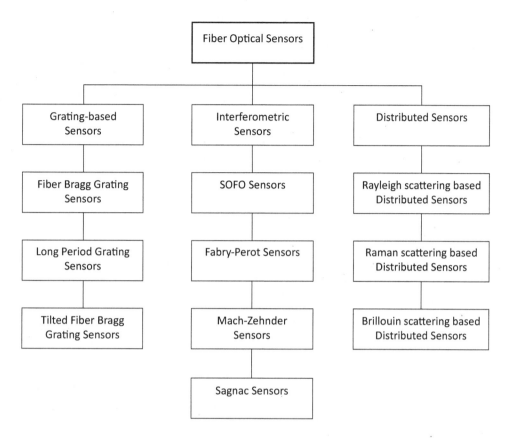

Figure 7.2 Overview of fiber optic sensor technologies
Source: Adapted from Guo et al. (2011).

López-Higuera et al., 2005; Barbosa et al., 2008; Peairs et al., 2009; Rodrigues et al., 2010). In this chapter, the main focus will be on the distributed sensors, the field where the authors have focused their research.

Furthermore, standard sensor instrumentation practice is normally centered on the choice of a limited and relatively small number of points that are supposed to be illustrative of the structural behavior (Glisic and Inaudi, 2012), using in this way discrete or point sensors. Nevertheless, for a large-scale structure the required number of this type of sensors to obtain a complete strain profile information of the monitored structure can grow rapidly.

An additional advantage of DOFS over standard point sensors when applied to load testing is the possibility to obtain the deflection profile and from this the real behavior of support conditions in the tested structure. In addition, the calculation of the deflected shape can be obtained by a single fiber optic cable and does not require a stationary point of reference, as is the case for displacement transducers (Villalba et al., 2012; Casas et al., 2014; Regier and Hoult, 2014; Brault et al., 2017; Nurmi et al., 2017; Secanell, 2017).

In this way, DOFS provide an advantage over point sensors for global strain measurements by providing thousands of sensing points, allowing for the mapping of strain distributions in mainly one dimension (the direction of the fiber), but also in two or even three dimensions. Therefore, experimental data can be used to obtain the global response of a structure to a certain load level, rather than having to use extrapolation techniques from a few point measurements by using a tailored numerical model.

Although still limited real applications were carried out, it is clear that DOFS present a real potential and feasibility for their application in future load tests. Consequently, they will be explained in great detail in the following sections.

7.1.2   Distributed optical fiber sensors (DOFS)

As it is easily comprehensible, DOFS share most of the advantages of other OFS. However, in contrast with the latter, they provide the possibility of measuring strain and temperature evolution both over time and space in a truly distributed way, virtually instrumenting every cross section along the length of the fiber attached to the monitored structural element.

Moreover, an additional advantage when using DOFS is related with the reduced necessity of connecting cables in order to transfer the acquired data from the sensors to the reading units in opposition to the otherwise hefty number of required connecting cables when using discrete or point sensors. As mentioned above, this is even more critical for large-scale structures, where these cables can cover extensive lengths in the range of kilometers (miles). This distinctive advantage permits OFS technology to be more cost-effective in the long term while opening a wide range of important applications in instrumentation of large structures.

Another way of achieving distributed sensing measurements is by multiplexing various discrete sensors, usually FBG, in a single optical fiber cable in what is called quasi-distributed sensing. In fact, up until 2014, two-thirds of the performed SHM projects where optical fiber sensors were deployed used quasi-distributed FBG sensors (Ferdinand, 2014), making it the most popular way of attaining almost spatially continuous optical fiber measurements. Notwithstanding, by using this technique, the instrumentation presents a predetermined finite number of sensors and respective sensing locations. Furthermore, quasi-distributed FBG sensing is based on the wavelength division multiplexing (WDM) technique, which usually limits the number of sensors that can be multiplexed to less than 100 (López-Higuera, 1998).

There is a wide variety of sensors that can be used to instrument a bridge or other structure for load testing or SHM, but only those based on optical fiber technology are able to achieve fully integrated, quasi-distributed, and truly distributed spatial measurements along extensive lengths and with reliable accuracy (Lopez-Higuera et al., 2011).

Optical fiber measurements are obtained through the modulation of the emitted light scattered signal by the external strain and temperature changes to the physical medium where the sensor is attached. When measuring the variation of this modulated signal, distributed fiber sensing is achieved. Hence when talking about optical fiber sensing, the phenomenon of scattering has to be addressed as well.

7.1.3   Scattering in optical fibers

The scattering phenomenon is at the source of all distributed optical fiber sensing technology and is fundamentally defined as the interaction between the emitted light and the optical medium where it is being reflected. Within this interaction, three different scattering processes can occur: Raman, Brillouin, and Rayleigh scattering (Boyd, 2008).

When an electromagnetic wave is emitted into an optical fiber, its propagation through the medium interacts with the constituent atoms and molecules and the electric field induces a time-dependent polarization dipole. This induced dipole generates a secondary electromagnetic wave, and this is called light scattering (Bao and Chen, 2012).

If the physical medium where this scattering takes place is homogeneous, only a forward scattering is possible. Nonetheless, due to the many variations in its density and composition, an optical fiber is in reality a non-homogeneous medium, which means that through the interaction of the emitted light with the fiber, scattering can occur in all directions including the opposite direction of the produced light beam (i.e. generating backscattering).

It is through the analysis of this back-propagating light that the inherent fiber properties are obtained. In this way, any external intervention such as strain or temperature variation is detected through the correlation with the variation of the backscattering components. Figure 7.3 presents the different scattering bandwidths, where it is possible to observe the lower intensity of the Raman and Brillouin spectral bands when compared to the Rayleigh one.

As also observed in Figure 7.3, Raman scattering is highly dependent on the temperature variation on the fiber but not on the strain. Hence, Raman-based DOFS have been researched and practiced in a wide range of applications (Cheng and Xie, 2004; Abalde-Cela et al., 2010; Oakley et al., 2011), but not so much in the bridge engineering field.

On the other hand, Brillouin scattering is inherently dependent on the fiber density, being in this way correlated with any temperature and strain variation. This is the base of all the explored Brillouin-based DOFS (Kurashima et al., 1990; Bao et al., 1995; Zeng et al., 2002; Bastianini et al., 2007; Ravet et al., 2009; Motil et al., 2015).

There are two main Brillouin-based DOFS techniques that have been used in the past few years. First, there is the Brillouin optical time domain reflectometry (BOTDR) technique, which uses the spontaneous Brillouin scattering and was the first to be used for strain and temperature measurements in civil engineering infrastructure. Then, there is the Brillouin optical time domain analysis (BOTDA) technique, which is based on stimulated Brillouin scattering amplifying the usually very weak Brillouin backscattering through the use of counter-propagating lasers (Bao and Chen, 2012; Galindez-Jamioy and López-Higuera, 2012; Leung et al., 2013).

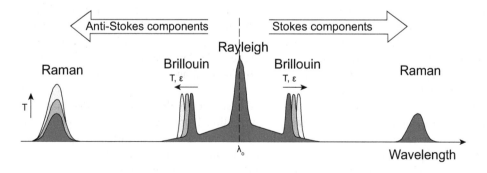

Figure 7.3 Wavelengths of the backscattered radiation (Rodriguez et al., 2014)

Both BOTDR and BOTDA are characterized for allowing distributed sensing measurements over extensive lengths, making them suitable for global monitoring of large-scale structures (Uchida et al., 2015). Nevertheless, both present a limited spatial resolution of 1 m (3.28 ft), which is not ideal for some applications such as damage detection.

On the other hand, Rayleigh scattering, as a quasi-elastic or linear phenomenon, is by itself independent of almost any external physical field, as seen in Figure 7.3. However, the Rayleigh backscattered profile of a specific optical fiber is a result of the heterogeneous reflective index of the latter (i.e. the imperfections), which is distributed randomly along the entire length of the fiber, establishing a fingerprint of each optical fiber as a result of its manufacturing process (Grave et al., 2015). Hence when subjected to an external stimulus, the Rayleigh backscatter pattern presents a spectral shift that is then used to calculate the strain or temperature variations along the length of the fiber by comparing with its unaltered reference state.

This technique is named Rayleigh based optical frequency domain reflectometry (OFDR). The most popular distributed OFS system is the optical backscatter reflectometer (OBR). Due to the random fluctuation of the refraction index that gives origin to the Rayleigh scattering, a comparison can be made with the use of fiber optic cable with a series of Bragg gratings with random yet definable periods (Froggatt and Moore, 1998; Hoult et al., 2014). By scanning the frequency with the OFDR technique, the spectral response of each equivalent FBG is obtained as well as high spatial resolution measurements of strain and temperature variations.

7.1.4 State of the art of fiber optic sensors in load testing

As mentioned before, optical fiber technology has been one of the most popular topics with civil engineering structural health monitoring in the past two decades, and in this way, successful applications of these sensors in load testing are found in the literature. This is truer for the use of discrete OFS when comparing to DOFS.

Some of the first applications of OFS sensors in bridge monitoring were conducted in Canada starting in 1993, where six bridges were equipped with the use of fiber optic sensing systems: Beddington Trail Bridge (Alberta), Salmon River Bridge (Nova Scotia), Taylor Bridge (Manitoba), Crowchild Trail Bridge (Alberta), Joffre Bridge (Quebec), and Confederation Bridge (New Brunswick) (Tennyson et al., 2001).

The Beddington Trail Bridge is a two-span, continuous skew bridge of 22.83 m (24.97 yd) and 19.23 m (21.03 yd) spans. Each span consists of 13 bulb-tee section, precast, prestressed concrete girders (Wu and Abe, 2003). In this first instrumented bridge, 18 out of the 20 FBG sensors were still in operation five years after their installation. These sensors had initially the objective of monitoring the stress relaxation in the steel and carbon fiber reinforced polymer (CFRP) tendons. Afterwards, with the use of an upgraded interrogator which improved the achievable strain resolution, the dynamic responses of a low-speed truck were measured and used to approximately estimate the weight of the truck and its driving direction (Maaskant et al., 1997).

The Taylor Bridge was the most instrumented case study with the use of 63 FBG sensors and two multi-Bragg sensors, being that 26 conventional strain gages were also deployed for comparison purposes. From the correctly sealed strain gages, 60% of them experienced malfunction due to excessive moisture originated in the steam curing process, whereas the OFS continued to perform well without showing any signs of being affected by similar malfunctions.

Although not technically deployed for the purpose of bridge load testing, some applications, especially the ones where OFS embedded solutions were used, are of interest in the scope of this publication.

For example, during the construction of the 117 m (127.95 yd) concrete arch span of the Siggenthal Bridge over the Limmart River in Baden in 2000, 58 long-gage OFS, with gage lengths ranging from 3 m (3.28 yd) to 5 m (5.47 yd), were embedded in pairs close to the top and bottom surfaces of its concrete arch slab. The purpose of these sensors was to measure the deformation of arch segments during both the construction and in-service periods (Inaudi et al., 2002).

The Vaux viaduct in Switzerland was also successfully monitored with the use of a total 12 FBG sensors. This structure is a steel-concrete composite bridge with a total length of 950 m (1038.93 yd) composed of two parallel bridges, the north one with 13 spans and the south with 14 spans (Navarro et al., 2000). Here the sensors were deployed in the interior walls of a box girder section during the push and pull stage during construction in order to measure the produced strain (Vohra et al., 2000). Moreover, on the concrete highway Versoix bridge in Switzerland, 104 total OFS transducers were installed during the requalification and widening of this bridge (Vurpillot et al., 1997).

OFS have also been applied in steel bridges load tests, such as the case of a circular pedestrian steel bridge in Aveiro, Portugal (Barbosa et al., 2008). In this case, 32 strain FBG transducers and eight temperature FBG transducers were welded to the bridge structure in a star configuration in eight branches. This installation occurred after the bridge construction, and all sensors were protected with waterproof sealant tape. The data collected by the DOFS during two load tests served to validate a developed numerical model used in the structural design by verifying that this model reproduced the real behavior of the bridge.

In the UK, Kerrouche et al. (2008) used FBG-based sensors embedded in CRFP composite rods incorporated into grooves used in the reinforcement of a reinforced concrete bridge while testing the bridge to destruction. The applied optical fiber sensors were able to measure in-situ, exceptionally high strains showing the feasibility of the use of these sensors even outside of their normal working parameters.

The same group (Kerrouche et al., 2009) then developed a relatively cheap and effective sensing system using a compact FBG-based monitoring system incorporating a scanning Fabry-Perot filter. The performance of this system was assessed and validated afterwards through both laboratory experiments and field tests in a real bridge.

In Korea, Chung et al. (2008) conducted an experimental study on the use of long-gage optical fiber sensors in order to estimate the deflection of prestressed concrete bridges. These results were compared with results using conventional strain gages and linear displacement transducers (LVDT) presenting the same level of accuracy.

More recently Torres Górriz et al. (2017) compared the performance between point and long-gage FBG-based strain sensors during static and dynamic load tests conducted on a concrete railway bridge. While the results confirm the predetermined idea that long-gage sensors are more suitable in the monitoring of heterogeneous materials, there are indeed some particular cases where both sensor types provide the same accurate strain measurements.

Regarding the use of DOFS in real-world bridge load test applications, there are relatively few examples. One of the performed applications was the use of a Brillouin optical time domain reflectometry on a diagnostic load test of Bridge A6358 on US-54 over the Osage River in Osage Beach, Missouri, USA (Matta et al., 2008). Here a 1.16 km (0.72 mi) optical fiber circuit was deployed onto the girders for strain measurement and thermal compensation. These experimental strain-measured profiles were integrated into vertical displacements, which presented a good agreement with total station measurements.

In 2012, in order to maximize the potential of OFS sensors use in a bridge monitoring strain test, Zhao et al. (2012) implemented a multiscale fiber optical sensing network with both FBG- and BOTDA-based sensors. In this way, the extended measurement range capability of DOFS and the high sensitivity achieved by FBG sensors were explored simultaneously. This combination allowed for the easy measurements of the overall stress condition of the structure, obtaining full information with local and distributed high precision.

Finally, Regier and Hoult (2014) deployed an OBR based DOFS in the load test of a reinforced concrete Black River Bridge outside of the town of Madoc, Ontario, Canada. The results provided by the DOFS presented a good agreement with instrumented strain gages also validating its results. Moreover, the strain data measured by the DOS were used to calculate deflections through double integration. These results compared well with measurements obtained by displacement transducer sensors.

A more complete state of the art on the use of DOFS in civil engineering applications on load testing and also general SHM systems is presented in Rodriguez et al. (2015a) and Barrias et al. (2016).

7.1.5 Advantages and disadvantages of fiber optic sensors versus other sensors for load testing

When comparing with the more traditional mechanical and electrical sensors, optical fiber sensors possess some unique advantages, which makes them a better solution for most of the civil engineering SHM applications and also load testing.

Some of these advantages are related to their immunity to electromagnetic interference (EMI). Furthermore, these sensors can also provide reliable measurements within a relatively wide temperature range and present large bandwidth and high sensitivity. Moreover, due to their small size and light weight, they are easy to carry and install on the monitored elements. Finally, they can also be embedded or simply bonded to a concrete or metallic support, which expands the possible applications of this technology (Casas and Cruz, 2003; Lopez-Higuera et al., 2011; Udd and Spillman, 2011; Ye et al., 2014).

On the other hand, with the use of electrical sensors, in some cases it is not possible to discriminate physical data signals from the noise produced by EMI. Additionally, these sensors can only be used as discrete or point sensors, restricting their use to one specific location while requiring long connection cables, limiting their application in large-scale structures (Li et al., 2004).

Nevertheless, between 1980 and 2000, the introduction of fiber optic sensors technology in the markets was somewhat slow while competing with the more conventional and established sensors. This was essentially due to the relatively higher costs associated with a limited number of suitable components. Since then, this situation was reversed, with even better expectations for the future (Udd and Spillman, 2011). With increasing use of the optical components driven by the commercial and telecommunication and optoelectronic market, their production costs have been steadily declining while the quality and the variety of choices are boosted, as seen in Figure 7.4.

For all these reasons, the authors feel confident on the increasing use of this type of sensors on structural load testing applications. This conviction is extended through the presentation of the personal experience acquired on the past few years on the use of fiber optic

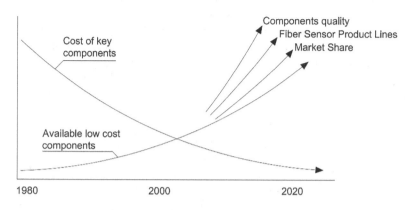

Figure 7.4 Trends for fiber optic sensors
Source: Adapted from Udd and Spillman (2011).

sensors, more specifically in their distributed form (DOFS) on SHM as well as load testing applications.

7.2 Distributed optical fibers in load testing

7.2.1 *Introduction*

OBR-based DOFS present an extremely high potential in the monitoring of bridge structural elements during load tests. Due to their distributed capability, extensive lengths of the structure can be monitored with both strain and temperature data. Furthermore, an additional advantage is that this technology does not require any ground reference in order to measure deflections and can be deployed both by attachment to the concrete, steel, or composite surfaces or in embedded solutions.

In the next sections, the experience obtained by the authors on the use of this technology during structural load tests is described. Initially, the performance of the system was assessed in controlled laboratory environments. Afterwards, with the experience and conclusions obtained, real-world experiences were conducted on load tests of two bridges in the area of Barcelona, where different challenges and conditions were dealt with and assessed.

In all these applications, the used OBR system was the commercial ODiSI-A model system from LUNA Technologies. This system enables fully distributed strain and temperature measurements with sub-cm spatial resolution and a ±2 με and ±0.2°C resolution. The maximum sensing length is 50 m (54 yd) and the strain measurement range is ±13,000 με.

7.2.2 *Experiences in laboratory: validation of the system*

7.2.2.1 *Bending tests of concrete slabs*

A reinforced concrete slab with a 5.60 m (6.12 yd) span length, 1.60 m (1.75 yd) width, and 0.285 m (11.22 in.) cross-section depth was load tested up to failure. The slab was simply supported at both ends and the loading was applied using an actuator "MTS" of 1 MN (225 kip) capacity "P" in the midspan of the slab (Fig. 7.5).

As seen in Figure 7.5, four different segments of OBR sensors were deployed: two on the top surface and two on the bottom surface. The used optical fiber had a polyimide coating and a 50 m (54.68 yd) length and was of a single mode type.

One important step when implementing DOFS in structural elements is related to the deployed adhesive and subsequent bonding. Hence before the deployment of the fiber,

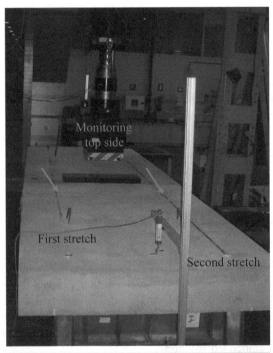

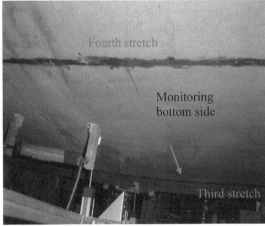

Figure 7.5 Instrumented slab and OBR sensors location
Source: Villalba and Casas (2012).

all predicted bonding areas were smoothed by mechanical means and cleaned and wiped of any residual grease through the use of a commercial cleaning solvent. Subsequently, the OBR sensor was bonded using a one-part component ethyl cyanoacrylate adhesive with low viscosity.

Aside from the DOFS, the reinforced concrete slab was also instrumented with dynamic strain gages in some reinforcing bars. Additionally, a total of five LVDTs were also deployed in order to measure the deflection at the center (300 mm [11.81 in.] stroke length range) and end sides (50 mm [1.97 in.] stroke length range) of the slab. Finally, three magnetic transducers, Tempsonics E-Series Model ER sensors with 50–75 mm (1.97–2.95 in.) of stroke length range, were used to monitor the joint opening at the construction joint of the slab. All of this is summarized and depicted in Figure 7.6. In Figure 7.5, one of the joint opening sensors is also shown. In Figure 7.6, the circled numbers refer to the position from the beginning of the fiber and indicate the points where the fiber is bent to pass from one stretch to the next. The numbers indicate the distances in centimeters from the edges of the slab to the closest fiber stretch.

The mechanical properties of the concrete and reinforcement are detailed in Table 7.1. With this information, the actuator load to induce cracking was calculated.

The ultimate load capacity was 255.15 kN (57.36 kip), which was 12.30% higher than the expected and theoretically calculated load of 227.20 kN (51.08 kip). Since the OBR sensor was adhered on both sides of the slab longitudinally, the strain fields of these positions and their respective variations were measured and examined.

In this way, in Figure 7.7, as an example the measured strain evolution on the deployed third stretch of the fiber is represented for different load stages. Strain peaks are identified which correspond to crack formations. Therefore, it is possible to observe how in this experiment DOFS technology was able to detect and track crack formations both for low load levels as well as severe cracking conditions. Additionally, by superimposing different load levels, it is possible to observe that the position where the peaks are located due to cracking remain stable. The peaks were observed in the DOFS far in advance of when they could be seen by visual inspection. In this sense, the use of this type of sensor in load tests where loads are high (proof load testing) can be really useful, as this may anticipate the occurrence of cracking just when it is forming, and therefore the test can be stopped before irreversible damage is produced in the structure.

In Figures 7.8 and 7.9, the behavior of both the third and fourth stretch (at the bottom surface of the slab) at the 50 kN (11 kip) and 110 kN (25 kip) load levels are specifically followed. As it is observable in these figures, taken for a load level of 110 kN (25 kip), the detected crack positions by the OBR system present a good correlation with the visually observed positions on the surface of the concrete slab. When a crack is formed, a strain increase is produced at the sensor which is picked up by the frequency shift of the DOFS. Additionally, as it was expected, it is possible to see that the majority of the produced cracks are located at midspan.

On the stretches of the fiber that are bonded on the upper surface of the reinforced concrete slab, the OBR system was also able to measure and follow the compressive strain evolution, as seen in Figure 7.10. It is observable that the measurements do not present noise perturbations and that the obtained strains are relatively low for the initial load levels. Only when close to the failure load does the DOFS detect peaks, indicating the formation of cracking due to excessive compression.

Once the strain profile evolution on the concrete surface is obtained with the OBR system, it is possible to create a formulation capable of calculating an average crack width. This is

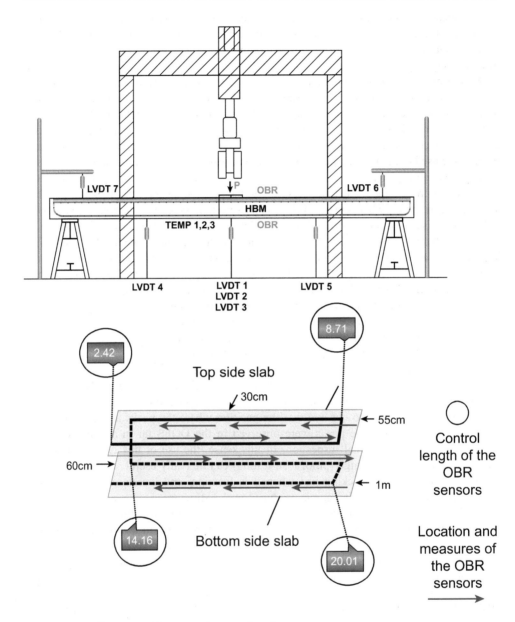

Figure 7.6 Experimental setup and sensor location

Source: Villalba and Casas (2012).

based on the integration of the obtained distributed experimental strains along a length L which corresponds to the zone where cracking is observed. It is obtained as the distance between the points where the strain in the concrete reaches the maximum tensile strain. This formulation is depicted in Figure 7.11 and is fully explained in Rodríguez et al. (2015b).

Table 7.1 Concrete slab material properties and characteristics. Conversion: 1 MPa = 145 psi, 1 kNm = 0.74 kip-ft, 1 kN = 0.225 kip

Yield limit of reinforcing bar (MPa)	Concrete compressive strength (MPa)	Additive	Cement	f_{ckm} (MPa)	E (MPa)	M_{crack} (kNm)	P_{crack} (kN)
500	35	Glenium C-355	1 52.5 R	51.31	33147.63	89.721	43.30

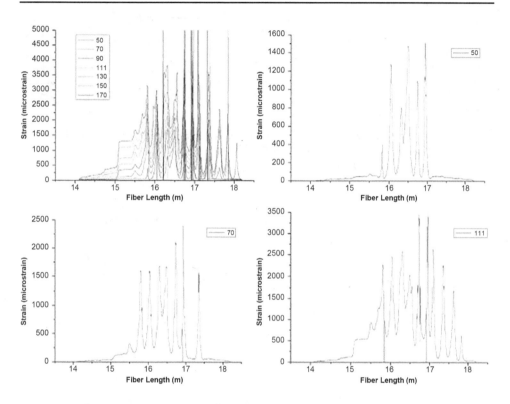

Figure 7.7 Strain along the third stretch (bottom side) of the fiber for increasing load level (50–170 kN). Upper-left: superposition of all load levels. Rest: 50, 70, and 111 kN load levels. Conversion: 1 m = 3.3 ft, 1 kN = 0.225 kip

Source: Villalba and Casas (2012).

In this calculation, in order to obtain an average strain (ε_{mean}) in the cracked length (L) the strain distribution (ε_{OBR}) is characterized in comparison to the maximum tensile strain of the concrete (ε_{fct}) and integrated as shown in Equation 7.1.

$$\varepsilon_{mean} = \frac{1}{L} \int_{0}^{L} \varepsilon_{OBR} \, dL \tag{7.1}$$

This average strain includes both the deformation due to concrete cracking and, starting here, the deformation due to the cracks in the cracked area. In this way,

Stretch "Length"	Peak Location in optical fiber Length (m)	Peak Location related to origin of stretch (m)	Micro strain (με) (50kN load level)	Micro strain (με) (110kN load level)
3	15.513	1.953	-	800
3	15.818	2.258	400	2250
3	16.046	2.486	1300	2450
3	16.318	2.758	800	2590
3	16.492	2.932	1480	2040
3	16.551	2.991	-	1550
3	16.745	3.185	1090	3450
3	16.942	3.382	1500	3500
3	17.085	3.525	-	2650
3	17.355	3.795	-	2240
3	17.626	4.066	-	1700
3	17.830	4.270	-	675

Figure 7.8 Peaks location in optical fiber and measured strain. Crack location (third stretch).
Conversion: 1 m = 3.3 ft, 1 kN = 0.225 kip

Source: Villalba and Casas (2012).

Stretch "Length"	Peak Location in optical fiber Length (m)	Peak Location related to origin of stretch (m)	Micro strain (με) (50kN load level)	Micro strain (με) (110kN load level)
4	20.522	3.588	-	800
4	20.694	3.416	270	4400
4	20.909	2.931	-	4460
4	21.123	2.987	400	2250
4	21.288	2.822	1450	2600
4	21.436	2.674	-	2150
4	21.562	2.548	3260	4910
4	21.735	2.375	-	2000
4	21.851	2.259	1650	2700
4	22.064	2.046	1270	2500
4	22.357	1.735	-	4500
4	22.940	1.170	-	350

Figure 7.9 Peaks location in optical fiber and measured strain. Crack location (fourth stretch).
Conversion: 1 m = 3.3 ft, 1 kN = 0.225 kip

Source: Villalba and Casas (2012).

knowing ε_{mean}, ε_{fct}, and L, the sum of the widths of all cracks (Σw) can be calculated with Equation 7.2.

$$\varepsilon_{mean} = \varepsilon_{fct} + \frac{\Sigma w}{L} \qquad (7.2)$$

An average crack width is then obtained with the use of Equation 7.3, where N corresponds to total number of cracks, which can be obtained by summing the number of the

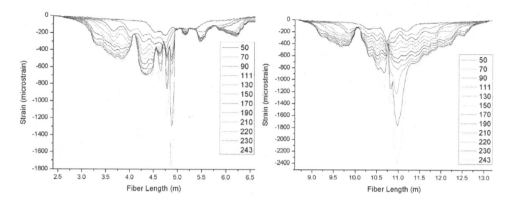

Figure 7.10 Strain along the first (left) and second stretch (right) of the fiber for increasing load level. Upper surface of the slab. Conversion: 1 m = 3.3 ft

Source: Villalba and Casas (2012).

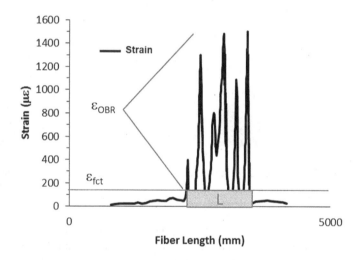

Figure 7.11 Average crack width parameters definition. Conversion: 1 mm = 0.04 in.

peaks detected by the OBR system.

$$w_{mean} = \frac{\Sigma w}{N} \tag{7.3}$$

This methodology can be used for different load levels. The comparison between this calculated average crack width and the one obtained experimentally at midspan with the use of crack transducers is described in Table 7.2.

The crack width transducers were located in two different points of the cross section, one near the edge of the slab and the other in the central part (Fig. 7.12). For this reason, the arithmetic mean of the two transducers was calculated to better represent the

Table 7.2 Comparison of obtained crack widths at midspan

Load (kN [kip])	Crack width transducer 1 (mm [in.])	Crack width transducer 2 (mm [in.])	Arithmetic mean (mm [in.])	OBR stretch 3 (mm [in.])	OBR stretch 4 (mm [in.])
50 [11]	0.058 [0.0023]	0.099 [0.0039]	0.079 [0.0031]	0.062 [0.0024]	0.065 [0.0026]
70 [16]	0.077 [0.0030]	0.154 [0.0061]	0.116 [0.0046]	0.112 [0.0044]	0.101 [0.0040]
90 [20]	0.105 [0.0041]	0.125 [0.0049]	0.115 [0.0045]	0.149 [0.0059]	0.127 [0.0050]
110 [25]	0.166 [0.0065]	0.147 [0.0058]	0.157 [0.0062]	0.190 [0.0075]	0.163 [0.0064]
130 [29]	0.296 [0.0117]	0.200 [0.0079]	0.248 [0.0098]	0.237 [0.0093]	0.209 [0.0082]
150 [34]	0.370 [0.0146]	0.267 [0.0105]	0.319 [0.0126]	0.298 [0.0117]	0.246 [0.0097]
170 [38]	0.439 [0.0173]	0.337 [0.0133]	0.388 [0.0153]	0.354 [0.0139]	0.213 [0.0084]

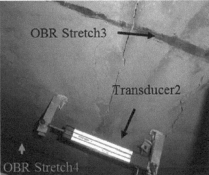

Figure 7.12 Cracks transducers' location. Left – near the edge of slab; Right – center of slab
Source: Rodríguez et al. (2015b).

overall crack width. The results show a good agreement between the method based on the measurements of the OBR system and the ones obtained by the transducers.

With this validation test it was possible to study the performance of this technology on the detection and localization of the developed cracks during the load increase. Through the comparison of the measured strain peaks with visual inspection, a good agreement was verified. This experience opens the possibility of using DOFS in proof load testing in order to avoid the occurrence of cracking. A complete description of the test is available in Villalba and Casas (2012).

7.2.2.2 Shear tests of partially prestressed concrete beams

A subsequent laboratory experiment was conducted again at the Structural Technology Laboratory of the Technical University of Catalonia (UPC-BarcelonaTECH), where DOFS were deployed for testing of two partially prestressed concrete (PPC) beams.

These PPC members are generally designed allowing for controlled cracking under service loading conditions. Therefore, the monitoring of the crack formation and evolution in these elements assumes an important challenge.

In this case, the focus of the tests was on the structural behavior under predominant shear loads. This adds additional complexity when compared with pure bending behavior where cracks are developed orthogonally to the beam axis, whereas in the case of shear, the unknown variable of the inclination of the cracking pattern is added.

The monitoring of shear crack pattern is usually achieved through the use of discrete and traditional instrumentation with limited results. Strain rosettes and digital microscopes have been the most used set of instrumentation in shear laboratory tests in order to measure crack widths, shear crack angles, and shear sliding displacements (Zakaria et al., 2009). Nevertheless, if these sensors are not located exactly at the point where cracks are going to be initiated, due to their discrete nature they are not able to effectively track the developed crack pattern formation. In opposition, and thanks to its distributed measuring capabilities, DOFS become greatly attractive tools for this kind of application.

In this laboratory experiment, two different PPC beams were tested under a three-point shear setup. The distance between the load position and the support was 1.80 m in order to generate a shear failure of the beams. Their respective mechanical properties are described in Table 7.3. Their cross-section shape, general dimensions, and the location of the load are depicted in Figure 7.13.

For the detection and tracing of the shear crack pattern using DOFS and the OBR system, an optical fiber cable arrangement was proposed by instrumenting the beams with a two-dimensional DOFS grid. As shown in Figure 7.14, in beam I1 two DOFS were used orthogonally to each other, one for horizontal segments and the other for vertical segments, whereas in beam I2 only one fiber was deployed for the entire two-dimensional grid.

In order to better assess the strain evolution, the results were divided in the different stretches illustrated in Figure 7.14, where horizontal stretches were named A, B, C, and D and the vertical stretches were numbered 1 through 10. Unfortunately, DOFS2, which

Table 7.3 PPC beams concrete properties. Conversion: 1 MPa = 145 psi

Specimen	f_{cm} (MPa)	f_{ct} (MPa)	E (MPa)	ε_{fct} με
I1	32.5	4.6	36440	126
I2	29.35	4.15	27264	152

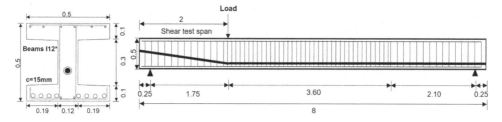

Figure 7.13 Cross section and longitudinal layout of PPC beams (dimensions in m). Conversion: 1 m = 3.3 ft

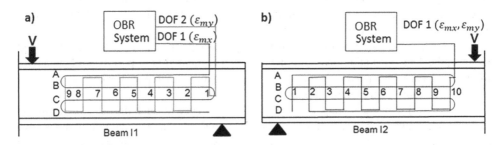

Figure 7.14 Proposed DOFS grids for shear crack monitoring in beams I1 (a) and I2 (b)

Source: Rodríguez et al. (2016).

was bonded to beam I1, failed to work properly during the experimental test and in this way, only horizontal measurements were obtained and analyzed for beam I1.

Notwithstanding, the measurements from DOFS 1 worked in a satisfactory way and important structural data was analyzed. The strain peaks which corresponded to the formation of cracking were detected by these horizontal measurements even before detected with the naked eye.

Some of the results for the horizontal measurements in beam I2 are illustrated in Figure 7.15. In this image, the concrete tensile strength of beam I2 (152 $\mu\varepsilon$) is plotted in order to better assess the crack formation by identifying the locations where the developed strain surpasses this value. In this case, it is possible to observe that the first crack originated with an 87 kN (20 kip) load at the bottom stretch followed by other cracks at higher loads.

Vertical strains distributions were also measured in this beam for the different load levels. Both stretches 2 and 8 are portrayed in Figure 7.16 for different levels of the test load.

From Figure 7.16 it is possible to see in left plots that for the initial load stages the concrete tensile strength was not exceed and in this way no diagonal cracks due to shear effects are developed. This conveys that the observed crack identified in Figure 7.15 at stretch D for 87 kN (20 kip) is purely due to bending.

Nevertheless, for higher load levels, starting at the 127 kN (29 kip) mark, different peaks were detected which correspond to developed diagonal cracks at the different web height levels. By pinpointing the locations of these peaks (i.e. where the measured strains surpass the concrete tensile capacity), a pattern of the shear cracks can be tracked.

Although in beam I1 only horizontal measurements were obtained, it was still possible to trace the developed shear cracking pattern in some parts of the beam as illustrated in Figure 7.17 (right) for a load of 262 kN (59 kip). This pattern shows a good agreement with the visually observed pattern (left).

The same exercise was conducted for beam I2 but this time having access to both the horizontal and vertical strains distributions and peaks. This is depicted in Figure 7.18 for a load level of 244 kN (55 kip). Again the shear crack pattern obtained through DOFS presents a good correlation with the shear crack pattern observed during the test.

In the same way as it was conducted for the cracks in bending, here an algorithm for the calculation of the average crack width based on the results obtained in the horizontal segments of the DOFS (x) and the vertical segments (y) was derived as depicted in Figure 7.7. A full explanation can be found in Rodriguez (2017).

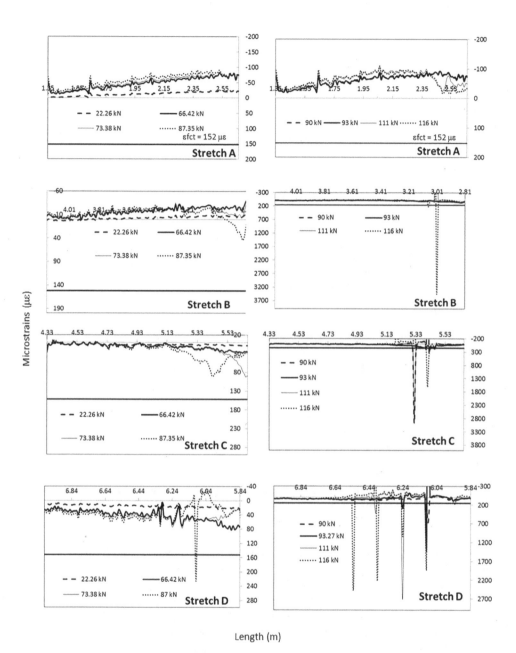

Figure 7.15 Horizontal strain distribution and peaks identification along DOFS 1 in beam 12. Conversion: 1 m = 3.3 ft, 1 kN = 0.225 kip

Source: Rodríguez et al. (2016).

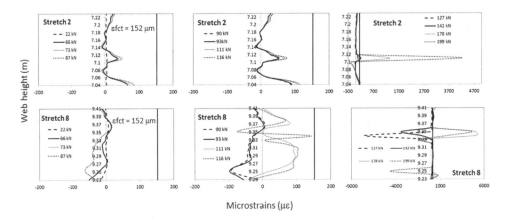

Figure 7.16 Vertical strain distribution and peaks identification along the DOFS in beam I2. Conversion: I kN = 0.225 kip

Source: Rodríguez et al. (2016).

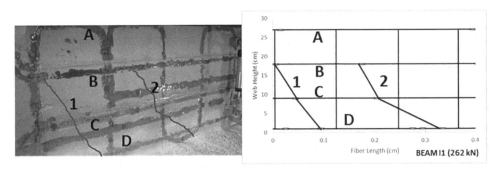

Figure 7.17 Correlation between shear crack pattern and detected peaks by OBR system at 262 kN (59 kip) in beam II. Conversion: I cm = 0.4 in.

Source: Rodríguez et al. (2016).

These beams were also instrumented with electrical transducers performing a strain rosette (Fig. 7.20). In this way, it was possible to compare the results obtained from both set of sensors as shown in Table 7.4.

A fairly good agreement can be seen between the results obtained from the DOFS sensors and the rosettes for load levels of 203 kN (45.6 kip) and 213 kN (47.9 kip). It should be pointed out that the technique of rosette sensor is not very accurate.

With this experiment, it was possible to assess the performance of DOFS technology in shear load testing and conclude that it is feasible to use this technology to monitor the shear structural behavior in reinforced concrete structures. High spatial resolution measurements of spatially continuous strain data at different loading levels were obtained allowing for the detection, location, and quantification of flexural and shear cracks. Therefore, the application of this type of sensor in the case of shear load testing is even more relevant than in the case of bending, as early detection of shear cracking and stoppage of the load increase is of paramount importance due to the brittle mode of failure.

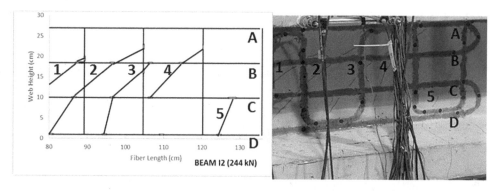

Figure 7.18 Shear crack pattern and detected peaks by OBR system at 244 kN (55 kip) in beam 12. Conversion: 1 cm = 0.4 in.

Source: Rodríguez et al. (2016).

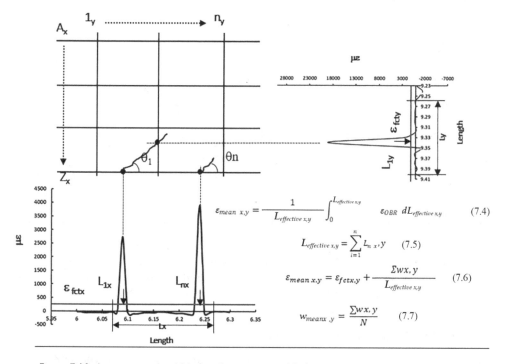

$$\varepsilon_{mean\ x,y} = \frac{1}{L_{effective\ x,y}} \int_0^{L_{effective\ x,y}} \varepsilon_{OBR}\ dL_{effective\ x,y} \qquad (7.4)$$

$$L_{effective\ x,y} = \sum_{i=1}^{n} L_{n\ x,y} \qquad (7.5)$$

$$\varepsilon_{mean\ x,y} = \varepsilon_{fctx,y} + \frac{\Sigma wx, y}{L_{effective\ x,y}} \qquad (7.6)$$

$$w_{meanx\ ,y} = \frac{\Sigma wx, y}{N} \qquad (7.7)$$

Figure 7.19 Average crack width for shear structural behavior

7.2.3   Application of DOFS in real structures

By now, there is still a relatively small amount of available examples using Rayleigh OFDR-based DOFS in real-world bridges, and not all them are perfectly related to load tests, being more the experiences oriented to SHM.

In the following sections, two different applications of this technology in real-world bridges are described, where different obstacles and challenges had to be assessed and dealt with,

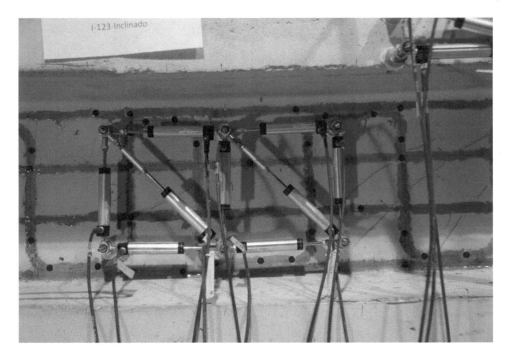

Figure 7.20 Electrical transducers in rosette shape for crack width measurement

Table 7.4 Comparison of the obtained crack widths in beam 12

Load (kN [kip])	Crack width (mm [in.])	
	Electrical transducer	OBR
203 [45.6]	0.24 [0.0094]	0.20 [0.0079]
213 [47.9]	0.27 [0.0106]	0.17 [0.0067]

leading to a better understanding of the capabilities and limitations of DOFS in bridge load testing. The first example is a diagnostic load test, and the second example describes the monitoring carried out during the rehabilitation of an existing bridge, and from its analysis, several conclusions about the performance of DOFS in load testing can be derived.

7.2.3.1 *San Cugat bridge in Barcelona*

The first application of the OBR system in a real bridge by the authors was conducted back in 2010, on a load test of a newly built viaduct on the BP-1413 road at Cerdanyola del Vallés in the Barcelona area, Spain. The structure is composed by five prestressed precast box beams with 25 m (27.34 yd) span length and 14 m (15.31 yd) width deck, performing a total longitudinal length of 125 m (136.7 yd). The beams are simply supported on double columns as seen in Figures 7.21 and 7.22.

For the diagnostic load, apart from the standard instrumentation, the setup consisted on the use of three sensors, DOFS 1 and DOFS 2 with 25 m (27.34 yd) length each and DOFS 3 with a length of 50 m (54.68 yd). The first two were positioned longitudinally along the

Figure 7.21 Viaduct in road BP-1413 in Cerdanyola del Vallés (Barcelona, Spain)
Source: Casas et al. (2014).

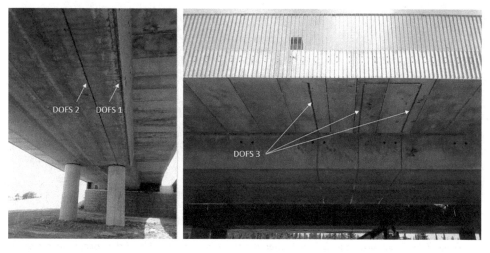

Figure 7.22 Instrumented DOFS in tested bridge
Source: Casas et al. (2014).

bottom flange of one beam of the central span to measure the strain due to the longitudinal bending, and the latter was placed orthogonally to the beam in the cantilever, as depicted in Figure 7.22, to measure the strains in the cantilever due to the transversal local bending moment.

Regarding the bonding of DOFS 3, special care was taken in the points where the fiber passes from the bottom flange of the beam to the web and from the web to the cantilever. At these critical points, the fiber was not fully bonded to prevent its local bending in order to obtain the optimal results in the remainder of the sensor.

In this way, except for the described points, the entire length of the DOFS was bonded to the bridge through the use of an epoxy adhesive, allowing for the monitoring of the developed strains under the load test. This bridge was also instrumented to measure deflections with conventional fleximeters sensors (LVDT) at the nearest points to the supports and midspan.

The main goals of the test were to (for the first time) assess the feasibility and performance of the OBR system on the monitoring of a real concrete bridge and to compare the data measured by this technology with both theoretical and other experimental values obtained by the LVDTs.

The load test consisted on two different load scenarios. The first one was conducted by the crossing of a 400 kN (90 kip) truck alone, while the second scenario was performed under the normal traffic in the bridge plus the additional load of the 400 kN (90 kip) truck.

7.2.3.1.1 Reading strains under 400 kN truck

When crossing the bridge from one end to another, the 400 kN (90 kip) truck induces the development of positive strains due to tension in the instrumented longitudinally DOFS. In Figure 7.23, for instance, it is possible to see the measured strain at the moment when the

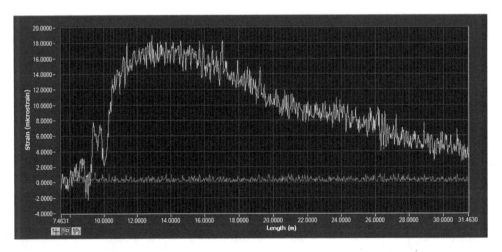

Figure 7.23 Strain profile in DOFS 2 when the truck is located at 1/4 of the span. Conversion: 1 m = 3.3 ft

Source: Casas et al. (2014).

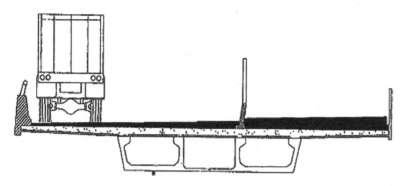

Figure 7.24 Position of the truck
Source: Casas et al. (2014).

truck is located in the first quarter position of the span in a centered position, generating a strain around 17 μɛ.

After the complete passage of the truck, the measured strain returns to values close to zero corresponding to the horizontal line in Figure 7.23, and the baseline of the experimental campaign, which is related to the static value of the bridge self-weight.

7.2.3.1.2   *Reading strains under normal traffic and 400 kN static load*

In the second load scenario, the 400 kN (90 kip) truck load was located at the midspan on the interior lane, and at the same time regular traffic was crossing on the other lane in the same direction (see Fig. 7.24). The data acquisition was performed with a sampling acquisition of one reading per minute of the three DOFS and a spatial resolution of 12 mm (0.47 in.). The first 10 minutes of measurements corresponded to the calibration and stabilizing period, after which a maximum increase of strain of 45 μɛ was obtained in the sensor DOFS 2.

From the strain measured along the entire length of the fiber, which corresponds to the entire length of the beam, and by taking into account the boundary conditions on the supports, it was possible to calculate the deflection at midspan with a value of 3.45 mm (0.136 in.) through the use of double integration of the curvature obtained by DOFS 2. The experimental distributed curvature was obtained, assuming that the point of zero strain (neutral axis) corresponds to the center of gravity of the cross section, obtained from the bridge drawings. The curvature at any point is obtained as the ratio between the strain measured by the fiber and the distance from the fiber to the neutral axis. The LVDT located at this point gave a value of 3.71 mm (0.146 in.), while the numerical model, consisting of a grillage of one-dimensional beam elements representing the beam plus slab in the longitudinal direction and the slab in the transverse direction, reported a value of 3.42 mm (0.135 in.). As presented in Table 7.5, a very good agreement was achieved.

Regarding the sensor DOFS 3, which was the one installed transversely to the deck of the bridge, the main goal was to obtain the deflection value at the end of the cantilever through double integration when the truck was placed as shown in Figure 7.24.

Table 7.5 Deflection at midspan (mm)

Sensor	DOFS 2	DOFS I	Fleximeters	Numerical Model
Deflection mm [in.]	3.45 [0.1358]	3.62 [0.1425]	3.71 [0.1461]	3.42 [0.1347]

In this case, it was important to pinpoint the exact length of the fiber which corresponded to the cantilever. This could be made through the software of the sensor interrogator, which provided the correct distance between any point of the fiber and the connector.

In this way, the analyzed cantilevers corresponded to the fiber length between 34 m and 37 m (37.18 yd and 40.46 yd) and 65 m and 68 m (71.09 yd and 74.37 yd).

When comparing the strain profile along the cantilever obtained from the two different sections of the sensor DOFS 3, it was realized that they present, as expected, very close values. With the measured strain at the bottom of the fiber and assuming the position of the neutral axis (possible due to the elastic range) and by again applying a double integration, where the two constants of this calculation are obtained through the conditions of a zero curvature at the end of the cantilever and a zero deflection at the fixed end, a deflection at the end of the cantilever of 0.26 mm (0.01 in.) was obtained. Again, to obtain the curvature profile along the cantilever, the point of zero strain was assumed the centroid of the corresponding cross section.

From the results reported here, it seems feasible to obtain the deflection profile of the bridge or the deflection at the points of interest by the use of the DOFS technology in a bridge load test. This can be achieved thanks to the high accuracy and spatial resolution provided by this technology. More detailed information about this load test is available in Villalba et al. (2012) and Casas et al. (2014).

7.2.3.2 Sarajevo Bridge in Barcelona

The second application of this technology on a real-world structure was performed on the Sarajevo Bridge located at one of the main entrances by highway to the city of Barcelona, Spain (Fig. 7.25).

This structure is composed of two simply supported spans of 36 m (39.37 yd) and 50 m (54.68 yd), where each span presents three box girder prestressed concrete beams which are connected through an upper reinforced concrete slab (Figs. 7.26 and 7.27).

This structure presented deficient conditions for pedestrian traffic, so the bridge authorities decided to widen the deck and improve the overall bridge aesthetics by adding overhead metal protections for the pedestrians. Naturally, this rehabilitation induced structural load changes which had to be monitored and analyzed during the works.

As said before, due to the fact that this structure is located at one of the road entrances to the city of Barcelona with higher traffic volume, the possibility of closing the bridge surroundings for this rehabilitation was not an option. In this situation, DOFS technology provided unique advantages by not requiring ground level reference and having the possibility of obtaining a great number of strain points data with the use of a minimal number of sensors.

As a result, it was decided to instrument one of the box girder beams with two 50 m (54.68 yd) DOFS placed at each side of its inner bottom slab at the junction with the beam web, as seen in Figure 7.28. This position was chosen since it was the location with the expected highest strain increments and possible cracking.

Figure 7.25 Sarajevo Bridge in Barcelona, Spain

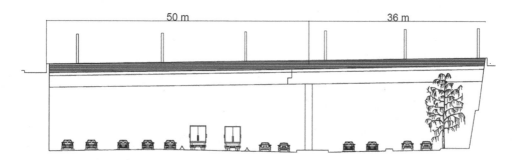

Figure 7.26 Side view of Sarajevo bridge. Conversion: 1 m = 3.3 ft
Source: Barrias et al. (2018).

In opposition to what was presented for the San Cugat bridge, in this case the sensors were placed inside the box girder, providing a better environmental protection of the sensors and an easier access for installation and operation. This was important as the monitoring system had to be in service for at least one year.

Of the 50 m (54.68 yd), only 36 m (39.37 yd) were bonded adjusting to the length of the instrumented viaduct span. A spatial resolution of 1 cm (0.39 in.) was used, which meant that each DOFS was measuring a total of 3600 points along the structure. The measurements were made sparsely along the monitoring period and are described in Table 7.6.

Figure 7.27 Bottom perspective of the bridge slab's box girders
Source: Barrias et al. (2018).

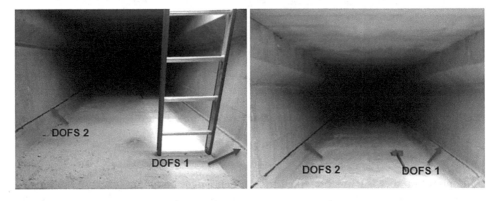

Figure 7.28 Deployed OBR system
Source: Barrias et al. (2018).

During these monitoring events the DOFS measurements were acquired every 5 minutes and from this data, the critical values (maximum and minimum) were assessed and used to produce envelope response graphs. While between June 29th and November 4th 2015 the measurements were made alternatively between DOFS 1 and 2, after the latter date only

Table 7.6 Summary of the monitoring events dates

Date	Description
29/06/2015	2 hrs 30 min measurement – DOFS 1 only
16/07/2015	3 hrs 20 min measurement – DOFS 1 only
06/08/2015	7 hrs measurement – DOFS 2 only
15/09/2015	7 hrs measurement – DOFS 1 only
01/10/2015	5 hrs measurement – DOFS 2 only
02/10/2015	1 hrs 30 min measurement – DOFS 2 only*
09/10/2015	5 hrs 30 min measurement – DOFS 1 only
04/11/2015	4 hrs measurement – DOFS 1** and DOFS 2
10/12/2015	6 hrs 30 min measurement – DOFS 2 only
22/12/2015	3 hrs measurement – DOFS 2 only
18/01/2016	5 hrs 40 min measurement – DOFS 2 only
19/01/2016	2 hrs 30 min measurement – DOFS 2 only
20/01/2016	3 hrs measurement – DOFS 2 only
18/02/2016	7 hrs 40 min measurement – DOFS 2 only

*The short measurement duration was due to the rupture of the cable that provided electrical power to the monitoring system.
**Measuring of a single 5-min event with DOFS 1.

data from DOFS 2 was acquired since the sensor DOFS 1 ceased to work properly. This was due to the puncture and following irreparable damage done to this sensor while works were being carried out in its whereabouts inside the box girder.

Notwithstanding, due to the long monitoring period of this application, one factor that had to be considered and dealt with was the temperature influence in the readings of the DOFS. Both the refractive index of the backscattered light and the materials which compose these sensors are dependent of these temperature changes, so compensation of its effect on the monitoring output was required. For this reason, the final part of the DOFS was not bonded to the concrete, creating an unbonded loop segment of the fiber located at the end, so only the strains due to temperature changes were measured there (Fig. 7.29). The algorithm used for temperature compensation is fully described in Barrias et al. (2018).

After the application of the thermal compensation method to the average measured strains of each segment from both DOFS during their monitoring operation period, the values represented in Table 7.7 and Table 7.8 were obtained.

The values in Table 7.7 and Table 7.8 are plotted in Figures 7.30 and 7.31 in order to better visualize the pure mechanical measured strain distribution along the monitored bridge span.

On both DOFS measurements it is observable that the measured strains in segment S1 are relatively smaller than the ones verified in the remaining segments. This is due to the proximity to the span support system and its elastomeric bearings. Moreover, it is observable that the mean mechanical strain distribution is essentially uniform in the remainder of the beam and with increasing compression as surrounding temperature decreases.

In conclusion, the applied loads on the deck due to the bridge rehabilitation produced very small strain variations along the bottom part of the bridge span, and in this way the main source of the observed strain variation was the uniform shortening of the box girder associated to the decrease of temperature from summer to winter. In fact, the measured strain difference between August and February was on the order of 500 $\mu\varepsilon$ (Fig. 7.31),

Figure 7.29 Unbonded loop of the DOFS
Source: Barrias et al. (2018).

Table 7.7 Average strain in different parts of DOFS 1 after temperature compensation (microstrain)

Segment ID	29/06/2015	16/07/2015	15/09/2015	09/10/2015	04/11/2015
S0	10.4	113.5	−389.3	−546.3	−670.7
S1	−2.8	45.3	−56.1	−104.1	−160.4
S2	13.4	23.0	−168.3	−230.8	−281.8
S3	17.0	27.8	−172.1	−246.5	−267.2
S4	18.2	29.9	−172.3	−254.5	−291.1

Table 7.8 Average strain in different parts of DOFS 2 after temperature compensation (microstrain)

Segment ID	06/08/ 2015	01/10/ 2015	04/11/ 2015	10/12/ 2015	22/12/ 2015	18/01/ 2016	19/01/ 2016	20/01/ 2016	18/02/ 2016
S0	14.1	−707.7	−771.0	−1168.5	−1177.5	−1393.4	−1383.2	−1397.4	−1353.6
S1	5.5	−235.0	−238.7	−379.1	−383.6	−499.5	−456.2	−453.5	−448.8
S2	17.1	−286.0	−271.2	−443.8	−446.2	−539.0	−536.5	−534.8	−504.8
S3	19.8	−284.9	−253.0	−425.6	−41.9	−528.7	−524.4	−524.1	−480.4
S4	20.7	−311.1	−287.0	−477.4	−479.3	−565.2	−571.7	−571.6	−532.3

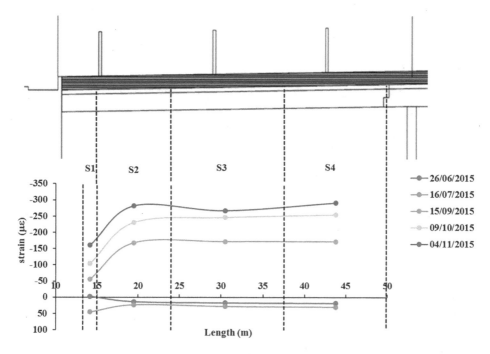

Figure 7.30 Mean mechanical distribution for DOFS 1. Conversion: 1 m = 3.3 ft
Source: Barrias et al. (2018).

which is very plausible for a structure of this typology and the climate of Barcelona. A complete description of the bridge tests can be found in Barrias et al. (2018).

7.2.3.3   *Lessons learned from the field tests*

The experiences obtained in the application of the field tests in the two real bridges show the feasibility of deploying the fiber optic sensors in-situ, both for the case of a diagnostic static test after construction and in the case of continuous structural health monitoring during the execution of a rehabilitation work. In the latter, because of the temperature change in the bridge site, the correction of the temperature effects can be done without problem as shown in the second bridge. In both cases, the bonding of the fiber optic sensor in the concrete surface was easily done with a standard bonding agent (epoxy).

However, it is important to smooth and clean the surface before bonding in order to get a perfect installation. As a concrete surface is a rather rough surface, this gives feasibility of a good bonding in the case of other bridge materials as steel and composite materials with much smoother finish. In fact, DOFS have been used in steel bridges and composite structures. The application to timber and masonry bridges is not yet available. However, the application to other masonry structures, as in the case of a masonry vault in a hospital building, as presented in Barrias et al. (2016) and Barrias et al. (2018), provides clear evidence that the deployment in masonry bridges is certainly feasible.

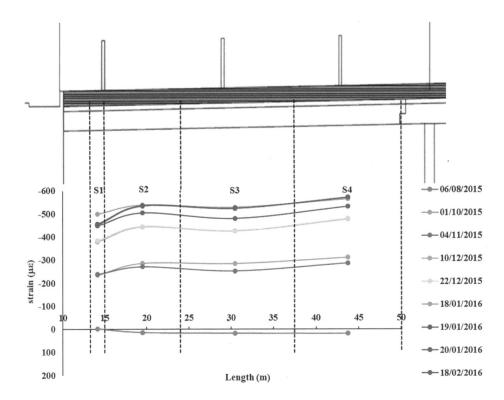

Figure 7.31 Mean mechanical distribution for DOFS 2. Conversion: I m = 3.3 ft
Source: Barrias et al. (2018).

7.3   Conclusions

In this chapter, the use of fiber optic sensors was described, especially in their distributed form in load testing. Initially, an overall introduction to optical fiber sensing was carried out, where the specific type of distributed optical fiber sensors (DOFS) were differentiated from the remainder while explaining their physical background, main field of applications, and their advantages and disadvantages.

This was then followed by the presentation of four different tests conducted by the authors using OBR-based DOFS in concrete elements and bridges during specific conditions of testing. Two of these tests were conducted in the controlled laboratory environment and the other two in real bridges with actual environmental conditions.

From the validation load tests carried out in the laboratory, the following conclusions can be drawn that are relevant for the implementation of DOFS in load testing:

1   It is possible with the use of a fiber in one direction or a grid in perpendicular directions to detect and locate imminent cracking in a reinforced or pretensioned concrete element. This result is relevant for diagnostic load testing but even more important to take into account in the design of a proof load test

2   The distributed strain profile obtained by two longitudinal fibers located at different heights of the element cross section allows to define a distributed profile of curvatures,

which after double integration with the corresponding boundary conditions, permit the calculation of the deflection profile and the assessment of the actual support conditions. This result is of interest when in the load test the deployment of displacement transducers to measure the vertical deflection becomes unrealistic and when there exist important discrepancies between the experimental results from the test and the expected ones calculated with a theoretical or numerical model.

The examples presented from the load tests in two real bridges allow the following conclusions:

1   The possibility of obtaining the deflection profile from the strain profile that was checked in laboratory specimens is also verified in the case of real-world bridges. The deflections were obtained both in the longitudinal and transversal direction with the use of a minimal number of sensors and without any reference to the ground level. A good agreement was achieved when comparing these results with the ones obtained through a theoretical numerical model and instrumented fleximeters. The possibility to obtain the full deflected shape by using DOFS and not only the deflection in discrete points, which is the case with the standard sensors, is relevant in the case of load testing as it may give guidance about the correct performance of the support conditions (bearings).
2   The additional challenge of the temperature influence on the measurements from a load test and the need of temperature effect compensation in the experimental results can be easily solved in the case of using DOFS and the OBR system in the monitoring. This is of relevance in the case when the results of two load tests carried out in different environmental conditions are compared in order to conclude about the correct performance of the structure after some time in service (diagnostic load test).

Overall, with the experimental campaigns presented, the OBR-based DOFS technology has proven its feasibility and application in the load test of bridges and many other structures, being one of the most promising tools for future applications of this type. With the decrease of the price of manufacturing components, the total cost of use of these sensors will only decrease and become an even more appealing and sought-out solution in the near future.

Acknowledgments

The authors want to knowledge the financial support provided by the Spanish Ministry of Economy and Innovation through research projects BIA2013–47290-R, BIA2012–36848 and FEDER (European Regional Development Funds), and also funding from the European Union's Horizon 2020 research and innovation program under the Marie Sklodowska-Curie grant agreement No. 642453.

References

Abalde-Cela, S., et al. (2010) Surface-enhanced Raman scattering biomedical applications of plasmonic colloidal particles. *J. R. Soc. Interface*, (7), S435–S450.

Bao, X., et al. (1995) Experimental and theoretical studies on a distributed temperature sensor based on Brillouin scattering. *Journal of Lightwave Technology*, 13(7), 1340–1348. doi: 10.1109/50.400678.

Bao, X. & Chen, L. (2012) Recent progress in distributed fiber optic sensors. *Sensors (Basel)*, 12(7), 8601–8639. doi: 10.3390/s120708601.

Barbosa, C., et al. (2008) Weldable fibre Bragg grating sensors for steel bridge monitoring. *Measurement Science and Technology*, 19(12), 125305. doi: 10.1088/0957-0233/19/12/125305.

Barrias, A., Casas, J. & Villalba, S. (2016) A review of distributed optical fiber sensors for civil engineering applications. *Sensors*. Multidisciplinary Digital Publishing Institute, 16(5), 748. doi: 10.3390/s16050748.

Barrias, A., et al. (2018) Application of distributed optical fiber sensors for the health monitoring of two real structures in Barcelona. *Structure and Infrastructure Engineering*, 14(7), 967–985. doi: 10.1080/15732479.2018.1438479.

Bastianini, F., et al. (2007) Overview of recent bridge monitoring applications using distributed Brillouin fiber optic sensors. *Journal of Nondestructive Testing*. Citeseer, 12(9), 269–276.

Boyd, R. (2008) *Nonlinear Optics*, 3rd edition. Academic Press, Rochester, NY.

Brault, A., Nurmi, S. & Hoult, N. A. (2017) Distributed deflection measurement of reinforced concrete elements using fibre optic sensors. *39th IABSE Symposium: Engineering the Future, 21–23 September*. Vancouver, Canada, pp. 1469–1477.

Casas, J. R. & Cruz, P. J. S. (2003) Fiber optic sensors for bridge monitoring. *Journal of Bridge Engineering*, 8(6), 362–373. doi: 10.1061/(ASCE)1084-0702(2003)8:6(362).

Casas, J. R., Villalba, S. & Villalba, V. (2014) Management and safety of existing concrete structures via optical fiber distributed sensing. *Maintenance and Safety of Aging Infrastructure: Structures and Infrastructures Book Series*, 10, 217–245. doi: 10.1201/b17073-9.

Cheng, J.-X. & Xie, X. S. (2004) Coherent anti-stokes Raman scattering: Instrumentation, theory, and applications. *Journal of Physical Chemistry*, B, 108, 827–840.

Chung, W., et al. (2008) Deflection estimation of a full scale prestressed concrete girder using long-gauge fiber optic sensors. *Construction and Building Materials*. Elsevier, 22(3), 394–401.

Del Grosso, A., et al. (2001) Monitoring of bridges and concrete structures with fibre optic sensors in Europe. *IABSE Symposium Reports*, 84(6), 15–22. Available from: www.ingentaconnect.com/content/iabse/report/2001/00000084/00000006/art00003.

Ferdinand, P. (2014) The evolution of optical fiber sensors technologies during the 35 last years and their applications in structure health monitoring. *EWSHM-7th European Workshop on Structural Health Monitoring*, July 8–11, Nantes, France.

Froggatt, M. & Moore, J. (1998) High-spatial-resolution distributed strain measurement in optical fiber with rayleigh scatter. *Applied Optics*, 37(10), 1735–1740. doi: 10.1364/AO.37.001735.

Galindez-Jamioy, C. A. & López-Higuera, J. M. (2012) Brillouin distributed fiber sensors: An overview and applications. *Journal of Sensors*. Hindawi Publishing Corporation, 2012, 1–17. doi: 10.1155/2012/204121.

Gholamzadeh, B. & Nabovati, H. (2008) Fiber optic sensors. *International Journal of Electrical, Computer, Energetic, Electronic and Communication Engineering*, 2(6), 1107–1117. Available from: http://onlinelibrary.wiley.com/doi/10.1002/9781118014103.fmatter/summary.

Glisic, B. & Inaudi, D. (2012) Development of method for in-service crack detection based on distributed fiber optic sensors. *Structural Health Monitoring*, 11(2), 161–171. doi: 10.1177/1475921711414233.

Grave, J. H. L., Håheim, M. L. & Echtermeyer, A. T. (2015) Measuring changing strain fields in composites with distributed fiber-optic sensing using the optical backscatter reflectometer. *Composites Part B: Engineering*. Elsevier, 74, 138–146.

Guo, H., et al. (2011) Fiber optic sensors for structural health monitoring of air platforms. *Sensors*, 11(12), 3687–3705. doi: 10.3390/s110403687.

Gupta, B. D. (2006) *Fiber Optic Sensors: Principles and Applications*. New India Publishing, New Delhi.

Habel, W. R. & Krebber, K. (2011) Fiber-optic sensor applications in civil and geotechnical engineering. *Photonic Sensors*, 1(3), 268–280. doi: 10.1007/s13320-011-0011-x.

Hoult, N. A., Ekim, O. & Regier, R. (2014) 'Damage/deterioration detection for steel structures using distributed fiber optic strain sensors. *Journal of Engineering Mechanics*, 140(1), 4014097. doi: 10.1061/(ASCE)EM.1943-7889.0000812.

Inaudi, D., et al. (2002) Monitoring of a concrete arch bridge during construction. *SPIE's 9th Annual International Symposium on Smart Structures and Materials*. International Society for Optics and Photonics, 17–21 March 2002, San Diego, CA. pp. 146–153.

Kerrouche, A., et al. (2008) Strain measurement on a rail bridge loaded to failure using a fiber Bragg grating-based distributed sensor system. *IEEE Sensors Journal*. IEEE, 8(12), 2059–2065.

Kerrouche, A., et al. (2009) Design and in-the-field performance evaluation of compact FBG sensor system for structural health monitoring applications. *Sensors and Actuators A: Physical*. Elsevier, 151(2), 107–112.

Kurashima, T., Horiguchi, T. & Tateda, M. (1990) Distributed-temperature sensing using stimulated Brillouin scattering in optical silica fibers. *Optics Letters*, 15(18), 1038–1040. doi: 10.1364/OL.15.001038.

Leung, C. K. Y., et al. (2013) Review: Optical fiber sensors for civil engineering applications. *Materials and Structures*, 48(4), 871–906. doi: 10.1617/s11527-013-0201-7.

Li, H.-N., Li, D.-S. & Song, G.-B. (2004) Recent applications of fiber optic sensors to health monitoring in civil engineering. *Engineering Structures*. Elsevier, 26(11), 1647–1657.

López-Higuera, J. M. (1998) *Optical Sensors*. Universidad de Cantabria, Cantabria, Spain.

López-Higuera, J. M., et al. (2005) Fiber optic civil structure monitoring system. *Optical Engineering*, 44(4 artigo n° 44401), 1–10. doi: 10.1117/1.1882392.

Lopez-Higuera, J. M., et al. (2011) Fiber optic sensors in structural health monitoring. *Journal of Lightwave Technology*, 29(4), 587–608. doi: 10.1109/JLT.2011.2106479.

Maaskant, R., et al. (1997) Fiber-optic Bragg grating sensors for bridge monitoring. *Cement and Concrete Composites*, 19, 21–33.

Matta, F., et al. (2008) Distributed strain measurement in steel bridge with fiber optic sensors: Validation through diagnostic load test. *Journal of Performance of Constructed Facilities*, 22(4), 264–273. doi: 10.1061/(ASCE)0887-3828(2008)22:4(264).

Motil, A., Bergman, A. & Tur, M. (2015) State of the art of Brillouin fiber-optic distributed sensing. *Optics and Laser Technology*. Elsevier, 78, 1–23. doi: 10.1016/j.optlastec.2015.09.013.

Navarro, M. G., Lebet, J.-P. & Beylouné, R. (2000) Launching of the Vaux viaduct. *Structural Engineering International*. International Association for Bridge and Structural Engineering, 10(1), 16–18.

Nurmi, S., Mauti, G. & Hoult, N. A. (2017) Distributed sensing to assess the impact of support conditions on slab behaviour. *39th IABSE Symposium: Engineering the Future, 21–23 September*. Vancouver, Canada, pp. 1478–1485.

Oakley, L. H., et al. (2011) Identification of organic materials in historical oil paintings using correlated extractionless surface-enhanced Raman scattering and fluorescence microscopy. *Analytical Chemistry*, (83), 3986–3989.

Peairs, D. M., et al. (2009) Fiber optic monitoring of structural composites using optical backscatter reflectometry. *Proceedings of the 41st International SAMPE Technical Conference*, Wichita, KS, USA, pp. 19–22.

Ravet, F., et al. (2009) Submillimeter crack detection with Brillouin-based fiber-optic sensors. *IEEE Sensors Journal*, 9(11), 1391–1396. doi: 10.1109/JSEN.2009.2019325.

Regier, R. & Hoult, N. A. (2014) Distributed strain behavior of a reinforced concrete bridge: Case study. *Journal of Bridge Engineering*, 19(12), 5014007. doi: 10.1061/(ASCE)BE.1943-5592.0000637.

Rodrigues, C., Felix, C. & Figueiras, J. (2010) Fiber-optic-based displacement transducer to measure bridge deflections. *Structural Health Monitoring*, 10(2), 147–156. doi: 10.1177/1475921710373289.

Rodriguez, G. (2017) *Monitoring of Concrete Structures by Distributed Fiber Optic Sensors*. Ph.D. Thesis (in Spanish), UPC-BarcelonaTECH, Barcelona, Catalonia, Spain.

Rodriguez, G., Casas, J. R. & Villalba, S. (2014) Assessing cracking characteristics of concrete structures by distributed optical fiber and non-linear finite element modelling to cite this version. *7th European Workshop on Structural Health Monitoring*. Nantes.

Rodriguez, G., Casas, J. R. & Villalba, S. (2015a) SHM by DOFS in civil engineering: A review. *Structural Monitoring and Maintenance*, 2(4), 357–382. doi: 10.12989/smm.2015.2.4.357.

Rodríguez, G., Casas, J. R. & Villalba, S. (2015b) Cracking assessment in concrete structures by distributed optical fiber. *Smart Materials and Structures*. IOP Publishing, 24(3), 35005.

Rodríguez, G., et al. (2016) Monitoring of shear cracking in partially prestressed concrete beams by distributed optical fiber sensors. *Proceedings 8th International Conference on Bridge Maintenance, Safety and Management*, IABMAS, June 26–30, Foz do Iguaçu, Brazil.

Secanell, A. (2017) *Distributed Deflection of Post-Tensioned Beam Based on Distributed Fiber Optic Sensors: Application to Bridges*. Master Thesis (in Catalan). UPC-BarcelonaTECH.

Tennyson, R. S., et al. (2001) Structural health monitoring of innovative bridges in Canada with fiber optic sensors. *Smart Materials Structures*. Manitoba, Canada, 10(3), 560–573.

Torres Górriz, B., Rinaudo, P. & García, C. (2017) Comparison between point and long-gage FBG-based strain sensors during a railway bridge load test. *Strain*. Wiley Online Library, 53(4).

Uchida, S., Levenberg, E. & Klar, A. (2015) On-specimen strain measurement with fiber optic distributed sensing. *Measurement*. Elsevier, 60, 104–113.

Udd, E. & Spillman Jr, W. B. (2011) *Fiber Optic Sensors: An Introduction for Engineers and Scientists*. 2nd edition. John Wiley & Sons, Hoboken, NJ.

Villalba, S. & Casas, J. R. (2012) Application of optical fiber distributed sensing to health monitoring of concrete structures. *Mechanical Systems and Signal Processing*. Elsevier, 39(1), 441–451.

Villalba, V., Casas, J. R. & Villalba, S. (2012) Application of OBR fiber optic technology in structural health monitoring of Can Fatjó Viaduct (Cerdanyola de Vallés-Spain). *VI International Conference on Bridge Maintenance, Safety and Management*. Stresa, Italy.

Vohra, S., et al. (2000) Distributed strain monitoring with arrays of fiber Bragg grating sensors on an in-construction steel box-girder bridge. *IEICE Transactions on Electronics*. The Institute of Electronics, Information and Communication Engineers, 83(3), 454–461.

Vurpillot, S., et al. (1997) Bridge spatial displacement monitoring with 100 fiber optic deformation sensors: Sensors network and preliminary results. *Smart Structures and Materials' 97*. International Society for Optics and Photonics, San Diego, CA. pp. 51–57.

Wu, Z. & Abe, M. (2003) *Structural Health Monitoring and Intelligent Infrastructure: Proceedings of the First International Conference on Structural Health Monitoring and Intelligent Infrastructure, 13–15 November 2003, Tokyo, Japan*. Taylor & Francis.

Ye, X. W., Su, Y. H. & Han, J. P. (2014) Structural health monitoring of civil infrastructure using optical fiber sensing technology: A comprehensive review. *The Scientific World Journal*, 2014, 652329. doi: 10.1155/2014/652329.

Zakaria, M., et al. (2009) Experimental investigation on shear cracking behavior in reinforced concrete beams with shear reinforcement. 7(1), 79–96.

Zeng, X., et al. (2002) Strain measurement in a concrete beam by use of the Brillouin-scattering-based distributed fiber sensor with single-mode fibers embedded in glass fiber reinforced polymer rods and bonded to steel reinforcing bars. *Applied Optics*, 41(24), 5105–5114. doi: 10.1364/AO.41.005105.

Zhao, X., et al. (2012) Application of multiscale fiber optical sensing network based on Brillouin and fiber Bragg grating sensing techniques on concrete structures. *International Journal of Distributed Sensor Networks*. Hindawi Publishing Corporation, 2012.

Chapter 8

Deflection Measurement on Bridges by Radar Techniques

Carmelo Gentile

Abstract

Recent advances in radar techniques and systems have favored the development of microwave interferometers, which are suitable for the non-contact measurement of deflections on large structures. The main characteristic of the new radar systems, entirely designed and developed by Italian researchers, is the possibility of simultaneously measuring the deflection of several points on a large structure with high accuracy, in static or dynamic conditions. In this chapter, the main radar techniques adopted in microwave remote sensing are described, and advantages and potential issues of these techniques are addressed and discussed. Subsequently, the application of microwave remote sensing in live load static and ambient vibration tests performed on full-scale bridges is presented in order to demonstrate the reliability and accuracy of the measurement technique. Furthermore, the simplicity of use of the radar technology is exemplified in practical cases, where the access with conventional techniques is uneasy or even hazardous, such as the stay cables of cable-stayed bridges.

8.1 Introduction

In many countries, live load static tests are mandatory as reception tests before the opening of new bridges or after the strengthening of existing ones, but these tests are hardly performed as routine tests to assess the structural condition of a bridge in service since the measurement of deflections of few target points generally requires very long traffic shutdown. Consequently, dynamic testing in operational conditions or ambient vibration testing (AVT) has become the main experimental method available for assessing the structural condition of bridges and full-scale structures in service, and hundreds of bridges and large structures have been tested worldwide in operational conditions.

AVT has the advantage of being rather inexpensive because no artificial excitation is necessary, so the experimental investigation can be performed without interfering with the normal use of the structure. On the other hand, AVT of large structures is generally performed by using accelerometers or other contact sensors that need to be wired and conveniently mounted on the structure. Within this context, the use of innovative non-contact systems for the measurement of structural response is very attractive and especially applications of laser-based systems (see e.g. Cunha and Caetano, 1999; Kaito et al. 2005) are reported in the literature. Other investigations suggest the application of global positioning systems (GPS) (Nickitopoulou et al., 2006; Meng et al., 2007) or image analysis and vision systems (Lee et al., 2006; Caetano et al., 2011). More recently, the combined use of high-

DOI: https://doi.org/10.1201/9780429265969

resolution radar waveforms (Wehner, 1995) and microwave interferometry (Henderson and Lewis, 1998) has led to the development of an innovative technology (Pieraccini et al. 2004), which simultaneously measures the deflection of several points on a large structure with high accuracy. Using a radar to simultaneously measure the displacement of several points on a large structure involves two main steps:

• Acquiring consecutive radar images of the structure at an appropriate sampling rate. Each radar image represents a distance map of the intensity of radar echoes coming from the reflecting targets: for example, each discontinuity of a structure (such as the "corner zones" corresponding to the intersection of girders and crossbeams in the deck of bridges) represents a good reflecting target, so that reflecting zones act as a series of virtual sensors.
• Evaluating the displacement of each target using the phase variation of the backscattered microwaves coming from each target at different times.

The practical implementation of the above principles in a sensor prototype was carried out by the Italian company IDS (Ingegneria Dei Sistemi, Pisa, Italy). Before the industrial production of the sensor, named IBIS-S, joint research started between IDS and the Politecnico di Milano (under the responsibility of the author), mainly aimed at validating the results of the IBIS-S interferometer and at assessing the equipment performances in live load and ambient vibration tests of full-scale bridges (see e.g. Gentile and Bernardini, 2008; Gentile, 2010; Gentile and Bernardini, 2010a, 2010b).

The microwave interferometer is capable of simultaneously measuring the displacement response of several points of a structure with high accuracy (i.e. the displacement resolution is lower than 0.02 mm [0.8×10^{-3} in]; Gentile and Bernardini, 2010a) and its use in static or dynamic tests exhibits several advantages:

• Remote sensing can be performed without the need of accessing the structure to install sensors or optical targets. In fact, each discontinuity of a structure – such as the "corner zones" corresponding to the intersection of girders and crossbeams in the decks of bridges – represents a potential source of reflection of the electromagnetic waves generated by the radar; hence an echo can be generated and the corner zones act as a series of virtual sensors. When the bridge structure is not sufficiently reflective to electromagnetic waves or when displacement of specific points on the structure must be measured, simple passive radar reflectors can be quickly and easily fixed (Gentile and Bernardini, 2008).
• Performing live load tests requires that the traffic has to be shut down only for the time strictly necessary for the test.
• The experimental procedure is especially suitable to structural elements very difficult to access by using conventional techniques, such as stay cables (Gentile, 2010; Luzi et al., 2014; Gentile and Cabboi, 2015).
• The equipment is quick and easy to install and can be used both during day and night and in almost all weather conditions.

In the first part of this chapter, the main radar principles and techniques adopted in microwave remote sensing are described, and advantages and potential issues of these techniques are addressed and discussed. The basic concepts are first introduced with reference to simple radar scenarios and subsequently extended to more complex scenarios. Subsequently, the results of past and recent full-scale tests are presented and discussed, in order to demonstrate (a) the long-term stability, required for actual employment in testing full-scale structures; (b) the accuracy of microwave remote sensing through the comparison between the data acquired from conventional sensors and the radar interferometer; and (c) the simplicity of use of the radar technology, especially when applied to the measurement of deflection of structural elements and structures very difficult to access by using conventional techniques, such as the stay cables in cable-stayed structures.

8.2 Radar technology and the microwave interferometer

As previously pointed out, two main steps are required to simultaneously measure the displacement of several points on a large structure by using radar: (1) acquiring consecutive radar images, where different points of the structure are individually observable, at an appropriate sampling rate; and (2) using the phase variation of the backscattered microwaves coming from each target point at different times to evaluate the displacement.

The latter task is in principle very simple through the microwave interferometry (see e.g. Henderson and Lewis, 1998). For example, let us consider the single degree of freedom system shown in Figure 8.1 and a radar emitting a sinusoidal wave. If the target does not move, the phase angle of the radar echo does not change in time; on the other hand, if the mass is vibrating, the received echoes obtained at different times exhibit phase differences, which are proportional to the displacement along the direction of wave propagation. Hence, the displacement d_{LOS} along the radar line of sight is simply computed from the phase shift $\Delta\varphi$ as:

$$d_{LOS} = -\frac{\lambda}{4\pi}\Delta\varphi \qquad (8.1)$$

where λ is the wavelength of the electromagnetic signal.

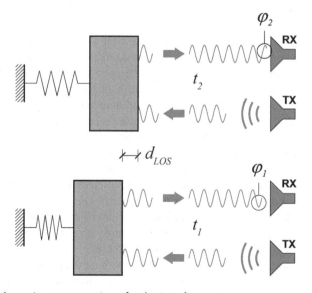

Figure 8.1 Schematic representation of radar interferometry

Nevertheless, the application of microwave interferometry (8.1) to real structures implies to acquire "images" of the structure at an appropriate sampling rate, with several points on the structure being individually observable in each image. Two or more target points, illuminated by the radar, are individually detectable if they produce different echoes. The *range resolution* or *distance resolution* Δr refers to the minimum separation that can be observed along the radar line of sight. The range resolution area is called range bin. Among the radar waveforms (Wehner, 1995) providing high range resolution, the more usual is probably the short pulse. The shorter the pulse, the more precise is the measurement of the range because the range resolution Δr is related to the pulse duration τ by the following:

$$\Delta r = \frac{c\tau}{2} \tag{8.2}$$

where c is the speed of light in free space. For a pulse of duration τ, the time-bandwidth product satisfies the equality $\tau B = 1$ (see e.g. Marple, 1987), where B is the bandwidth (i.e. the width of the range of emitted frequencies). Hence, the range resolution Δr may be expressed as:

$$\Delta r = \frac{c}{2B} \tag{8.3}$$

Equations 8.2 and 8.3 show that a better range resolution (corresponding to a smaller numerical value of Δr) can be obtained either by decreasing τ or increasing B. Instead of using short-time pulses, the stepped frequency continuous wave (SF-CW) technique (Wehner, 1995) can be adopted to increase B. SF-CW radars exhibit a large bandwidth by linearly increasing the frequency of successive pulses in discrete steps, as shown in Figure 8.2. A SF-CW radar has a narrow instantaneous bandwidth (corresponding to individual pulse) and a large effective bandwidth $B = (N - 1)\Delta f$ is obtained, emitting a burst of N electromagnetic pulses (tones), whose frequencies are increased from tone to tone by a constant frequency increment Δf. It should be noticed that a SF-CW radar emits one burst of N tones at each sample time interval (Fig. 8.2b).

By taking the inverse discrete Fourier transform (IDFT) of the received signal sampled at N discrete frequencies, the response is reconstructed in the time domain of the radar: each complex sample in this domain represents the echo from a range (distance) interval of length c/2B. The synthetic profile, or range profile, of the radar echoes is then obtained by calculating the magnitude of the IDFT of acquired vector samples. The range profile is simply a one-dimensional map of the intensity of radar echoes in function of the distance of the objects generating the echoes themselves; in other words, it represents a 1D map of the scattering objects versus their distances.

The concept of range profile is better illustrated in Figure 8.3, where an ideal range profile is shown, as obtained when the radar beam illuminates a series of targets at different distances and different angles from the axis of the system. The peaks in the amplitude of the IDFT at each time interval (Fig. 8.3) identify the position of the targets detected in the scenario, whereas the phase difference between two consecutive IDFTs provides the targets' deflection through Equation 8.1. Figure 8.3 shows the angle of transmission covered by the main lobe of the antenna in the horizontal plane, with all the points inside the shadowed area of Figure 8.3 being observable from the sensor. It is to be noticed that a radar sensor transmits electromagnetic waves also in the vertical plane and that different transmission

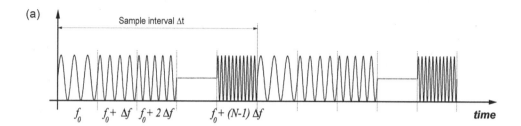

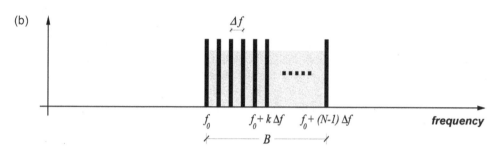

Figure 8.2 Representation of a stepped frequency continuous waveform in (a) time and (b) frequency domain

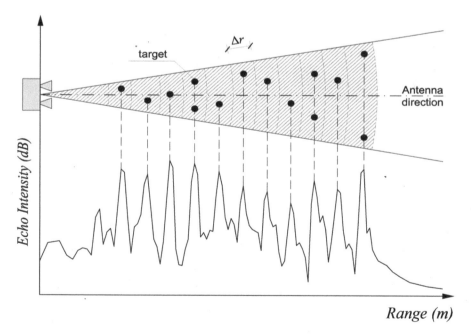

Figure 8.3 Idealization of a radar image profile (range profile)

angles in the vertical and horizontal plane could be obtained by using different antennas. Figure 8.3 also shows that the radar has only 1D imaging capabilities, that is, different targets can be individually detected if they are placed at different distances from the radar. Consequently, measurement errors may arise from the multiplicity of contributions to the same range bin, coming from different points placed at the same distance from the radar but not lying on the same axis (Gentile and Bernardini, 2008, 2010a; Rödelsperger et al., 2010).

In order to provide a simple example of a real range profile, let us refer to the cable-stayed bridge shown in Figure 8.4a, that crosses the river Oglio between the towns of Bordolano and Quinzano, about 70 km (43.5 mi) from Milan (Gentile, 2010; Benedettini and Gentile, 2011). The deflection response of the two arrays of forestays to wind and traffic excitation was acquired by positioning the microwave interferometer at the base of the upstream-side and downstream-side tower, respectively, as shown in Figure 8.4b. Since the position of the sensor is inclined upward, the only targets encountered along the path of the electromagnetic waves are the stays themselves (Fig. 8.4b). Therefore, the range profile exhibits well-defined peaks, which correspond to the reflecting targets and clearly identify the position of the cables, as it is shown in Figure 8.4c for the forestays on the downstream side. The inspection of Figure 8.4b and 8.4c reveals that the peaks in the range profile (indicated as S_{1D}, S_{2D}, and S_{3D} in Figure 8.4c occur exactly at the expected distances from the sensor (Fig. 8.4b).

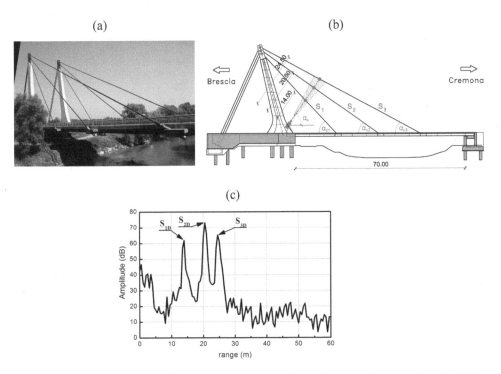

Figure 8.4 (a) View of the cable-stayed bridge between Bordolano and Quinzano (Gentile, 2010); (b) elevation view of the bridge and radar position in the test of forestays (dimensions in m); (c) range profile of the test scenario on downstream side. Conversion: 1 m = 3.3 ft

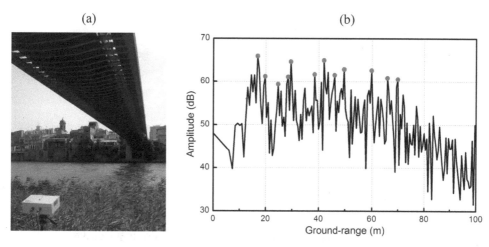

Figure 8.5 (a) Typical scenario corresponding to a bridge deck and (b) relevant (radar image) range profile. Conversion: 1 m = 3.3 ft

Figure 8.5 refers to a different and more complex scenario: the underside of the deck of the suspension bridge crossing the river Ebro at Amposta, Spain (Gentile and Luzi, 2014), where the reflecting targets consist of equally spaced steel transverse crossbeams (Fig. 8.5a). As shown in Figure 8.5b, the corresponding range profile (radar image) exhibits several equally spaced peaks marking the relative (projected) distance from the sensor of the transverse crossbeams and identifying with excellent accuracy the position of the reflecting crossbeams.

It should be noticed that the radar sensor can simultaneously measure the displacement response of the reflecting points detected in the scenario and corresponding to the peaks of the range profile. On the other hand, the interferometric technique, represented by Equation 8.1, provides a measurement of the displacement along the line of sight of all the range bins of the structure illuminated by the antenna beam; hence, the evaluation of the actual displacement requires the knowledge of the direction of motion. For many bridges (simple or continuous spans, frame or truss bridges), the displacement under traffic loads can be assumed as vertical and it can be easily evaluated by making straightforward geometric projections, as shown in Figure 8.6.

The radar technology, based on the combined use of SF-CW and microwave interferometry, was implemented in the industrially engineered microwave interferometer (IDS, IBIS-S system) used in this work. The radar equipment (Fig. 8.7) consists of a sensor module, a control PC, and a power supply unit.

The sensor module is a coherent radar (i.e. a radar preserving the phase information of the received signal) generating, transmitting, and receiving the electromagnetic signals to be processed in order to provide the deflection measurements. This unit, weighing 12 kg (1.5 lbs), includes two horn antennas for transmission and reception of the electromagnetic waves and is installed on a tripod equipped with a rotating head (Fig. 8.7), allowing the sensor to be orientated in the desired direction. The equipment radiates at a central frequency of 17.20 GHz, so that the radar is classified as Ku-band (see e.g. Skolnik, 1990),

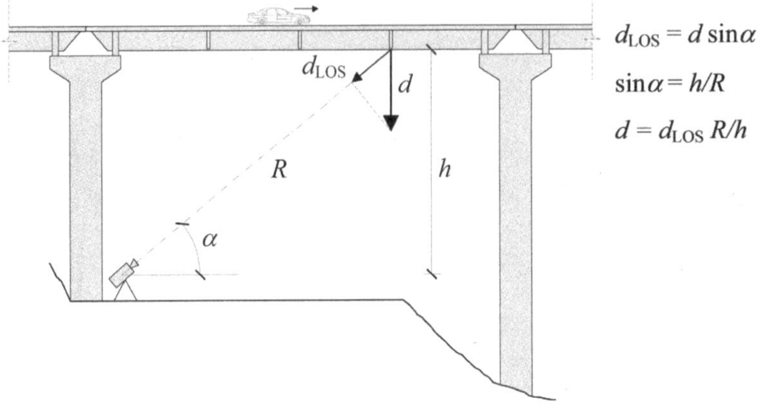

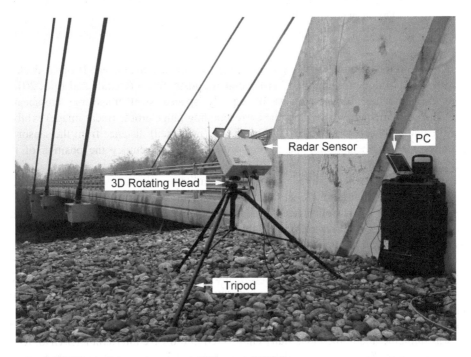

$$d_{LOS} = d \sin\alpha$$
$$\sin\alpha = h/R$$
$$d = d_{LOS} R/h$$

Figure 8.6 Displacement along the line of sight vs. actual (i.e. vertical) displacement

Figure 8.7 View of the radar sensor (IDS, model IBIS-S)

according to the standard radar frequency letter-band nomenclature from IEEE Standard 521–1984.

The sensor unit is connected to the control PC by means of a standard USB 2.0 interface; the control PC is provided with specific software for the management of the system and is used to configure the acquisition parameters, to manage and store the measurements, and to show in real time the range profile and the displacement of some points corresponding to peaks of the range profile.

The power supply unit, consisting of 12V battery packs, provides power to the system for 5–6 hours.

The interferometric radar was designed to provide:

- Minimum range resolution Δr of 0.50 m (19.68 in.), so that two targets can still be detected individually if their relative distance is greater than or equal to 0.50 m (19.68 in.).
- Maximum sampling frequency of the scenario f_{sample} of 200 Hz, which is an excellent performance since the significant frequency content of displacement time-histories is generally in the frequency range 0–20 Hz for a civil engineering structure. In addition, sampling interval $\Delta t = 0.005$ s is in principle well suitable to provide a good waveform definition of the acquired signals.

As a consequence of the radar techniques implemented in the sensor, the maximum operating distance depends on f_{sample} and Δr (Gentile and Bernardini, 2010b). The dependence of sampling rate on the maximum distance is shown in Figure 8.8, for three different values of the range resolution. The inspection of Figure 8.8 reveals that, for a range resolution of 0.50 m (19.68 in.), the sampling rate drops off for distances greater than 150.0 m (492.0 ft) while, for a range resolution of 1.0 m (39.37 in.), the sampling rate starts to decrease for distances greater than 300.0 m (984.0 ft) and reaches the value of 35 Hz for a range of 1000.0 m (3280.8 ft).

It is further noticed that the radar technology provides various advantages including independence of daylight and weather, portability and quick setup time (about 10 minutes). On the other hand, obvious issues and sources of uncertainties are related to the 1D imaging capabilities, not always easy localization of measurement points (geo-referencing of target points), and relative displacements in line of sight only.

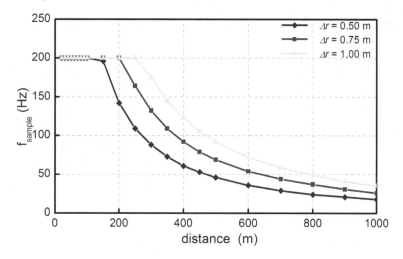

Figure 8.8 Sampling rate vs. maximum distance for three different values of the range resolution Δr. Conversion: 1 m = 3.3 ft

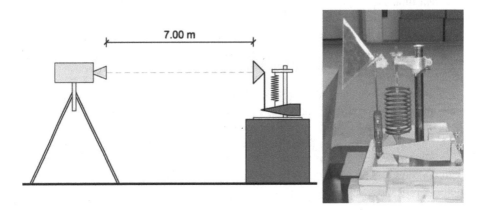

Figure 8.9 Mass-spring system tested in laboratory and test setup. Conversion: 1 m = 3.3 ft

8.3 Accuracy and validation of the radar technique

8.3.1 *Laboratory test*

Unlike other non-contact techniques of deflection measurement that are characterized by an accuracy generally ranging between 1.0–4.0 mm (0.04–1.6 in.) (image-based techniques) and 1.0 cm (0.39 in.) (GPS), sub-millimetric accuracy has in principle to be expected from the design specification on the components of the microwave interferometer. This performance was verified in various laboratory tests before using the radar in the field on full-scale structures. Among these tests, the free-vibration response of a simple mass-spring system was measured (Gentile and Bernardini, 2010a).

The test setup was arranged by installing the mass-spring system, modified by adding a small, light-passive radar reflector (corner reflector) in front of the radar sensor at a distance of 7.0 m (22.97 ft). Figure 8.9 shows a sketch of the test setup and a photograph of the oscillator equipped with the corner reflector. The control PC of the sensor was configured to measure targets up to a distance of 50.0 m (164 ft) and with a scenario sampling frequency of 50 Hz.

Figure 8.10a shows the free-damped displacement measured by the radar sensor in 1000 s of observation and the measured time history corresponded perfectly to what was expected for a lightly damped single-degree-of-freedom system. In order to better illustrate the characteristics of the measured response, Figure 8.10b and 8.10c show temporal zooms of the displacement time-histories in the low-amplitude range. Figure 8.10b clearly shows that the damped harmonic motion is very well described when its amplitude ranges between 0.1 mm (3.937×10^{-3} in.) and 0.2 mm (7.784×10^{-3} in.). A similar performance appears in Figure 8.10c, corresponding to the end of the free-damped motion: the measurement seems to exhibit excellent quality until the amplitude of the displacement exceeds 0.01–0.02 mm (3.937–7.784×10^{-4} in.).

8.3.2 *Comparison with position transducer data*

Taking profit of the static load tests performed on a viaduct, a few deflection time series were collected by using the microwave sensor and cable extension position transducer.

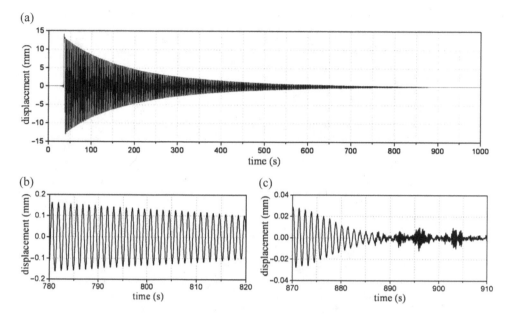

Figure 8.10 (a) Radar-based measurement of the displacement of a mass-spring system in laboratory test; (b) zoom of the measured displacement in time interval 780–820 s; (c) zoom of the measured displacement in time interval 870–910 s. Conversion: 1 mm = 0.04 in.

Figure 8.11a shows the acquisition geometry adopted on site to measure the vertical displacement at the center of one of the viaduct spans by using the conventional sensor and the radar (used in this case as a single-point interferometer).

Although the focus of the test program was the measurement of vertical deflections of the bridge under the live load provided by heavy trucks of known weight and geometry located at selected points of the structure, few vehicle running tests were performed, using one truck with a load of 40 tons (88 kip) and a running speed of 30 km/h (18.64 mph) and 40 km/h (24.85 mph). Figure 8.11b shows an example of comparison between the displacement time-histories simultaneously measured by the different sensors. Figure 8.11b shows, on the one hand, that the two time series are almost indistinguishable and clearly demonstrates the reliability and accuracy of microwave remote sensing. On the other hand, Figure 8.11b clearly suggests the use of the radar survey for the simple and effective evaluation of the dynamic effects of traffic on railway and road bridges (see e.g. Paultre et al., 1992). To this regard, a further experimental validation was carried out within the study of the historic San Michele bridge (1889), again by using the radar sensor as a single-point interferometer (Gentile and Saisi, 2011).

8.4   Static and dynamic tests of a steel-composite bridge

The microwave interferometer was first used on site during the ambient vibration-based assessment of a reinforced concrete bridge crossing the river Adda (Gentile and Bernardini, 2008, 2010a) and the stability of the radar in long-term functioning was accurately verified.

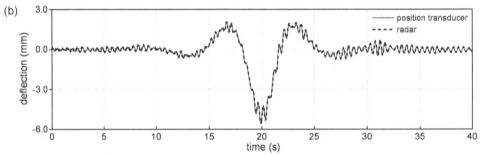

Figure 8.11 (a) Radar and position transducer setup in the field; (b) typical comparison between displacement time series. Conversion: 1 mm = 0.04 in.

In these tests, ambient vibration response of the bridges was measured by simultaneously using the radar and conventional accelerometers. In order to provide accurate comparison between the signals acquired from the different sensors, simple passive radar reflectors or "corner reflectors" were placed as close as possible to the accelerometers. Time-histories, resonant frequencies, and mode shapes provided by the radar sensor turned out to be as accurate as those obtained with traditional accelerometers.

Subsequently, in order to assess the performance of the non-contact radar technique without the use of corner reflectors, extensive static and dynamic tests were carried out on some spans of a steel-composite bridge. Steel and steel-composite bridges are much more reflective to electromagnetic waves than the concrete ones; furthermore, the deck generally includes a large number of corner zones, provided by the intersection of girders and cross-beams. In both static and dynamic tests, experimental data were collected by simultaneously using the radar technique and more conventional techniques, with validation purposes.

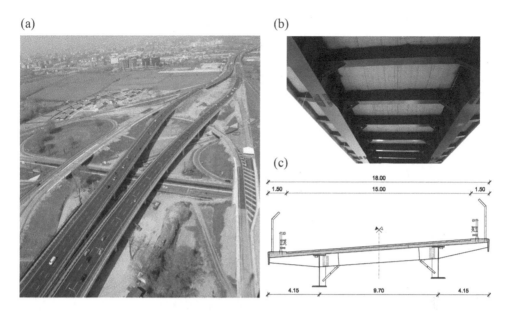

Figure 8.12 Viaduct passing over Forlanini Avenue (Milan, Italy): (a) aerial view; (b) bottom view of the bridge deck; (c) cross section (dimensions in m). Conversion: 1 m = 3.3 ft

8.4.1 Description of the bridge

The investigated bridge belongs to a motorway intersection recently completed in the neighborhood of Forlanini Avenue, Milan, which is the main road linking the city center to the city airport of Linate. The infrastructure, shown in Figure 8.12a, includes two viaducts passing over Forlanini Avenue. The south-side viaduct is a continuous span steel-concrete composite bridge consisting of eight spans; the intermediate spans are generally 50.0 m (164 ft) long while the end spans are 38.0 m (124.67 ft) long, for a total length of 376.0 m (1233.6 ft). The structure consists of ladder beams with cantilevers so that the cross section (Fig. 8.12b and 8.12c) is characterized by two main longitudinal girders with transverse crossbeams, at a longitudinal spacing of 4.17 m (13.68 ft). The crossbeams are extended beyond the girder to form cantilevers spanning 4.15 m (13.62 ft). The wide flanges support a reinforced concrete slab 25.0 cm (9.84 in.) thick. The total width of the deck is 18.0 m (59.0 ft) for three traffic lanes and two lateral emergency lanes.

8.4.2 Load test: experimental procedures and radar results

As usual in the reception test of bridges, the focus of the test program was the measurement of vertical deflections of the bridge under live load. Vehicles of known weight were located at selected points of the viaduct and vertical deflections were measured at the center of the girders of loaded spans by using traditional LVDT Schaewitz extensometers.

The source of the live load for the test program was a series of 12 two-axle trucks weighing between 415 kN (93.3 kip) and 440 kN (98.9 kip). The test vehicles were placed according to four different arrangements to provide four live load cases. For all live load configurations, the test vehicles were positioned to simultaneously load two spans of the

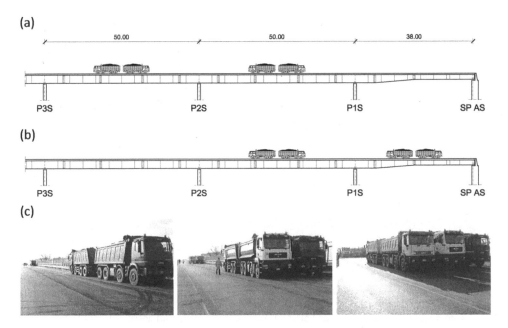

Figure 8.13 (a) Schematic of load condition LC1 (dimensions in m); (b) schematic of load condition LC2; (c) typical arrangement of test vehicles. Units in m. Conversion: 1 m = 3.3 ft

viaduct and longitudinally with the rear axle centered on the midspan (Fig. 8.13a and 8.13b). Since the deck is characterized by a significant transverse slope, the position of the vehicles was transversely non-centered between the two main girders in order to experimentally evaluate the torsion effects (Fig. 8.13c).

The results of two load configurations will be presented and discussed in the following. These two load cases, shown in Figure 8.13, are hereinafter referred to as LC1 (test vehicles loading the two spans between piers P3S and P1S on the west side of the structure, Fig. 8.13a) and LC2 (test vehicle loading the two end spans on the west side, Fig. 8.13b). In all load conditions, the radar has been configured to measure targets up to a distance of 150.0 m (492.0 ft), with a scenario sampling frequency of 10 Hz. Figure 8.14a shows the position of the extensometers on span P2S-P3S and IBIS-S sensor during load configuration LC1. Since the deck includes a large number of corner zones, provided by the intersection of girders and crossbeams, the exterior position of the microwave interferometer (Fig. 8.14a) has to be preferred in order to avoid the possible occurrence of multiple contributions to the same range bin coming from different reflective zones placed at the same distance from the radar. Figure 8.14b shows the range profile of the scenario detected in LC1, projected along the longitudinal axis of the bridge. The analysis of the results provided by the microwave interferometer begins with the inspection of the range profile; this inspection, performed on site, allows to verify that the sensor positioning provides a correct image of the scenario. The radar image of Figure 8.14b exhibits several peaks clearly marking the relative distance from the sensor of the transverse crossbeams reflecting the electromagnetic waves. It should be noticed that the peaks of Figure 8.14b identify with excellent accuracy the

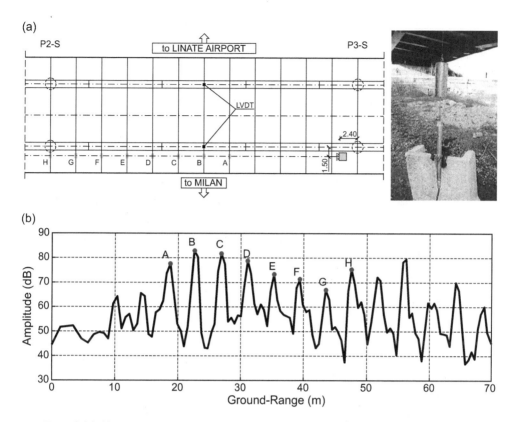

Figure 8.14 (a) Radar and LVDT position in load condition LCI; (b) ground-range profile in load condition LCI. Conversion: Im = 3.3 ft

crossbeams, provided that the distance between the radar and the axis of P3S pier (2.40 m [7.87 ft]; see Fig. 8.14a) is properly accounted for. It is further observed that the areas of the crossbeams corresponding to the peaks of Figure 8.14b are in principle placed along the radar's line of sight (as it is schematically indicated in Fig. 8.14a).

Figure 8.15 shows an example of displacement time-histories corresponding to different range bins and hence to different positions along the deck. It is observed that all deflections exhibit similar evolution and the time windows corresponding to·successive entrance and motion (150–400 s, 700–950 s, and 1050–1250 s) of the test vehicles along the bridge are clearly identified in Figure 8.15; in addition, as to be expected, deflection decreases from midspan (curve B in Fig. 8.15) to pier (curve H in Fig. 8.15). Figure 8.15 also compares the deflection obtained by the radar at midspan (region B of Fig. 8.14a) to the one directly measured by the neighboring extensometer; it has to be noticed that the radar-based measurement slightly exceeds the conventional measurement, conceivably as a consequence of the torsion behavior of the deck.

In load condition LC2, the radar position was moved as shown in Figure 8.16a in order to have significant echo from all the transverse cantilevers and to obtain the deflected elastic curve of the whole span P1S-P2S. The corresponding range profile (Fig. 8.16b) allows to clearly identify the crossbeams and confirms that all cantilevers of span P1S-

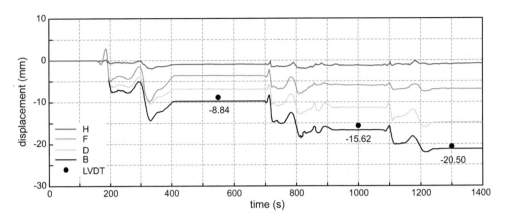

Figure 8.15 Load condition LCI: progress in time of the vertical displacements measured by the radar technique and comparison with the extensometer measurement. Conversion: 1 mm = 0.04 in.

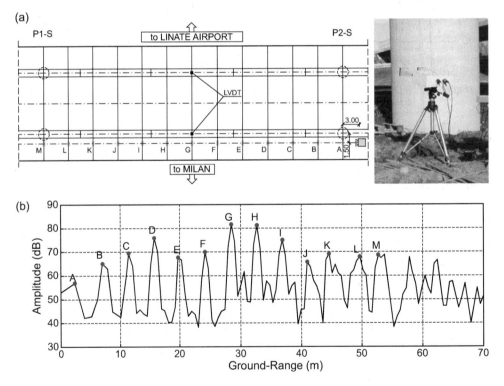

Figure 8.16 (a) Radar and LVDT position in load condition LC2; (b) ground-range profile in load condition LC2. Conversion: 1 m = 3.3 ft

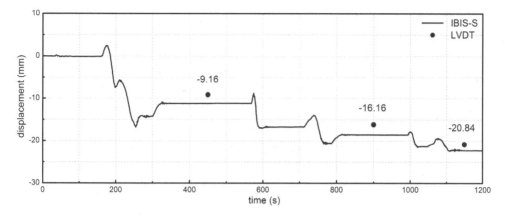

Figure 8.17 Load condition LC2: progress in time of the vertical displacement measured by the radar technique and comparison with the extensometer measurement. Conversion: 1 mm = 0.04 in.

P2S provide a sufficient echo. Again the progress in time of the deflections measured by the radar clearly corresponds to the different phases of the load condition (Fig. 8.17), and good agreement was found between the results provided by radar and extensometer; in this case, as shown in Figure 8.17, the difference between conventional and radar measurement is larger than in previous load condition LC1 (Fig. 8.16), due to the torsion behavior of the deck and the slightly curved geometry of the investigated span. The percentage difference between extensometer and radar measurement ranges from 6.96% to 18.14%.

8.4.3 Ambient vibration test: experimental procedures and radar results

Ambient vibration tests were carried out only on the span between piers P2S and P3S (Fig. 8.18a); during the tests, 10 WR-731A accelerometers (Fig. 8.18b and 8.18c) and the radar sensor were simultaneously used. Velocity responses were recorded by the conventional sensors to allow an easier comparison with the signals collected by the radar system.

Figure 8.18b shows a sketch of the accelerometer layout on span P2S-P3S and the position of IBIS-S sensor. The control PC of the microwave interferometer has been configured to detect targets in the distance interval 0–80.0 m (0–262.5 ft), with a scenario sampling frequency of 200 Hz. The analysis of the radar results first involved the inspection of the ground-range profile, shown in the correct image of the scenario.

The velocities recorded by the WR-731A sensors were compared to the velocities computed by deriving the displacement obtained from the IBIS-S sensor. An example of comparison is given in Figure 8.19, which refers to the signals simultaneously recorded in the central cross section (sensor A06 and zone F of the cantilever, Fig. 8.18b) over a time period of 10 s. Figure 8.19 clearly shows that the two series of data exhibit the same evolution in time and very similar shapes, with the radar time series being about 30% larger than the one recorded by the conventional sensor. Such differences in the amplitude, observed in all crossbeams over a 2260 s period, may be explained by recalling that

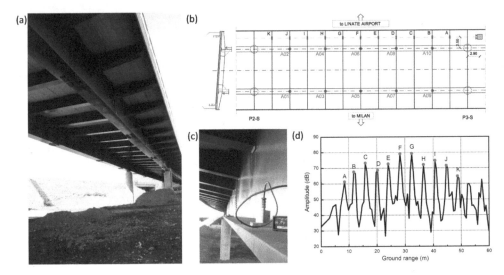

Figure 8.18 Viaduct passing over Forlanini Avenue: (a) view of the span subjected to ambient vibration test; (b) accelerometer layout and radar position during the tests (dimensions in m); (c) example of WR-731A accelerometer installed on the bridge; (d) typical ground-range profile obtained from the radar. Conversion: 1 m = 3.3 ft

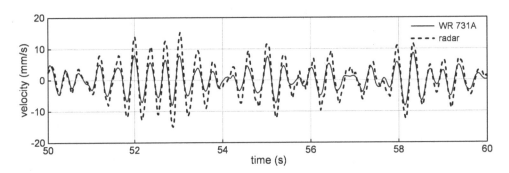

Figure 8.19 Example of comparison between the velocities recorded by conventional and radar sensors. Conversion: 1 mm/s = 0.04 in/s

conventional and radar measurements do not refer to the same points and are possibly related to the torsion of the deck. The OMA of simultaneously acquired data (over a time window of 2260 s) seems to confirm the above hypothesis. Thirteen normal modes were identified in the frequency range 0–9 Hz by applying the frequency domain decomposition technique (Brincker et al., 2001) to the velocity data recorded by the WR-731A sensors and to the displacement data provided by the IBIS-S sensor. The identified modes are dominant bending (B) modes and dominant torsion modes (T) of the deck; due to the transverse slope of the deck, deck torsion was never negligible, even under dominant bending.

Table 8.1 Viaduct passing over Forlanini Avenue. Modal parameters identified from conventional and radar data

Mode Identifier	f (Hz)		MAC
	WR-731A	IBIS-S	
B_1	1.904	1.905	0.996
B_2	2.075	2.076	0.992
B_3	2.368	2.345	0.989
B_4	2.686	2.663	0.997
B_5	3.052	3.029	0.998
B_6	3.369	3.371	0.995
T_1	3.638	3.640	0.995
B_7	3.833	3.835	0.998
T_2	6.055	6.058	0.994
B_8	6.348	6.327	0.989
B_9	6.641	6.645	0.985
B_{10}	6.982	6.987	0.976
T_3	7.251	7.280	0.984

Table 8.1 summarizes the modal parameters identified from the data acquired by using the two different measurement techniques. In more detail, Table 8.1 compares the corresponding mode shapes and scaled modal vectors through the well-known modal assurance criterion (MAC) (Allemang and Brown, 1983). The natural frequencies identified from radar data turned out to be practically equal to the ones from conventional data (with the frequency discrepancy being less than 1%, as shown in Table 8.1). Furthermore, also the modal displacements exhibit a very good agreement, with MAC values always very close to unity (provided that the modal amplitudes are scaled to their maximum values).

8.5   A challenging application: structural health monitoring of stay cables

Application of the radar technique to the measurement of cable vibrations seems especially promising in order to perform systematic dynamic assessment of stay cables in a simple and quick way. First, high accuracy has to be expected from radar-based measurements in terms of both natural frequencies and cable tensions (Gentile, 2010; Benedettini and Gentile, 2011). Furthermore, the radar technique provides the deflection time history, which could be used to directly evaluate the susceptibility of cables to large amplitude oscillations or the efficiency of devices (e.g. external dampers) adopted to prevent excessive vibrations. Finally, the microwave interferometry exhibits in principle some advantages with respect to other techniques of remote sensing, such as (a) its reliability in case of fog or rain and in almost all weather conditions; (b) higher precision of the measured deflections; and (c) the possibility of simultaneously measuring the response of several cables. In addition, the possible issues that may occur in the application of the radar technique to bridges and large structures (i.e. 1D imaging capabilities and a priori knowledge of the direction of motion), can hardly affect the survey of an array of cables. More specifically:

- As already shown in Figure 8.4, the typical position of the sensor in the survey of an array of cables is inclined upward; hence, the stay cables are the only targets encountered along the path of the electromagnetic waves, so that 1D imaging capability is perfectly adequate to the test scenario.

- It can be assumed that the in-plane motion of the cable is orthogonal to its axis, so that the actual deflection d can be expressed as:

$$d = \frac{d_{\text{LOS}}}{\cos[\pi/2 - (\alpha_c + \alpha_s)]} \tag{8.4}$$

where α_c and α_s are the slope of the cable and of the sensor. In other words, the prior knowledge of the direction of motion is available for cables, so that it is possible to evaluate the actual displacement from the radial one.

Dynamic measurements on stay cables are often aimed at identifying the local natural frequencies. In order to evaluate the reliability and the accuracy of microwave remote sensing, the radar technique was first applied to few stays of a cable-stayed bridge crossing the river Adda and to the forestays of the bridge between Bordolano and Quinzano (Fig. 8.4, Gentile, 2010; Benedettini and Gentile, 2011). Subsequently, two series of extensive measurements were performed in operational conditions on all stay cables of the curved cable-stayed bridge (Fig. 8.20) erected in the commercial harbor of Porto Marghera, Venice, Italy (Gentile and Cabboi, 2015), by simultaneously using accelerometers and microwave remote sensing.

The cable-stayed bridge belongs to a viaduct, including six spans of 42 m + 105 m + 126 m + 30 m + 42 m + 42 m (137.8 ft + 344.5 ft + 413.4 ft + 98.4 ft + 137.8 ft + 137.8 ft), that generally curves with a radius of 175 m (574.15 ft). The two longer curved bridge parts consist of an inclined concrete tower, single-plane cables, and a composite deck. The curved deck has a centerline length of 231 m (757.9 ft), with two different side spans and nine cables supporting each side span. The cast-in-place inclined tower (Fig. 8.20) is a visually memorable landmark and played a determining role in the conceptual and executive design of the bridge (Fig. 8.21). The tower is about 75 m (246 ft) high

Figure 8.20 Views of the cable-stayed bridge in Porto Marghera (Venice, Italy)

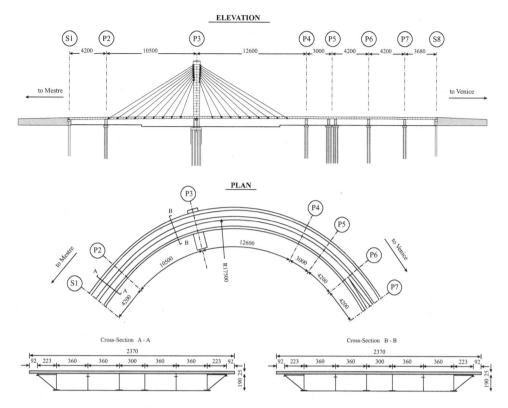

Figure 8.21 Elevation, plan, and typical cross-sections of the cable-stayed bridge in Porto Marghera. Units: cm. Conversion: 1 cm ≐ 0.4 in.

and is characterized by a complex geometric layout, where both the base and the height of the triangular cross section are varying along the inclined longitudinal axis.

Two ambient vibration tests were performed, in July 2010 and April 2011, on all stay cables of the bridge with the objective of investigating the repeatability of radar survey with structural health monitoring (SHM) purposes. The dynamic tests were carried out on one array of cables at time and simultaneously using accelerometers and radar interferometer. For each array and for each test, 3000 s of radar and accelerometer data were acquired at a rate of 200 Hz.

Figure 8.22a shows the accelerometers and radar position in the test of the stay cables on Mestre side. The radar was placed at the cross section of the deck that is vertically supported by the basement of the tower and inclined 55° upward (Fig. 8.22b); a similar setup was adopted in testing the array of stay cables of the opposite (Venice) side of the bridge. Figure 8.22a also shows the angle of transmission covered by the main lobe of the antenna in the vertical plane, with all the points inside the shadowed area of Figure 8.22a being observable from the sensor. It is worth underlining that the sensor transmits electromagnetic waves also in the horizontal plane and that different transmission angles in the vertical and horizontal plane could be obtained by using different antennas. Due to the cone-shaped emission of the sensor, notwithstanding the slightly spatial arrangement of the cables of each array, all the cables are clearly detected and identified in the

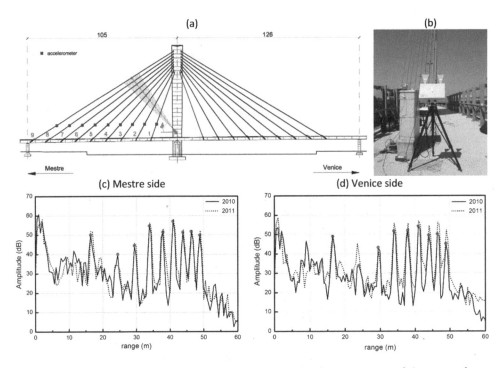

Figure 8.22 (a) Accelerometers and radar position in the dynamic survey of the array of stay cables on Mestre side; (b) view of the radar interferometer on site; (c–d) range profile of the test scenarios detected in 2010 and 2011. Conversion: 1 m = 3.3 ft

range profiles (Fig. 8.22c and 8.22d). The range profiles of the test scenarios in the two tests are presented in Figure 8.22c and 8.22d. Since the test scenario on the two sides was practically the same in the two tests, the radar image profiles are very similar and each range profile exhibits nine well-defined peaks occurring at the expected distance from the sensor and clearly identifying the position in range of the cables.

Figure 8.23 shows the auto-spectral densities (ASD) of the ambient responses acquired, by using the two measurement systems, on the longer cables of both arrays during the first test (July 2010). Although the ASDs of Figure 8.23 are associated to different mechanical quantities measured (displacement and acceleration) and to different points of the cables, the spectral plots clearly highlight that (1) a large number of local resonant frequencies are identified from radar data and these natural frequencies are in excellent agreement with the ones obtained from accelerometer; and (2) usually, the number of frequencies identified from radar data is large enough to establish if the cables behave as a taut string or deviate from a taut string (so that accurate estimate of the cable tensions can be retrieved from the identified natural frequencies as well).

It is worth underlining that similar results, in terms of number and agreement of natural frequencies, have been obtained for all the stay cables of the two arrays in both tests, with the exception of the two shorter ones (cables 1–2 in Fig. 8.22a). For the shorter stays, the radar technique detected the lower 3–4 local natural frequencies only,

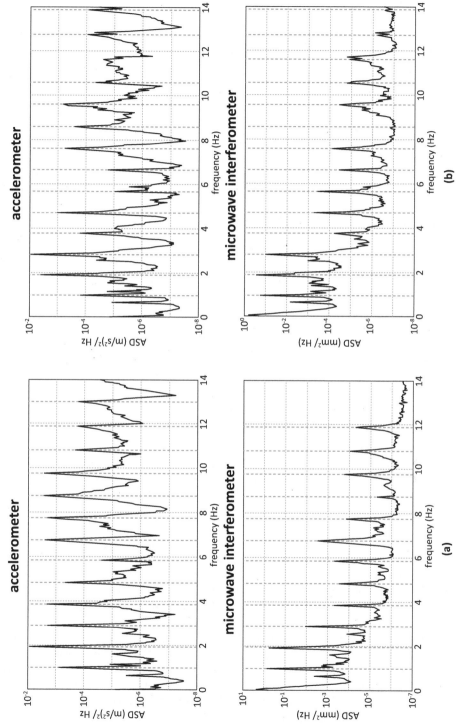

Figure 8.23 Auto-spectra (ASD) of acceleration and displacement data measured in 2010 on the longer stays of (a) Mestre side and (b) Venice side. Conversion: 1 m/s² = 3.3 ft/s², 1 mm² = 0.0016 in.²

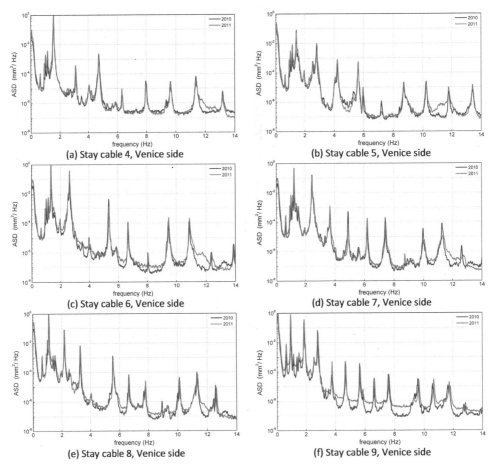

Figure 8.24 Auto-spectra (ASD) of the displacement data measured in 2010 and 2011 on stays 4–9, Venice side. Conversion: 1 mm² = 0.0016 in.²

whereas the accelerometers provided a larger number of cable frequencies. This specific aspect, first observed in the test of July 2010, was subsequently investigated in April 2011 and the main reason of the different performance of the two measurement techniques on the shorter stays seems related to the different electromagnetic reflectivity of the protective sheath.

Figure 8.24 shows the ASDs of the deflection response of stay cables 4–9 on Venice side. The spectral plots in Figure 8.24 are a synthesis of the frequency content present on those cables in the tests performed in 2010 and 2011, and allowed the identification of several local resonant frequencies in the frequency range of 0–15 Hz. The inspection of the ASDs in Figure 8.24 reveals that the local natural frequencies of cables 4–6 are practically equal in the two tests, whereas the frequencies of the longer cables 7–9 tend to slightly increase in the second test.

8.6   Summary

8.6.1   *Advantages and disadvantages of microwave remote sensing of deflections*

An innovative radar technology, developed for remote measurements of deflections on bridges and large structures in both static and dynamic conditions, has been presented and discussed in this chapter. The remote sensing technology, based on the combined use of microwave interferometry and high-resolution waveforms, has been implemented in a portable equipment exhibiting various advantages with respect to contact sensors and, at the same time, providing a direct measurement of displacements, which is of utmost importance in the in-service monitoring of structures.

The advantages of microwave remote sensing from the practical standpoint include (a) the portability of the sensor and quick setup time (so that, for instance, performing load tests on a bridge requires only limited road traffic disruption); (b) independence of daylight and weather; (c) the simplicity and safety of use, especially when the radar sensor is applied to survey structures and structural elements that are very difficult to access by using the conventional procedures. The technical advantages of the radar technique involve the simultaneous monitoring of the deflection of several targets within the sensor applicable distance, the excellent displacement resolution (of the order of 0.02 mm [0.8×10^{-3} in.]) even at large distances from the equipment, the high spatial resolution (i.e. the space sampling of the target points on the structure is of the order of 0.50 m [19.68 in.]) and the possibility of working in both static and dynamic condition (indeed the quasi-static measurement can be seen as a dynamic measurement with a low sampling frequency).

On the other hand, typical issues are related to the 1D imaging capabilities, not always easy localization of measurement points (geo-referencing of target points), and relative displacements in line of sight only. When a radar-based survey is performed, these issues have to be considered and properly addressed through a careful planning of the experimental survey.

The second part of the chapter is aimed at exemplifying the practical use of the radar interferometer in dynamic tests of full-scale bridges and includes the extensive comparison with conventional measurement techniques. When the tests are performed with appropriate care and planning, the evidence for the following conclusions arises: (1) the radar-based measurements (in terms of deflection and velocity) generally exhibit excellent agreement with the ones recorded by conventional sensors; (2) an excellent agreement was found between the modal parameters (resonant frequencies and mode shapes) identified from radar data and from data obtained with conventional accelerometers; and (3) the radar technique turns out to be especially suitable to vibration survey and SHM of stay cables, since it allows to simultaneously measure the dynamic response of several stay cables and provides measurements of high level of accuracy in terms of identification of natural frequencies.

8.6.2   *Recommendations for practice*

As previously pointed out, the main issue in the application of the radar technique on real structures are related to (a) the 1D imaging capabilities of the equipment, meaning that it is not possible to distinguish the position and the displacement of two target points which are placed at the same distance from the sensor (Fig. 8.3); and (b) the relative displacements are measured along the line of sight, implying that the actual displacement can be evaluated if and only if its direction is known a priori. From a practical standpoint, accurate and reliable measurements might be obtained using microwave remote sensing bearing in mind the previous possible issues as well as the information available in real time from the software controlling the data acquisition: the range profile and the deflection time series corresponding to pre-selected points of the investigated structure (i.e. to pre-selected peaks of the range profile).

Since the geometric arrangement of the structure and of the radar position plays an important role in the measurement procedures, an accurate planning of the experimental campaign is mandatory and involves the preliminary selection of the positions available on site for the sensor installation and the prediction of the reflecting points illuminated by the radar beam, so that the expected range profile is approximately established before the test. The good agreement of expected and actual range profiles usually guarantees the success of the tests, provided that the deflection amplitudes is sufficiently larger than the sensor resolution.

8.6.3   *Future developments*

In the author's opinion, the practical application and developments of radar interferometry in the field of civil engineering structures has been so far limited by the high cost of the radar technology and by the need of coupling structural and electronics skills to obtain good results in real tests. On the other hand, the number of application seems to be slowly increasing and involves both rather flexible structures (such as buildings, bridges, footbridges, and guyed masts) and stiff structures (such as chimneys and historic towers).

Beyond the increase of case studies on different classes of civil engineering structures, probably the most promising and easy development of the current radar technology is its use in continuous monitoring (quasi-static or dynamic). This development is conceptually straightforward as it implies mainly the modification of the software codes available for the data acquisition. At present, the data acquisition is performed by storing the radar echoes collected from all the points illuminated by the radar beam, so that the corresponding data files are very large and not suitable even for a few days of continuous monitoring. The continuously acquired radar echoes should be processed in real time to extract the deflection time series and much smaller files, containing only the deflection of pre-selected reflecting points, should be stored.

Acknowledgments

The support of IDS (Ingegneria Dei Sistemi, Pisa, Italy) in supplying the IBIS-S radar sensor employed in the tests is gratefully acknowledged. Sincere thanks are due to M. Antico and M. Cucchi (VIBLAB, Politecnico di Milano) for their assistance in conducting the field tests.

References

Allemang, R. J. & Brown, D. L. (1983) Correlation coefficient for modal vector analysis. *Proceedings of the 1st International Modal Analysis Conference, IMAC-I*, Society for Experimental Mechanics, Inc., Bethel, CT.

Benedettini, F. & Gentile, C. (2011) Operational modal testing and FE model tuning of a cable-stayed bridge. *Engineering Structures*, 33(6), 2063–2073.

Brincker, R., Zhang, L. & Andersen, P. (2001) Modal identification of output-only systems using frequency domain de-composition. *Smart Materials and Structures*, 10(3), 441–445.

Caetano, E., Silva, S. & Bateira, J. (2011) A vision system for vibration monitoring of civil engineering structures. *Experimental Techniques*, 35(4), 74–82.

Cunha, A. & Caetano, E. (1999) Dynamic measurements on stay cables of cable-stayed bridges using an interferometry laser system. *Experimental Techniques*, 23(3), 38–43.

Gentile, C. (2010) Deflection measurement on vibrating stay cables by non-contact microwave interferometer. *Nondestructive Testing & Evaluation International*, 43(3), 231–240.

Gentile, C. & Bernardini, G. (2008) Output-only modal identification of a reinforced concrete bridge from radar-based measurements. *Nondestructive Testing & Evaluation International*, 41(7), 544–553.

Gentile, C. & Bernardini, G. (2010a) An interferometric radar for non-contact measurement of deflections on civil engineering structures: Laboratory and full-scale tests. *Structure & Infrastructure Engineering*, 6(5), 521–534.

Gentile, C. & Bernardini, G. (2010b) Radar-based measurement of deflections on bridges and large structures. *European Journal of Environmental and Civil Engineering*, 14(4), 495–516.

Gentile, C. & Cabboi, A. (2015) Vibration-based structural health monitoring of stay cables by microwave remote sensing. *Smart Structures and Systems*, 16(2), 263–280.

Gentile, C. & Luzi, G. (2014) Radar-based dynamic testing of the cable-suspended bridge crossing the Ebro River at Amposta, Spain. *Proceedings of the 11th International Conference on Vibration Measurements by Laser and Noncontact Techniques, AIVELA 2014*, AIP Conference Proceedings 1600. pp. 180–189.

Gentile, C. & Saisi, A. (2011) Ambient vibration testing and condition assessment of the Paderno iron arch bridge (1889). *Construction and Building Materials*, 25(9), 3709–3720.

Henderson, F. M. & Lewis, A. J. (eds.) (1998) *Manual of Remote Sensing: Principles and Applications of Imaging Radar*. Wiley & Sons, New York, NY.

Kaito, K., Abe, M. & Fujino, Y. (2005) Development of a non-contact scanning vibration measurement system for real-scale structures. *Structure & Infrastructure Engineering*, 1(3), 189–205.

Lee, J. J. & Shinozuka, M. (2006) A vision-based system for remote sensing of bridge displacement. *Nondestructive Testing & Evaluation International*, 39(5), 425–431.

Luzi, G., Crosetto, M. & Cuevas-González, M. (2014) A radar-based monitoring of the Collserola Tower (Barcelona). *Mechanical Systems and Signal Processing*, 49, 234–248.

Marple, S. L., Jr. (1987) *Digital Spectral Analysis with Applications*. Prentice-Hall, Englewood Cliffs, NJ.

Meng, X., Dodson, A. H. & Roberts, G. W. (2007) Detecting bridge dynamics with GPS and triaxial accelerometers. *Engineering Structures*, 29(11), 3178–3184.

Nickitopoulou, A., Protopsalti, K. & Stiros, S. (2006) Monitoring dynamic and quasi-static deformations of large flexible engineering structures with GPS: Accuracy, limitations and promises. *Engineering Structures*, 28(10), 1471–1482.

Paultre, P., Chaallal, O. & Proulx, J. (1992) Bridge dynamics and dynamic amplification factor: A review of analytical and experimental findings. *Canadian Journal of Civil Engineering*, 19, 260–278.

Pieraccini, M., Fratini, M., Parrini, F., Macaluso, G. & Atzeni, C. (2004) High-speed CW step-frequency coherent radar for dynamic monitoring of civil engineering structures. *Electrononics Letters*, 40(14), 907–908.

Rödelsperger, S., Läufer, G., Gerstenecker, C. & Becker, M. (2010) Monitoring of displacements with ground-based microwave interferometry: IBIS-S and IBIS-L. *Journal of Applied Geodesy*, 4(1), 51–54.

Skolnik, M. I. (ed.) (1990) *Radar Handbook*. McGraw-Hill, Boston, MA.

Wehner, D. R. (1995) *High-Resolution Radar*. Artech House, Norwood, MA.

Part IV

Load Testing in the Framework of Reliability-Based Decision-Making and Bridge Management Decisions

Chapter 9

Reliability-Based Analysis and Life-Cycle Management of Load Tests

Dan M. Frangopol, David Y. Yang, Eva O. L. Lantsoght, and Raphael D. J. M. Steenbergen

Abstract

This chapter reviews concepts related to the uncertainties associated with structures and how the results of load tests can be used to reduce these uncertainties. When an existing bridge is subjected to a load test, it is known that the capacity of the cross section is at least equal to the largest load effect that was successfully resisted. As such, the probability density function of the capacity can be truncated after the load test, and the reliability index can be recalculated. These concepts can be applied to determine the required target load for a proof load test to demonstrate that a structure fulfills a certain reliability index. Whereas the available methods focus on member strength and the evaluation of isolated members, a more appropriate approach for structures would be to consider the complete structure in this reliability-based approach. For this purpose, concepts of systems reliability are introduced. It is also interesting to place load testing decisions within the entire life cycle of a structure. A cost-optimization analysis can be used to determine the optimum time in the life cycle of the structure to carry out a load test.

9.1 Introduction

In this chapter, the effect of load testing on the reliability index is discussed. As the uncertainties play an important role in determining bridge performance, the role of load testing in reducing these uncertainties is an important aspect. The source of the uncertainties can be aleatoric (caused by the inherent randomness of a process) or epistemic (caused by imperfect knowledge) (Ang and Tang, 2007). The benefit of a successful load test is that the uncertainty associated with the capacity is reduced. Since during a proof load test relatively high loads are applied, which correspond to the factored live loads, the probability density function (PDF) of the capacity can be truncated after a proof load test at the level of the largest load effect achieved in the test.

The benefit of a diagnostic load test lies in reducing the uncertainties with regard to structural response in terms of, for example, transverse distribution, the effect of bearing restraints, and the contribution of secondary elements like parapets and barriers to the overall stiffness of the structure. A finer assessment can be carried out after a diagnostic load test, and the reliability analysis can be carried out based on the updated finite element model (Gokce et al., 2011). For this approach, no standard procedures have been developed yet. Closely related to the reduction of uncertainties in load tests is also the reduction of uncertainties on the live loads by using weigh-in-motion (WIM) measurements (Casas and Gómez, 2013).

DOI: https://doi.org/10.1201/9780429265969

The first topic that is discussed in this chapter is the influence of load testing on the reliability index. General concepts related to the determination of the probability of failure and reliability index before, during, and after a proof load test are summarized.

The second topic in this chapter deals with the application of the previously discussed reliability-based concepts to derive the required target proof load to demonstrate a certain reliability index. An example of application is added. This chapter deals with the basic principles of the effect of load testing on the reliability index of a given structure. The analysis of the effects of deterioration are discussed in Chapter 10.

The previously discussed approach deals only with the probabilistic analysis of a structural member. For an evaluation of the entire structure, it is necessary to consider concepts of system reliability. Where direct derivations and research results are not available, possibilities for future research are pointed out.

Zooming out even more brings us to the point of evaluating the structure from the perspective of its life cycle (Frangopol et al., 1997; Frangopol, 2011; Frangopol and Soliman, 2016; Frangopol et al., 2017). Cost-optimization techniques and time-dependent effects such as material degradation and deterioration can be used to evaluate which point in time during the life cycle of the structure would be the optimal moment for load testing and assessing the structure (Frangopol and Liu, 2007; Okasha and Frangopol, 2009; Barone and Frangopol, 2014a; Sabatino et al., 2015; Kim and Frangopol, 2017). These concepts fit in the philosophy of using life-cycle analysis to determine the optimal time for maintenance, repair, rehabilitation, and inspection of a given structure. An even further step would be to consider the bridge structure as part of an infrastructure network and determine the optimum point in time for load testing the structure based on a cost optimization that balances the economic, environmental, and social costs of the load test and benefits from the perspective of the entire infrastructure network (Liu and Frangopol, 2005, 2006a, 2006b; Bocchini and Frangopol, 2011, 2012, 2013; Dong et al., 2015).

9.2 Influence of load testing on reliability index

9.2.1 General principles

To determine the probability of failure P_f, the distribution functions of the resistance R and the loading S are necessary. Resistance or load effect can be approximated by different probability distributions. Examples of such well-defined distribution functions are uniform distribution, normal distribution, lognormal distribution, and extreme value distribution, among others. The mathematical expression of these distributions can be consulted in textbooks (Melchers, 1999). For structural engineering applications, recommendations for the choice of the type of distribution functions and governing parameters of the selected distribution functions are given in the JCSS Probabilistic Model Code (JCSS, 2001b). These recommendations, however, do not differentiate between newly designed structures and existing structures.

The next step in determining the probability of failure is the determination of the limit state function g. For structural applications, the limit state function g can be taken as the difference between the resistance R and the loading S:

$$g = R - S \tag{9.1}$$

When $g < 0$, the resistance is smaller than the applied loading, and failure occurs. The chance that $g < 0$ is called the probability of failure, P_f. This probability of failure P_f can be translated into the reliability index β:

$$\beta = \Phi^{-1}(1 - P_f) \tag{9.2}$$

where Φ^{-1} is the inverse normal distribution. Current design codes and codes for assessment have derived load and resistance factors based on a minimum required reliability index (target reliability index).

The expression of the probability of failure before the load test, P_{fb}, as shown in Figure 9.1a, is expressed based on the following convolution integral:

$$P_{fb} = \int_{-\infty}^{+\infty} (1 - F_s(r)) f_R(r) dr \tag{9.3}$$

In Equation 9.3, $F_s(r)$ is the cumulative distribution function (CDF) of the loading S and $f_R(r)$ is the probability density function (PDF) of the resistance R.

During a load test, the loading and resulting load effect S is not a random variable but a deterministic value of the applied load s_p. The distribution function of the loading, f_s, is thus replaced by the deterministic value s_p, as shown in Figure 9.1b. The probability of failure during the test, P_{fd}, is described by the cumulative distribution function (CDF) of the resistance F_R:

$$P_{fd} = F_R(s_p) \tag{9.4}$$

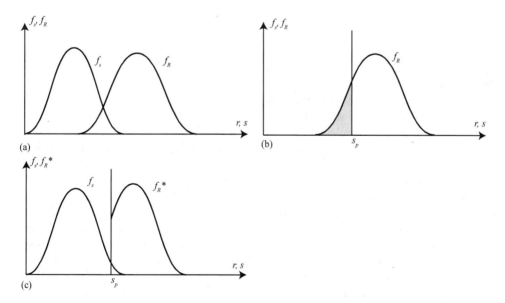

Figure 9.1 Determination of the probability of failure (a) before, (b) during, and (c) after a load test during which the target proof load s_p was applied. Highlighted region shows the probability of failure in Figure 9.1b

Source: Figure from Lantsoght et al. (2017c). Reprinted with permission.

After a successful load test, it is known that the resistance is at least equal to the maximum load effect that was achieved during the test, provided that no signs of distress were observed. The convolution integral from Equation 9.3 can be updated with the information that is obtained during the load test. As a result, the PDF of the resistance, f_R, can be updated into a truncated distribution function f_R^*, as shown in Figure 9.1c. The probability of failure after the load test, updated with the information from this test, P_{fa}, is then determined as

$$P_{fa} = \frac{1}{1 - F_R(s_p)} \int_{s_p}^{+\infty} (1 - F_s(r)) f_R(r) dr \tag{9.5}$$

The presented values for the probability of failure before, during, and after a load test are valid provided that there is no correlation between R and S. Solutions for cases where R and S are correlated, or for structures with quality problems, are available in the literature (Spaethe, 1994).

The previous considerations for P_{fa}, for which the value of P_{fa} is larger than P_{fb} as a result of the load test, are only valid for a successful load test. Another possible outcome of a load test is that P_{fa} is smaller than P_{fb}. This case occurs after a load test during which a stop criterion is exceeded. Exceeding a stop criterion means that further loading will result in irreversible damage to the structure or even collapse. As such, the load for which a stop criterion is reached can be considered the lower bound of the structural capacity. The outcome of the load test is then that a deterministic value of the capacity is found as the load effect caused by the load s_s for which a stop criterion is exceeded. When a stop criterion is exceeded during a load test, the probability of failure during the test $P_{fd} = 1$ with $s_s < s_p$ (see Fig. 9.2b). The probability of failure after the load test P_{fa} can then be calculated as:

$$P_{fa} = 1 - F_S(s_s) \tag{9.6}$$

where F_S is the CDF of the load effect and s_s is the load for which a stop criterion is exceeded, as shown in Figure 9.2c. The result is then that the reliability index after the test β_a is lower than that before the test β_b.

The magnitude of the load before (s_k), during (s_p), and after the load test (s_k) is shown in Figure 9.3a. The figure also shows the possibility for an increase of the loads with a factor ξ_p over time as a result of changes in traffic loads and intensities, for example, when a heavier truck type is permitted circulation. Figure 9.3a reflects the fact that in proof load tests, loads (s_p) are used that are higher than the characteristic live loads. The effect on the reliability index before (β_b), during (β_d), and after (β_a) a load test is shown in Figure 9.3b for the case when the target proof load s_p is applied and in Figure 9.3c for the case when a stop criterion is exceeded at s_s prior to reaching s_p. The value of β_b can be quantified with Equation 9.3. The value of β_d for a load test in which s_p is applied is quantified with Equation 9.4 and $\beta_d = 0$ when a stop criterion is reached at a load s_s. The value of β_a for a load test during which s_p was applied can be quantified according to Equation 9.5 and for a load test terminated at s_s according to Equation 9.6. If the applied target load s_p is large enough, the updated information after the test will result in a larger reliability index β_a after the load test than β_b before the load test. Since during the load test a higher load s_p is used than the characteristic live load s_k, the reliability index temporarily drops to β_d during the load test. Note for Figure 9.3b that a drop in β is shown. Whether β drops or not during the load test is subject to discussion: the applied load is higher than the traffic load, but the execution of

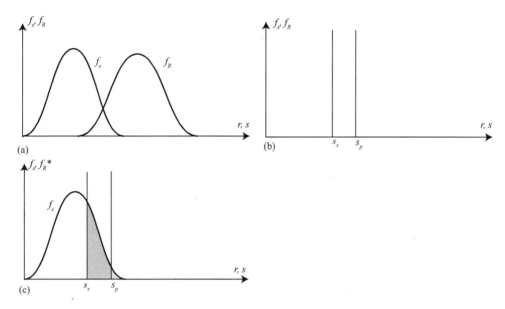

Figure 9.2 Determination of the probability of failure (a) before, (b) during, and (c) after a load test during which a stop criterion was exceeded for s_s. Highlighted region shows the probability of failure

the test itself is without uncertainty. If a stop criterion is exceeded during the test and a load s_s lower than s_p is applied, the lower bound for failure is found during the test, β_d becomes zero, and the value of β_a will be lower than β_b.

9.2.2 *Effect of degradation*

The effect of degradation is discussed in more detail in Chapter 10. Here, only a few basic concepts are reviewed. The concepts shown in Figure 9.3 do not take into account the effect of degradation. The resistance R decreases over time as a result of material degradation and deterioration. For computations, the probability density function (PDF) of the resistance f_R and the cumulative distribution function (CDF) F_R can be expressed as a function of the time t. The limit state function then becomes time-dependent (Frangopol and Kim, 2014). Degradation increases the probability of failure over time and decreases the reliability index over time. The effect of this reduction of the reliability index is shown in Figure 9.4a for a load test in which s_p was applied and in Figure 9.4b for a load test terminated at s_s.

For concrete bridges, the main inducer of service life reduction is corrosion of the reinforcement. Corrosion can affect both flexural reliability and shear reliability (Enright and Frangopol, 1998; Vatteri et al., 2016). Corrosion also reduces the probability of exceeding the serviceability requirements (Li et al., 2005).

For steel bridges, fatigue and fracture can cause structural failure. The results from inspections can be used to update estimates of the remaining service life by improving the calibration of the probabilistic degradation model (Righiniotis and Chryssanthopoulos, 2003). The failure probability is then determined based on conditional probabilities (Lukic and Cremona, 2001; Zhu and Frangopol, 2013).

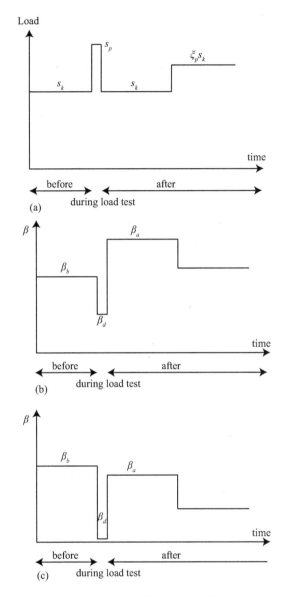

Figure 9.3 Change of reliability index before (β_b), during (β_d), and after load test (β_a), based on (Spaethe, 1994): (a) value of the load before (s_k), during (s_p), and after load test (s_k), including the effect for future increases in load with a factor ξ_p; (b) effect on reliability index before, during, and after load test when s_p is applied; and (c) effect on reliability index before, during, and after load test when the test is terminated at s_s

The idea of updating the effect of degradation after inspections can also be applied to load testing. The information from a load test can be used to update the estimate of the remaining service life. The concepts from updating the estimated service life based on data from structural health monitoring (Messervey and Frangopol, 2008; Messervey, 2009; Messervey

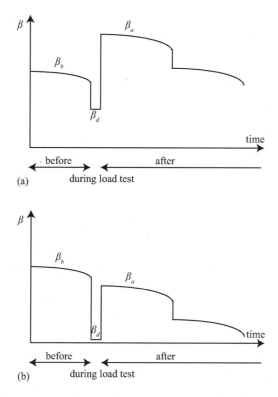

Figure 9.4 Change of reliability index before (β_b), during (β_d), and after load test (β_a), for applied loads shown in Figure 9.3a and taking into account degradation (a) effect on reliability index before, during, and after load test when s_p is applied; and (b) effect on reliability index before, during, and after load test when the test is terminated at s_s

et al., 2011) can be applied. The use of load testing data to update the estimate of the remaining service life is a topic that needs further research.

9.2.3 *Target reliability index and applied loads*

For existing structures, the target reliability index is lower than that of a structure in the design stage (Stewart et al., 2001; Steenbergen and Vrouwenvelder, 2010). The following factors determine the target reliability index for assessment: consequences of failure, reference period, remaining service life, relative cost of safety, and importance of the structure. When the maintenance and repair costs are large and the consequences of failure are minor, lower reliability indices are tolerated as assessment result. These values result from a cost optimization that considers the structural cost, the cost of damage, and the probability of failure, see Figure 9.5. For the loss of human life, a lower bound of $\beta = 2.5$ with a reference period of one year (Steenbergen and Vrouwenvelder, 2010) should be considered.

Stewart et al. (Stewart et al., 2001) suggest a target reliability index for a one-year reference period between 3.1 and 4.7. Target reliability indices have also been suggested as a function of the age of the bridge and its remaining lifetime (Koteš and Vican, 2013),

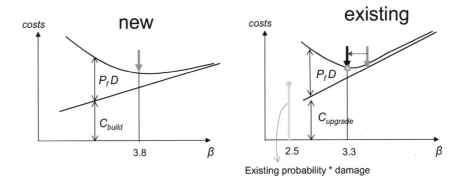

Figure 9.5 Resulting target reliability index after cost optimization

resulting in target indices between 2.692 (for a bridge of 90 years with a remaining lifetime of two years) and 3.773 (for a bridge of 10 years with a remaining lifetime of 90 years).

In Europe, the factored live load model is used (Olaszek et al., 2012) for proof load tests, and a target proof load is then applied that creates the same sectional shear or moment as the factored live load model. When the factored live load model is used to find the target proof load, it is tacitly assumed that the resulting reliability index and probability of failure after the test (if the test is successful) of the bridge have the same value as what the load factors are calibrated for.

In the Netherlands, target reliability indices and load factors have been formulated for existing structures in the Dutch code NEN 8700:2011 (Code Committee 351001, 2011). The application to highway bridges is provided in the guidelines for the assessment of bridges RBK (Rijkswaterstaat, 2013), and prescribes reliability indices between 3.1 (with a reference period of 15 years) and 3.6 (with a reference period of 30 years). In the United States, a target reliability index of 2.3 (with a reference period of five years; Šavor and Šavor Novak, 2015) was determined for rating at the operating level, and 3.5 for rating at the inventory level (for a lifetime reference period of 75 years; Šavor and Šavor Novak, 2015) (NCHRP, 1998).

9.3 Required target load for updating reliability index

9.3.1 *Principles*

The general expression for the limit state function is given in Equation 9.1. Depending on the goal of the load test, the expression for the limit state function can be expressed based on the failure mode that needs to be evaluated. For concrete bridges, typically bending moment and shear are evaluated, and the governing failure mode is further studied. For bending moment, the following limit state violation is found:

$$g = m_R - m_S < 0 \tag{9.7}$$

This limit state is expressed based on the bending moment capacity m_R and the sectional moment caused by the applied loads m_S, where m_R and m_S are random variables.

For shear, the following limit state violation can be used:

$$g = v_R - v_S < 0 \tag{9.8}$$

where v_R the shear capacity of the cross section under consideration and v_S the shear stress caused by the applied loads, with v_R and v_S random variables.

In order to determine the probability of failure, the distribution functions need to be determined. There are different approaches to determine the distribution functions of m_R, m_S, v_R, and v_S. When no information about the actual traffic is available and no distribution function of the live loads can be extrapolated from WIM measurements (Obrien et al., 2015), the load combination using the load model from the code is used. Another possibility, when no information about the actual traffic is available, is to take the traffic load models from *fib* Bulletin 80 (fib Task Group 3.1, 2016). In the case of a bridge-specific traffic load model, it is advised to use Monte Carlo simulations of traffic flow using WIM data over influence lines or fields of the bridge sections under consideration. Here, appropriate values for the statistical and model uncertainty should be taken into account. The distribution function of the resistance can be determined considering aleatoric uncertainties of material properties and epistemic uncertainties of structural models. The Probabilistic Model Code (Joint Committee on Structural Safety, 2001a, 2001b) can be used as a starting point to select the shape of the distribution function, the bias, and the coefficient of variation. However, this code makes no distinction between newly designed structures and existing structures that need to be assessed.

9.3.2 Example: viaduct De Beek – information about traffic is not available

9.3.2.1 Description of viaduct De Beek

Viaduct De Beek (see Fig. 9.6) (Lantsoght et al., 2017a, 2017d) is a reinforced concrete slab bridge over highway A67 in the province of Noord Brabant in the Netherlands. It has been in service since 1963. In 2015, the conclusion of an assessment was that posting or restricting the use of the viaduct is necessary (Willems et al., 2015) because the flexural capacity of the viaduct is insufficient. As a result, the use of the viaduct was restricted from two lanes to one lane (see Fig. 9.6b).

The geometry of viaduct De Beek is shown in Figure 9.7. The viaduct has four spans: two end spans with a length of 10.81 m (35.5 ft) and two central spans with a length of 15.40 m (50.5 ft). The total width is 9.94 m (32.6 ft) and the carriageway is 7.44 m (24.4 ft) wide, which facilitates two lanes of 3.5 m (11.5 m) width each way. Since 2015, the traffic restriction results in only one lane, which is facilitated by the use of barriers. The profile is parabolic in the longitudinal direction and varies from 470 mm (18.5 in.) to 870 mm (34.3 in.) (see Fig. 9.7b).

The concrete compressive strength was determined based on core samples. The characteristic concrete compressive strength was determined as $f_{ck} = 44.5$ MPa (6.5 ksi), and the design compressive strength was found as $f_{cd} = 30$ MPa (4.4 ksi). The properties of the steel were determined by sampling. It was found that the average yield strength of the steel was $f_{ym} = 291$ MPa (42.2 ksi), the tensile strength was $f_{tm} = 420$ MPa (60.9 ksi), and that the design yield strength can be assumed as $f_{yd} = 252$ MPa (36.6 ksi). Plain reinforcement bars were used. The reinforcement layout is shown in Figure 9.8. The thickness of the

Figure 9.6 Viaduct De Beek: (a) side view; (b) traffic restriction

asphalt layer was measured on core samples to lie between 50 mm (2.0 in.) and 75 mm (3.0 in.).

At the Dutch RBK Usage level (load factors derived for $\beta = 3.3$ for a reference period of 30 years) (Rijkswaterstaat, 2013), the Unity Check (ratio of factored acting moment to factored moment capacity) is determined as UC = 1.02 in the governing section, and at the

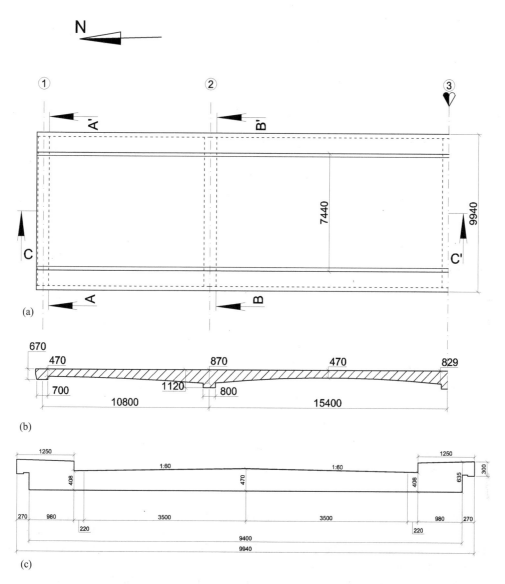

Figure 9.7 Geometry of viaduct De Beek: (a) top view; (b): side view at section C-C'; (c) cross section at section A-A'. Dimensions in mm. Conversion: 1 mm = 0.04 in.

Source: Reprinted with permission (Lantsoght et al., 2017d).

Eurocode ULS level (load factors derived for $\beta = 4.3$ for a reference period of 100 years), the maximum value is UC = 1.10. Since the Unity Checks are larger than 1, the assessment indicates that the section does not fulfill the requirements for the Eurocode ULS level based on an analytical assessment. This assessment is carried out by using a linear finite element model in which the load combination of the self-weight, superimposed dead load, and live loads from NEN-EN 1991–2:2003 (CEN, 2003) are applied. For shear, the Unity Check at

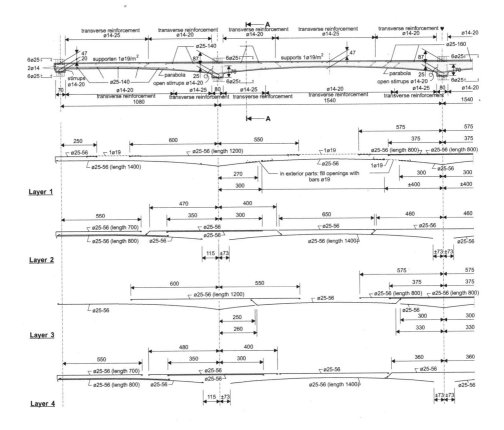

Figure 9.8 Reinforcement layout of viaduct De Beek

Source: Reprinted with permission from ASCE from (Lantsoght et al., 2017b).

the RBK Usage level is UC = 0.48, so that the failure mode of shear will not be further studied, as the bridge fulfills the requirements for shear for all safety levels. To verify if viaduct De Beek can carry the code-prescribed live loads for bending moment, a load test was carried out.

A limitation for the execution of the proof load test on viaduct De Beek was that only the first span could be tested. Testing the more critical second or third spans was not allowed, as these spans are above the highway. To safely test these spans, the highway would have to be closed, which was not permitted by the road authority.

9.3.2.2 Determination of required target load

Based on a traditional approach, as described in Chapter 1, the required proof load for bending moment and shear, applied on a single design tandem, are as given in Table 9.1. In this approach, the target proof load is determined as the load that results in the same sectional moment or shear as the factored load combination. For research purposes, the shear-critical position was also tested. These results are not discussed here. For more information

Table 9.1 Required proof load for bending $P_{load,bending}$ as determined based on the traditional approach for the different safety levels. Conversion: 1 kN = 0.225 kip

Safety level	$P_{load,bending}$ (kN)
Eurocode Ultimate Limit State	1656
RBK Design	1649
RBK Reconstruction	1427
RBK Usage	1373
RBK Disapproval	1369
Eurocode Serviceability Limit State	1070

Table 9.2 Considered safety levels, and reliability index and reference period associated with the load factors of the considered safety level

Safety level	β	Reference period	$\alpha\beta$
Eurocode Ultimate Limit State	4.3	100 years	3.44
RBK Design	4.3	100 years	3.44
RBK Reconstruction	3.6	30 years	2.88
RBK Usage	3.3	30 years	2.64
RBK Disapproval	3.1	15 years	2.48
Eurocode Serviceability Limit State	1.5	50 years	1.20

about these results, please refer to (Lantsoght et al., 2017a). The maximum load that was applied during the bending moment test was 1751 kN (394 kip) (including the weight of the equipment), which corresponds to the Eurocode ULS safety level, plus 6% extra.

To determine the target proof load based on the principles outlined before, the solution for Equation 9.5 in terms of s_p is sought that corresponds to the target reliability index β_a that needs to be proven with the applied proof load. The reliability index that is associated with each safety level used in the Netherlands is indicated in Table 9.2. In a proof load test, the probabilistic influence factor α_S for stochastic considerations with proof load testing can be taken as $\alpha_S = 0.8$. The target reliability indices that would result after a proof load test are then given in Table 9.2 as $\alpha\beta$.

Since no information about the traffic on the bridge is available, the analysis is carried out based on the bending moment capacity m_R and the occurring bending moment m_S caused by the load combination of the code. This load combination consists of the self-weight, the superimposed dead load, and the live loads consisting of a design truck in both lanes and the distributed live loads. It is necessary to consider the live loads in both lanes if the goal of the proof load test is to remove the current traffic restrictions. The bending moment m_S was determined by using a linear finite element model. The average value of the acting bending moment is determined by using all load factors as equal to 1, and results in $m_S = 385$ kNm/m (87 kip-ft/ft). The average value of the bending moment resistance is determined based on the mean values of the material parameters and is $m_R = 673$ kNm/m (151 kip-ft/ft).

To develop the probability density functions of the acting bending moment and the bending moment resistance, the recommendations of the JCSS Probabilistic Model Code (JCSS, 2001b) are followed. The shape of the functions is recommended to be lognormal. The recommendation for the bending moment resistance includes a mean of 1.2 and a coefficient of variation of 0.15. For the acting bending moment, the case of moments in plates is selected, for which the recommendations are to use a mean of 1.0 and a coefficient of variation of 20%. The resulting probability density functions are shown in Figure 9.9.

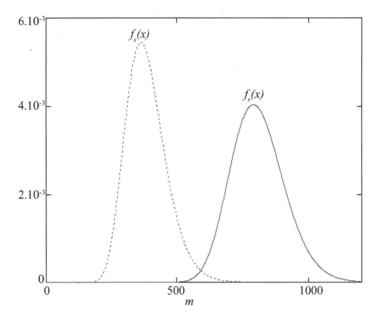

Figure 9.9 Resulting probability density functions for the acting bending moment m_s and the bending moment resistance m_R

The solution of Equation 9.3 for the PDFs in Figure 9.9 gives the reliability index prior to the proof load test, which equals $\beta_b = 3.02$. Considering this result, it would not be interesting to proof load test for any other safety level than the RBK Design level. The reliability index prior to load testing was $\beta_b = 3.02$. The goal of a proof load test is to increase this value. It can be seen in Table 9.2 that only the RBK Design level has a value of $\alpha\beta$ that is larger than 3.02, which is 3.44.

In a next step, the target proof load s_p is determined so that after the proof load test the requirements of the RBK Design safety level are fulfilled. The value of $\alpha\beta$ for this safety level can be read from Table 9.2 as 3.44. As such, the value of s_p is sought that results in a reliability index after proof load testing of $\beta_a = 3.44$. The required value to find $\beta_a = 3.44$ is a load that causes a bending moment of 655 kNm/m (147 kip-ft/ft). Reverse analysis then was used to determine that this load equals 1951 kN (439 kip). This required load is significantly larger than the load found by using the traditional approach. This result shows that when a certain reliability index needs to be proven during a proof load test, high loads are required. These high loads increase the risks for the personnel, structure, and traveling public during the test.

9.3.2.3 Discussion of results

The fact that the required load of 1951 kN (439 kip) for proving the target reliability index of 3.44 is larger than the load found with the traditional approach as 1656 kN (372 kip) can be explained by three reasons. The first reason is that no information about the traffic distribution is taken into account for this example. This reason is mitigated by the next example. The second reason is that the recommendations for developing the probability density function from the JCSS Probabilistic Model Code (JCSS, 2001b) prescribe rather large coefficients of variation. The third reason is that the recommendations from the JCSS Probabilistic Model Code (JCSS, 2001b) are general recommendations, and that these recommendations may need to be altered for the particularities related to existing structures.

A sensitivity analysis of the assumptions from the JCSS Probabilistic Model Code (JCSS, 2001b) was carried out (Lantsoght et al., 2017c). The value of the coefficient of variation on the bending moment resistance m_R varied between 5% and 15%. It can be argued that the coefficient of variation can be reduced from the recommended value of 15%, since the only variable is the yield strength of the steel. For modern steel types (JCSS, 2001b; Karmazinova and Melcher, 2012) the coefficient of variation of the yield strength of the steel is 7%. However, this value may not be representative of the variation on the steel that was used in the past and that can be found in many existing bridges. Data and recommendations for historically used rebar steel types in the Netherlands are not available. In the sensitivity analysis, a mean value of both 1.2 and of 1.0 was used for m_R. Additionally, the value of the coefficient of variation on the acting bending moment m_s was varied between 5% and the recommended value of 20%. The recommended value for moments in plates of a coefficient of variation of 20% can be considered rather large, and is significantly larger than the coefficient of variation of 5% for stresses in 3D models recommended by the JCSS Probabilistic Model Code. Therefore, this range of values was studied in the sensitivity analysis.

With the aforementioned ranges for the coefficients of variation of the acting bending moment and the bending moment capacity and for the mean value of the bending moment capacity, the convolution integrals of Equations 9.3, 9.4, and 9.5 are solved. The applied load during the proof load test is taken for all cases as 1751 kN (394 kip), the maximum load that was applied in the field. This load results in a bending moment of 597 kNm/m (134 kip-ft/ft), which is used for the value of s_p. In the sensitivity analyses it is found that the resulting reliability index after testing β_a varies between 2.85 and 6.66. This analysis thus shows that uniform recommendations for the required coefficient of variation need to be developed that are applicable to proof load testing and existing structures, so that a simplified reliability-based approach can be used for the determination of the target proof load for bridges where no WIM data are available. It must be noted here as well that the effect of carrying out the proof load test becomes smaller as the coefficients of variation (and thus, the uncertainties) are reduced. Another discussion that should be held is whether proof load tests should have as their goal to demonstrate a certain reliability index and probability of failure for a bridge, or if it is sufficient to know that a certain type of vehicle can pass safely, taking a safety margin based on a simple magnification factor as used in the *Manual for Bridge Evaluation* (AASHTO, 2016), into account.

9.3.3   Example: Halvemaans Bridge – information about traffic is modeled

9.3.3.1   Description of Halvemaans Bridge

The Halvemaans Bridge (see Fig. 9.10) (Fennis and Hordijk, 2014) is a single-span reinforced concrete slab bridge in the city of Alkmaar in the Netherlands. The bridge has been in service since 1939. An assessment led to the conclusion that the bridge does not fulfill the code requirements for bending moment when subjected to live load model 1 of NEN-EN 1991–2:2003 (CEN, 2003). The Halvemaans Bridge was subjected to a proof load test in the spring of 2014.

Figure 9.10 Halvemaans Bridge

The geometry of the Halvemaans Bridge is shown in Figure 9.11. The span length is 8.2 m (27 ft) and the slab has a skew angle of 22°. The total width is 7.5 m (25 ft) and the carriageway width is 5.5 m (18 ft) (see Fig. 9.12). The thickness of the concrete deck is 450 mm (18 in.) and the thickness is increased to 590 mm (23 in.) at the sides (see Fig. 9.12). The abutments are masonry walls with a thickness of 0.33 m (1 ft). Details of the reinforcement layout are given in Figure 9.12. It is assumed that the reinforcement lies parallel to the axis of the bridge. No core samples were taken to determine the concrete compressive strength, which was estimated to be $f_{cm} = 68$ MPa (9863 psi).

9.3.3.2 Determination of proof load

In the analysis for the determination of the exact value of the proof loading, the following aspects were taken into account. In these aspects the determination of the proof load differs from the determination of the "normal" design load for a bridge assessment using models and building codes.

1   In the event of a proof loading, the bridge is deterministically loaded by its self-weight, so there is no uncertainty about it. For the permanent action for this bridge, it is assumed that it will not change in the future. As a result, the contribution of the permanent load to the failure probability will be zero and need not be included in the proof loading value.

2   It is clear that if the bridge is proof loaded to its ultimate capacity, there is no uncertainty about the strength either. In a probabilistic analysis (the basis of the proof

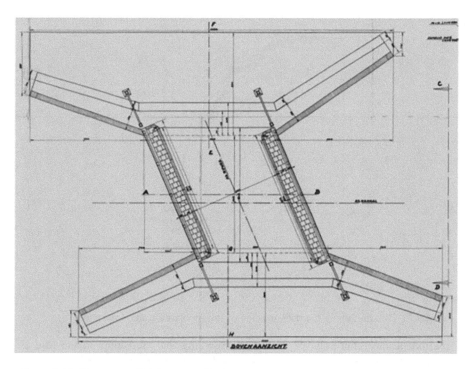

Figure 9.11 Geometry of Halvemaans Bridge, top view

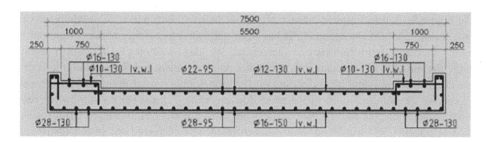

Figure 9.12 Cross section of Halvemaans Bridge

loading determination) all uncertainty will be on the load side. This means that from that perspective the load factor during proof loads is significantly larger than normally used for calculations. Differently formulated, for the loads S the probabilistic influence factor α_S is usually taken as 0.7 (ISO 2394 [ISO/TC 98/SC 2 Reliability of structures, 2015] and NEN-EN 1990:2002 [CEN, 2002]); in the case of a proof load α_S should be increased from 0.7 to 1.0, if through the proof loading we get to know the resistance R exactly. However, through proof loading, in general, we know that the real capacity is larger than the capacity needed to carry the proof loading. This leads to an increase in α_S. In this case $\alpha_S = 0.8$ was assumed.

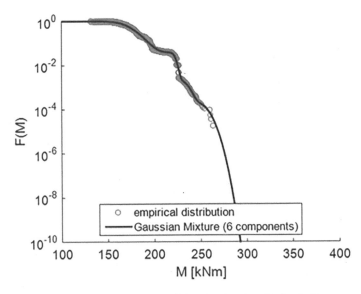

Figure 9.13 Distribution of the daily maxima of the simulated traffic load effect

3   The model uncertainty (JCSS, 2001a), representing the uncertainty in the load effect calculation, will be smaller than normally used. Using proof loading, namely, in general fewer calculations are done. This reduces the load factor if compared to normal structural calculations.

In Figure 9.13, the complementary cumulative distribution function of the daily maxima of the traffic load effects is shown; it has been generated using the influence lines of the bending moment in the midspan of the bridge. Weigh-in-motion (WIM) data is used to sample the traffic flow. The datapoints give the empirical distribution function; the continuous line is the fitted analytical distribution function which is used in the full probabilistic analysis for the determination of the proof loading value. The WIM measurements result from a bridge subjected to 2.5 million trucks per year. Since the Halvemaans Bridge is subjected to only 51,500 trucks per year, this difference is corrected for in the distribution function of the traffic load effect. Statistical uncertainty was included to account for the uncertainty in the extrapolated part of the distribution function.

The proof load that is applied to the Halvemaans Bridge in the field is a distributed load over the width of one lane (3 m = 9.8 ft) instead of over four wheel prints to facilitate execution. With this proof load, it should be demonstrated that the requirements of the repair level for consequences class 2 (CC2) of NEN 8700:2011 (Code Committee 351001, 2011) are fulfilled, namely a reliability index $\beta = 3.1$ for a reference period of 15 years. Using the reduction factor $\alpha_S = 0.8$ gives a target reliability index of $\alpha_S \beta = 2.5$ (reference period 15 years). The WIM measurements are used for simulations for a simply supported wide beam (representing the slab bridge) of 7 m (23 ft) length. The resulting bending moment at midspan of the bridge is translated into a distributed load q_{EUDL} over the entire span length. Probabilistic calculations showed that a load of $q_{EUDL} = 175$ kN/m (12 kip/ft), or a load that causes the same sectional moment at midspan in the bridge, should be applied during a proof load test. The proof load that causes the same sectional moment at midspan of the bridge is found to be 85 tons (94 short tons). The maximum load that was ultimately applied during the proof load test was 90 tons (99 short tons), which resulted in the conclusion that, based on the current knowledge regarding the interpretation of proof load test results for structural safety, the bridge fulfills the code requirements for CC2 at the repair level.

9.4 Systems reliability considerations

A bridge structure comprises various structural elements. The method described previously is mainly focused on element reliability. In order to ensure the safety of the entire bridge, system reliability analyses should be conducted (Frangopol, 2011). In general, system reliability of a bridge is governed by not only the element reliability but also a number of other factors including system arrangement, correlation of element failures, and post-failure behavior of elements, among others (Estes and Frangopol, 1999; Zhu and Frangopol, 2012; Saydam and Frangopol, 2013; Barone and Frangopol, 2014b; Zhu and Frangopol, 2014a, 2014b, 2015).

Ideally, the system model of a structure can be classified into one of the following four categories: (a) series systems, (b) parallel systems, (c) parallel-series systems, and (d) series-parallel systems. These four types of idealized systems are schematically represented by either reliability block diagrams or fault tree models (Rausand and Arnljot, 2004). Figure 9.14 shows schematically the examples of these four types of systems. In the following discussions, reliability block diagrams are used to represent systems whenever possible. It should be noted that the latter two system types (i.e. parallel-series and series-parallel systems) can be transformed to one another by conducting minimal cut or minimal path analyses of systems (Leemis, 1995). Therefore, the failure of a system can be represented by one of the following three cases:

1   For a series system with n components, the event of system failure (E) can be represented by

$$E \equiv \bigcup_{k=1}^{n} \{g_k(\mathbf{x}) < 0\} \qquad (9.9)$$

where $g_k(\mathbf{x}) < 0$ indicates the failure of element k.

2   For a parallel system with n components, the event of system failure can be represented by

$$E \equiv \bigcap_{k=1}^{n} \{g_k(\mathbf{x}) < 0\} \qquad (9.10)$$

3   For a series-parallel system with n minimal cut sets, each of which contains c_k elements in parallel $(k = 1, 2, \ldots n)$, the event of system failure can be represented by

$$E \equiv \bigcup_{k=1}^{n} \bigcap_{j=1}^{c_k} \{g_{kj}(\mathbf{x}) < 0\} \qquad (9.11)$$

where $g_{kj}(\mathbf{x}) < 0$ indicates the failure of element j in the kth minimal cut set of the system.

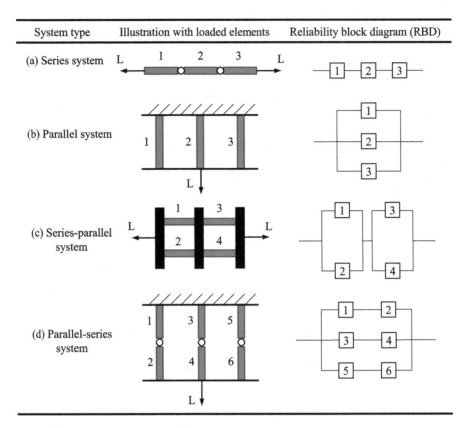

| System type | Illustration with loaded elements | Reliability block diagram (RBD) |

Figure 9.14 Idealized system models

The system model of a bridge can be created based on the judgment and experience of bridge engineers. For instance, if reliable end and center diaphragms exist in the superstructure, parallel subsystems can be used to represent that the failure of three adjacent girders is required for the failure of the entire bridge superstructure (Estes and Frangopol, 1999). As a result, the girder bridge shown in Figure 9.15 can be modeled as a series-parallel system considering 16 failure modes of different structural members including flexural failure of concrete deck [$g(1)$], shear failure of interior, exterior, and interior-exterior girders [$g(2)$, $g(4)$, and $g(6)$], flexural failure of interior, exterior, and interior-exterior girders [$g(3)$, $g(5)$, and $g(7)$], shear failure of pier caps [$g(8)$], pier cap failure under positive and negative moments [$g(9)$ and $g(10)$], crushing of top columns [$g(11)$], crushing of bottom columns [$g(12)$], one-way shear failure of footing [$g(13)$], two-way shear failure of footing [$g(14)$], flexural failure of footing [$g(15)$], and crushing of expansion bearing [$g(16)$] (Estes and Frangopol, 1999). Alternatively, finite element models can be used to analyze the effects of different failure modes to system failure (Imam et al., 2012; Saydam and Frangopol, 2013). For instance, Imai and Frangopol (2002) established a series-parallel system model for a suspension bridge (Honshu Shikoku Bridge, Japan) based on nonlinear finite element models.

Apart from defining the system model, another important factor for determining system reliability is the correlation among different failure modes (i.e. element failures) (Estes and

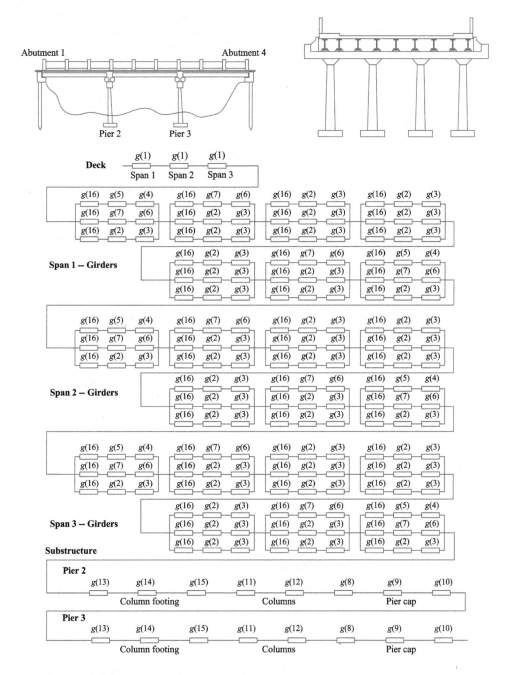

Figure 9.15 System model of a girder bridge

Source: Adapted from Estes and Frangopol (1999).

Frangopol, 1999; Zhu and Frangopol, 2012). For a bridge, the load-carrying capacities of different structural members are likely to be correlated due to their similar materials and construction process. Similarly, the loading effects in the structural members are very probable to be highly correlated, especially in the case of vehicle loads where the loading effects are caused by one or a set of heavy vehicles passing the bridge and are thus nearly fully correlated. Therefore, the element failures in a bridge system are also correlated. In general, direct evaluation of correlation among random variables representing the element failure is difficult. Nevertheless, this correlation can be implicitly considered by the coefficients of correlation between random variables (R's or S's) in different element limit state functions. The precise evaluation of system reliability usually requires carrying out Monte Carlo simulation, which is not always viable due to the low failure probabilities of civil engineering structures. Nevertheless, the system reliability with correlated failure modes can be approximated based on the reliability bounds of series, parallel, and series-parallel systems (Estes and Frangopol, 1998). Using this approach, RELSYS (RELiability of SYStems), a Fortran 77 computer program, was developed at the University of Colorado at Boulder based on the Ditlevsen bounds of system failure probabilities (Ditlevsen, 1979; Estes and Frangopol, 1998, 1999). The program is currently available at the Computational Laboratory for Life-Cycle Structural Engineering at Lehigh University. For systems with large numbers of elements (high-dimensional problems), narrower bounds of system failure probabilities can be used to improve the quality of the approximated system reliability (Song and Der Kiureghian, 2003; Song and Kang, 2009). For different system models, the system reliability can be lower (as in series systems) or higher (as in parallel systems) than the element reliability. In order to obtain a consistent level of safety for different bridges, the element target reliability should be adjusted according to the system model and the correlation condition. Target reliability indices of elements in different system models and with different correlation conditions have been calculated for typical bridge system models (Zhu and Frangopol, 2014b). The ultimate goal is to achieve a consistent level of reliability for different systems (e.g. the system reliability index $\beta_{sys} = 3.5$). These element target reliability indices can be used in load testing. Brittle or ductile behavior after element failure can lead to very different load redistribution paths within a system and thus affect the system reliability (Enright and Frangopol, 1999). The effect of post-failure behavior has been extensively studied primarily by using event tree models (Zhu and Frangopol, 2014a, 2015).

9.5   Life-cycle cost considerations

As mentioned previously, the reliability of a bridge decreases in its life cycle due to various deterioration mechanisms. Therefore, inspection, structural health monitoring, and/or timely maintenance actions must be performed to ensure structural safety in the service life (Frangopol and Soliman, 2016). All these actions will bring in additional life-cycle maintenance costs. Figure 9.16 shows the evolution of life-cycle performance in terms of reliability index as well as life-cycle maintenance cost under a generic deterioration process and multiple maintenance actions. In Figure 9.16, two types of maintenance actions are illustrated. If a maintenance action is proactive and implemented before the reliability threshold is reach, it is usually referred to as a preventative maintenance action (Frangopol et al., 1997). Alternatively, if a maintenance action is reactive as a result of

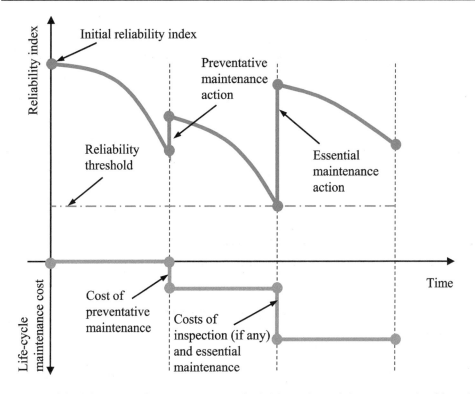

Figure 9.16 Life-cycle performance in terms of reliability index and the corresponding life-cycle maintenance cost

the violation of a prescribed reliability threshold, such an action is called an essential maintenance action (Frangopol et al., 1997). Usually, the maintenance cost of an essential action is higher than that of a preventative action. Figure 9.16 shows the growth of life-cycle cost with respect to the service time. In general, the life-cycle cost of a bridge can be expressed as (Frangopol et al., 1997)

$$C_{life} = C_{ini} + \sum_{i=1}^{N_r} \frac{C_{r,i}}{(1+r)^{t_i}} + \sum_{j=1}^{N_s} \frac{C_{s,j}}{(1+r)^{t_j}} + C_{fail} \tag{9.12}$$

where C_{ini} is the initial cost; N_r is the number of maintenance actions; t_i and $C_{r,i}$ are the time and the cost of the ith maintenance action, respectively; N_s is the number of inspection actions; t_j and $C_{s,j}$ are the time and the cost of the jth inspection action, respectively; r is the discount rate of money; and C_{fail} is the expected failure cost.

As discussed in Section 9.2, proof load testing provides useful insight into the load-carrying capacities of bridges. In essence, load testing plays a similar role as that of an inspection action. The information obtained from load testing can be used to adjust life-cycle maintenance plans. For the two cases mentioned in Section 9.2, Figure 9.17 shows schematically the corresponding changes in essential maintenance schedules and their associated life-cycle performance and life-cycle maintenance cost. Since load testing can induce additional cost into the total life-cycle cost, the time and technique used in the testing

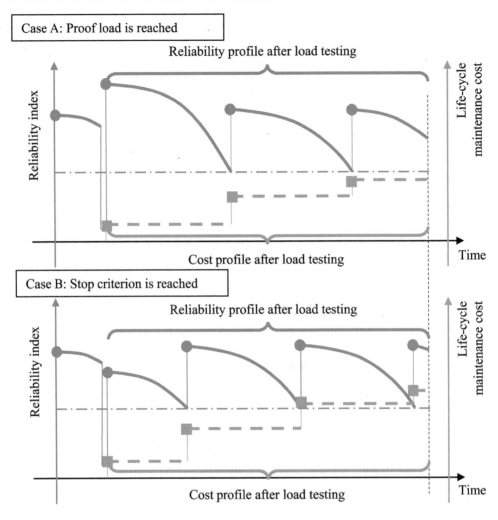

Figure 9.17 Life-cycle performance (in reliability index) and life-cycle maintenance cost after load testing based on different cases of testing results

should be optimized in the life cycle of a bridge so that the life-cycle cost can be minimized. In Equation 9.12, the initial cost C_{ini} is fixed in cases where the load testing is planned for an existing bridge. In addition, for reliability-based planning, the expected failure cost (i.e. the failure risk) can be excluded from the planning process based on the fact that the failure risk is considered tolerable as long as the reliability target is satisfied and that the decision-makers are indifferent to the further decrease of this tolerable risk. Otherwise, risk-based planning is needed. Therefore, minimization of total life-cycle cost in reliability-based load testing planning is equivalent to the minimization of life-cycle maintenance cost.

It should be noted that the occurrence of Case A (i.e. proof load is reached) or Case B (i.e. stop criterion is reached) in Figure 9.17 is not known a priori in the planning phase of

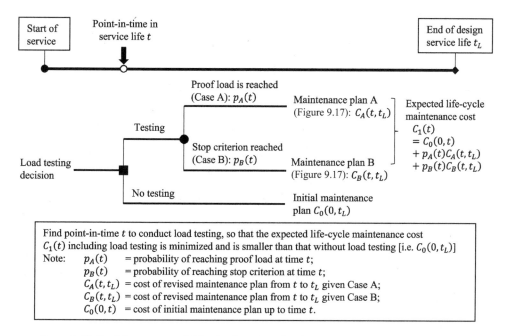

Figure 9.18 Decision-tree model for load testing planning

load testing. Therefore, optimization of load testing plans must be conducted in a prepos-terior manner (Ang and Tang, 1984). Such approaches have been used to optimize inspection plans of various civil and marine structures under different deterioration mechanisms (Kwon and Frangopol, 2011; Kim and Frangopol, 2012; Kim et al., 2013; Soliman et al., 2016). For load testing planning, an illustrative decision-tree model can be established as shown in Figure 9.18. The reliability-based load testing planning can be formulated as the following optimization problem:

Given

Bridge model, deterioration model, and models for preventative and essential main-tenance actions

Find

Time and technique of loading testing in the life cycle of the bridge

So that

The total life-cycle cost is minimized

Subjected to

(a) that the lowest reliability index of the bridge in its life cycle is higher than the reliability target;
(b) the budget for load testing;
(c) the budget for maintenance actions.

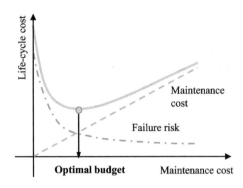

Figure 9.19 Determination of optimal budget for maintenance

The aforementioned reliability-based planning can be extended further to risk-based planning by considering the consequence of structural failure. Risk is herein defined as the product of failure probability and the associated failure consequences, as represented by the following equation:

$$R_f = p_f \cdot C_f \tag{9.13}$$

where R_f is the failure risk; p_f is the failure probability; and C_f is the failure consequences. The difference between reliability-based and risk-based planning is that the former uses a reliability target to implicitly regulate the tolerable risk, while the latter explicitly calculates the risk value and treats it as a part of the optimization objective or a separate objective overall. Usually the failure consequences are measured or converted to monetary value. For a bridge structure, the consequences of bridge failure include the direct cost of reconstruction and the social cost borne by traffic users in the form of economic losses due to extra travel time and extra travel distance (Decò and Frangopol, 2011).

The obtained risk value can be considered as a separate objective, thereby forming a bi-objective optimization that simultaneously minimizes the life-cycle failure risk and life-cycle maintenance cost. The results of this bi-objective problem can be presented in a Pareto front which represents the optimal compromises between these two objectives. Alternatively, the life-cycle failure risk can be considered as failure cost and be added to the total life-cycle cost. As expected, with increasing budgets for load testing and maintenance, the failure risk is likely to be reduced. Figure 9.19 shows the qualitative relation between budgets and risk. It can be seen from Figure 9.19 that by using risk-based planning, the optimal budget for reducing total life-cycle cost can be determined.

Bridge failure can have dire impacts in economic, social, and environmental terms. Not all these consequences can be expressed in monetary value. The utilities of economic, social, and environmental impacts in the decision-making process are largely governed by the risk perceptions and risk attitudes of stakeholders. Therefore, a sustainability-informed approach for risk assessment and risk-based planning has been advocated in recent studies (Dong et al., 2013; Bocchini et al., 2014; Sabatino et al., 2015; Liu et al., 2018). García-Segura et al. (2017) showed that life-cycle maintenance plans optimized based on economic or environmental objectives can hold very different outlooks due to the different time values of economic and environmental consequences. Sabatino et al. (2016) introduced multi-attribute utility theory (MAUT) into sustainability-informed risk-based planning to (a) convert consequences of different units to a consistent utility value in the range of (0, 1); and (b) combine economic, social, and environmental consequences to a sustainability utility value based on risk attitudes of decision makers. MAUT-based frameworks for life-cycle management have been proven to be an effective tool to harmonize the economic, social, and environmental aspects of sustainability (Dong et al., 2015; Sabatino et al., 2016; Yang and Frangopol, 2018).

9.6   Summary and conclusions

This chapter discusses load testing from the perspective of structural reliability. In a proof load test, the applied target proof load results in a certain load effect. If the structure can withstand the applied load without signs of distress (and the proof load test is successful), it has been shown experimentally that the structure has a capacity that is larger than or equal to the load effect caused by the applied load. As such, the probability density function of the resistance at the considered cross section can be truncated at the level of the load effect caused by the applied load, and the reliability index can be recalculated. Alternatively, the target proof load to demonstrate a certain reliability index can be found by applying these principles. This reliability index is lower for existing structures than for new structures. Taking into account as well the dimension of time and the effects of degradation, the optimum time for a load test can be determined.

This chapter contains two examples of proof load tests for which the target proof load was determined as a function of the target reliability index that the load test should demonstrate. For viaduct De Beek, the target proof load was determined a posteriori and for the Halvemaans Bridge, the target proof load was determined a priori. For viaduct De Beek, a flexure-critical reinforced concrete slab bridge, no traffic information is available. Recommendations from the JCSS Probabilistic Model Code are used to determine the probability density functions of the acting bending moment and the bending moment resistance. The example shows the need for recommendations on the assumptions for the coefficient of variation that can be used in simplified analyses for the determination of the target proof load when no traffic information is available. The example of the Halvemaans Bridge, a flexure-critical reinforced concrete slab bridge, shows that weigh-in-motion data can be used to develop a cumulative distribution function of the load effect, and how this information can be used to derive the target proof load. When traffic information is available or can be estimated with reasonable assumptions, the presented method can be followed to determine the target proof load.

The concepts in the first half of this chapter are based on an analysis at the element level, and based on a sectional analysis. It is more realistic to consider the bridge in its entirety and calculate the systems reliability, or to even consider the bridge as part of the entire infrastructure network by determining the measures (including load testing) that should be taken during the life cycle of the bridge to minimize the cost (economic, environmental, and social). To consider the bridge in its entirety, systems reliability considerations are required. The failure mode of the system needs to be determined by analyzing the system based on the system type (series, parallel, or a combination of series and parallel), based on judgement and experience of bridge engineers, or based on nonlinear finite element models. In principle, the correlation between failure modes (correlation of capacity for different failure modes and correlation of loading effects) should be taken into account. Such considerations require a large computational effort. When the systems reliability index is a constant and known value, depending on the structural system and correlation between failure modes, the target reliability index of each element can be determined. The target reliability index of the elements can then be used to determine the target proof load to be used during a load test to demonstrate the required systems reliability index.

When load testing is considered from a life-cycle perspective, the goal is to minimize the life-cycle cost of the structure by including actions such as maintenance, inspections, and load testing. Depending on the outcome of a load test, the life-cycle maintenance plan of a given structure may need to be adjusted. Life-cycle cost-optimization calculations can also be used to determine the optimal time during the life of a structure for a load test. This minimized cost should encompass the economic, environmental, and social cost.

References

AASHTO (2016) *The Manual for Bridge Evaluation with 2016 Interim Revisions*. American Association of State Highway and Transportation Officials (AASHTO), Washington, DC.

Ang, A. H.-S. & Tang, W. H. (1984) *Probability Concepts in Engineering Planning and Design, Vol. 2: Decision, Risk, and Reliability*. Wiley, New York, NY.

Ang, A. H.-S. & Tang, W. H. (2007) *Probability Concepts in Engineering: Emphasis on Applications in Civil & Environmental Engineering*. Wiley, Hoboken, NJ.

Barone, G. & Frangopol, D. M. (2014a) Life-cycle maintenance of deteriorating structures by multi-objective optimization involving reliability, risk, availability, hazard and cost. *Structural Safety*, 48, 40–50.

Barone, G. & Frangopol, D. M. (2014b) Reliability, risk and lifetime distributions as performance indicators for life-cycle maintenance of deteriorating structures. *Reliability Engineering and System Safety*, 123, 21–37.

Bocchini, P. & Frangopol, D. M. (2011) Generalized bridge network performance analysis with correlation and time-variant reliability. *Structural Safety*, 33(2), 155–164.

Bocchini, P. & Frangopol, D. M. (2012) Restoration of bridge networks after an earthquake: Multi-criteria intervention optimization. *Earthquake Spectra*, 28(2), 427–455.

Bocchini, P. & Frangopol, D. M. (2013) Connectivity-based optimal scheduling for maintenance of bridge networks. *Journal of Engineering Mechanics*, 139(6), 760–769.

Bocchini, P., Frangopol, D. M., Ummenhofer, T. & Zinke, T. (2014) Resilience and sustainability of civil infrastructure: Toward a unified approach. *Journal of Infrastructure Systems, ASCE*, 20(2), 04014004.

Casas, J. R. & Gómez, J. D. (2013) Load rating of highway bridges by proof-loading. *KSCE Journal of Civil Engineering*, 17, 556–567.

CEN (2002) *Eurocode – Basis of Structural Design, NEN-EN 1990:2002*. Comité Européen de Normalisation, Brussels, Belgium.

CEN (2003) *Eurocode 1: Actions on Structures, Part 2: Traffic Loads on Bridges, NEN-EN 1991–2:2003*. Comité Européen de Normalisation, Brussels, Belgium.

Code Committee 351001 (2011) *Assessment of Structural Safety of an Existing Structure at Repair or Unfit for Use: Basic Requirements, NEN 8700:2011 (in Dutch)*. Civil center for the execution of research and standard, Dutch Normalisation Institute, Delft, the Netherlands.

Decò, A. & Frangopol, D. M. (2011) Risk assessment of highway bridges under multiple hazards. *Journal of Risk Research*, 14(9), 1057–1089.

Ditlevsen, O. (1979) Narrow reliability bounds for structural systems. *Journal of Structural Mechanics*, 7(4), 453–472.

Dong, Y., Frangopol, D. M. & Saydam, D. (2013) Time-variant sustainability assessment of seismically vulnerable bridges subjected to multiple hazards. *Earthquake Engineering & Structural Dynamics*, 42, 1451–1467.

Dong, Y., Frangopol, D. M. & Sabatino, S. (2015) Optimizing bridge network retrofit planning based on cost-benefit evaluation and multi-attribute utility associated with sustainability. *Earthquake Spectra*, 31(4), 2255–2280.

Enright, M. P. & Frangopol, D. M. (1998) Probabilistic analysis of resistance degradation of reinforced concrete bridge beams under corrosion. *Engineering Structures*, 20, 960–971.

Enright, M. P. & Frangopol, D. M. (1999) Reliability-based condition assessment of deteriorating concrete bridges considering load redistribution. *Structural Safety*, 21(2), 159–195.

Estes, A. C. & Frangopol, D. M. (1998) RELSYS: A computer program for structural system reliability. *Structural Engineering and Mechanics*, 6(8), 901–919.

Estes, A. C. & Frangopol, D. M. (1999) Repair optimization of highway bridges using system reliability approach. *Journal of Structural Engineering*, 125(7), 766–775.

Fennis, S. A. A. M. & Hordijk, D. A. (2014) *Proof Loading Halvemaans Bridge Alkmaar* (in Dutch). Delft University of Technology, Delft, the Netherlands.

fib Task Group 3.1 (2016) *Partial Factor Methods for Existing Concrete Structures*. Fédération Internationale du Béton, Lausanne, Switzerland.

Frangopol, D. M. (2011) Life-cycle performance, management, and optimisation of structural systems under uncertainty: Accomplishments and challenges. *Structure and Infrastructure Engineering*, 7(6), 389–413.

Frangopol, D. M. & Kim, S. (2014) Chapter 18: Life-cycle analysis and optimization. In: Chen, W. F. & Duan, L. (eds.) *Bridge Engineering Handbook, Vol. 5: Construction and Maintenance*. CRC Press/Taylor & Francis Group, Boca Raton, FL.

Frangopol, D. M. & Liu, M. (2007) Maintenance and management of civil infrastructure based on condition, safety, optimization, and life-cycle cost. *Structure and Infrastructure Engineering*, 3(1), 29–41.

Frangopol, D. M. & Soliman, M. (2016) Life-cycle of structural systems: Recent achievements and future directions. *Structure and Infrastructure Engineering*, 12(1), 1–20.

Frangopol, D. M., Lin, K.-Y. & Estes, A. C. (1997) Life-cycle cost design of deteriorating structures. *Journal of Structural Engineering*, 123(10), 1390–1401.

Frangopol, D. M., Dong, Y. & Sabatino, S. (2017) Bridge life-cycle performance and cost: Analysis, prediction, optimisation and decision-making. *Structure and Infrastructure Engineering*, 13(10), 1239–1257.

García-Segura, T., Yepes, V., Frangopol, D. M. & Yang, D. Y. (2017) Lifetime reliability-based optimization of post-tensioned box-girder bridges. *Engineering Structures*, 145, 381–391.

Gokce, H. B., Catbas, F. N. & Frangopol, D. M. (2011) Evaluation of load rating and system reliability of movable bridge. *Transportation Research Record*, 2251, 114–122.

Imai, K. & Frangopol, D. M. (2002) System reliability of suspension bridges. *Structural Safety*, 24(2–4), 219–259.

Imam, B. M., Chryssanthopoulos, M. K. & Frangopol, D. M. (2012) Fatigue system reliability analysis of riveted railway bridge connections. *Structure and Infrastructure Engineering*, 8(10), 967–984.

ISO/TC 98/SC 2 Reliability of Structures (2015) ISO 2394:2015 *General Principles on Reliability for Structures*. International Organization for Standardization, Geneva, Switzerland.

JCSS (2001a) *JCSS Probabilistic Model Code, Part 3: Resistance Models*. Joint Committee On Structural Safety (JCSS), Kongens Lyngby, Denmark.

JCSS (2001b) *JCSS Probabilistic Model Code, Part 2: Load Models*. Joint Committee On Structural Safety (JCSS), Kongens Lyngby, Denmark.

Karmazinova, M. & Melcher, J. (2012) Influence of steel yield strength value on structural reliability. *Recent Researches in Environmental and Geological Sciences*, 441–446.

Kim, S. & Frangopol, D. M. (2012) Probabilistic bicriterion optimum inspection/monitoring planning: Applications to naval ships and bridges under fatigue. *Structure and Infrastructure Engineering*, 8(10), 912–927.

Kim, S. & Frangopol, D. M. (2017) Efficient multi-objective optimisation of probabilistic service life management. *Structure and Infrastructure Engineering*, 13(1), 147–159.

Kim, S., Frangopol, D. M. & Soliman, M. (2013) Generalized probabilistic framework for optimum inspection and maintenance planning. *Journal of Structural Engineering, ASCE*, 139, 435–447.

Koteš, P. & Vican, J. (2013) Recommended reliability levels for the evaluation of existing bridges according to Eurocodes. *Structural Engineering International*, 23, 411–417.

Kwon, K. & Frangopol, D. M. (2011) Bridge fatigue assessment and management using reliability-based crack growth and probability of detection models. *Probabilistic Engineering Mechanics*, 26(3), 471–480.

Lantsoght, E. O. L., Koekkoek, R. T., Yang, Y., Van Der Veen, C., Hordijk, D. A. & De Boer, A. (2017a) Proof load testing of the viaduct De Beek. *39th IABSE Symposium: Engineering the Future*. Vancouver, Canada.

Lantsoght, E. O. L., Koekkoek, R. T., Van Der Veen, C., Hordijk, D. A. & De Boer, A. (2017b) Pilot proof-load test on viaduct De Beek: Case study. *Journal of Bridge Engineering*, 22, 05017014.

Lantsoght, E. O. L., Van Der Veen, C., De Boer, A. & Hordijk, D. A. (2017c) Required proof load magnitude for probabilistic field assessment of viaduct De Beek. *Engineering Structures*, 148, 767–779.

Lantsoght, E. O. L., Van Der Veen, C., Hordijk, D. A. & De Boer, A. (2017d) Reliability index after proof load testing: Viaduct De Beek. *ESREL 2017*. Protoroz, Slovenia.

Leemis, L. M. (1995) *Reliability: Probabilistic Models and Statistical Methods*. Prentice-Hall, Inc., Upper Saddle River, NJ, USA.

Li, C. Q., Lawanwisut, W. & Zheng, J. J. (2005) Time-dependent reliability method to assess the serviceability of corrosion-affected concrete structures. *Journal of Structural Engineering-ASCE*, 131, 1674–1680.

Liu, M. & Frangopol, D. M. (2005) Time-dependent bridge network reliability: Novel approach. *Journal of Structural Engineering*, 131(2), 329–337.

Liu, M. & Frangopol, D. M. (2006a) Optimizing bridge network maintenance management under uncertainty with conflicting criteria: Life-cycle maintenance, failure, and user costs. *Journal of Structural Engineering*, 132(11), 1835–1845.

Liu, M. & Frangopol, D. M. (2006b) Probability-based bridge network performance evaluation. *Journal of Bridge Engineering*, 11(5), 633–641.

Liu, L., Frangopol, D. M., Mondoro, A. & Yang, D. Y. (2018) Sustainability-informed bridge ranking under scour based on transportation network performance and multi-attribute utility. *Journal of Bridge Engineering, ASCE* (in press).

Lukic, M. & Cremona, C. (2001) Probabilistic assessment of welded joints versus fatigue and fracture. *Journal of Structural Engineering-ASCE*, 127, 211–218.

Melchers, R. E. (1999) *Structural Reliability: Analysis and Prediction*. John Wiley, Chichester.

Messervey, T. B. (2009) *Integration of Structural Health Monitoring into the Design, Assessment, and Management of Civil Infrastructure*. Università di Pavia, Pavia, Italy.

Messervey, T. B. & Frangopol, D. M. (2008) Innovative treatment of monitoring data for reliability-based structural assessment. In: H.-M. Koh & D. M. Frangopol (eds.) *Proceedings of the Fourth International Conference on Bridge Maintenance, Safety, and Management, IABMAS'08*. CRC Press/Balkema, Taylor & Francis Group, Seoul, Korea. pp. 1466–1474.

Messervey, T. B., Frangopol, D. M. & Casciati, S. (2011) Application of the statistics of extremes to the reliability assessment and performance prediction of monitored highway bridges. *Structure and Infrastructure Engineering*, 7(1), 87–99.

NCHRP (1998) *Manual for Bridge Rating through Load Testing*. Transportation Research Board, Washington, DC.

Obrien, E. J., Schmidt, F., Hajializadeh, D., Zhou, X. Y., Enright, B., Caprani, C. C., Wilson, S. & Sheils, E. (2015) A review of probabilistic methods of assessment of load effects in bridges. *Structural Safety*, 53, 44–56.

Okasha, N. M. & Frangopol, D. M. (2009) Lifetime-oriented multi-objective optimization of structural maintenance considering system reliability, redundancy and life-cycle cost using GA. *Structural Safety*, 31(6), 460–474.

Olaszek, P., Świt, G. & Casas, J. R. (2012) Proof load testing supported by acoustic emission: An example of application. *IABMAS 2012*, Stresa, Lake Maggiore, Italy.

Rausand, M. & Arnljot, H. Ã. (2004) *System Reliability Theory: Models, Statistical Methods, and Applications*, 2nd edition. John Wiley & Sons, New York City, NY.

Righiniotis, T. D. & Chryssanthopoulos, M. K. (2003) Probabilistic fatigue analysis under constant amplitude loading. *Journal of Constructional Steel Research*, 59, 867–886.

Rijkswaterstaat (2013) *Guidelines Assessment Bridges: Assessment of Structural Safety of an Existing Bridge at Reconstruction, Usage and Disapproval* (in Dutch), RTD 1006:2013 1.1. Ministry of Infrastructure and the Environment, Utrecht, The Netherlands.

Sabatino, S., Frangopol, D. M. & Dong, Y. (2015) Sustainability-informed maintenance optimization of highway bridges considering multi-attribute utility and risk attitude. *Engineering Structures*, 102, 310–321.

Sabatino, S., Frangopol, D. M. & Dong, Y. (2016) Life cycle utility-informed maintenance planning based on lifetime functions: Optimum balancing of cost, failure consequences and performance benefit. *Structure and Infrastructure Engineering*, 12(7), 830–847.

Šavor, Z. & Šavor Novak, M. (2015) Procedures for reliability assessment of existing bridge. *Gradevinar*, 67, 557–572.

Saydam, D. & Frangopol, D. M. (2013) Applicability of simple expressions for bridge system reliability assessment. *Computers and Structures*, 114–115, 59–71.

Soliman, M., Frangopol, D. M. & Mondoro, A. (2016) A probabilistic approach for optimizing inspection, monitoring, and maintenance actions against fatigue of critical ship details. *Structural Safety*, 60, 91–101.

Song, J. & Der Kiureghian, A. (2003) Bounds on system reliability by linear programming. *Journal of Engineering Mechanics*, 129(6), 627–636.

Song, J. & Kang, W. H. (2009) System reliability and sensitivity under statistical dependence by matrix-based system reliability method. *Structural Safety*, 31(2), 148–156.

Spaethe, G. (1994) The effect of proof load testing on the safety of a structure (in German). *Bauingenieur*, 69, 459–468.

Steenbergen, R. D. J. M. & Vrouwenvelder, A. C. W. M. (2010) Safety philosophy for existing structures and partial factors for traffic loads on bridges. *Heron*, 55, 123–140.

Stewart, M. G., Rosowsky, D. V. & Val, D. V. (2001) Reliability-based bridge assessment using risk-ranking decision analysis. *Structural Safety*, 23, 397–405.

Vatteri, A. P., Balaji Rao, K. & Bharathan, A. M. (2016) Time-variant reliability analysis of RC bridge girders subjected to corrosion: Shear limit state. *Structural Concrete*, 17, 162–174.

Willems, M., Ruiter, P. B. D. & Heystek, A. P. (2015) *Inspection Report Object 51H-304-01* (in Dutch). Rijkswaterstaat, Utrecht, The Netherlands.

Yang, D. Y. & Frangopol, D. M. (2018) Bridging the gap between sustainability and resilience of civil infrastructure using lifetime resilience. In: P. Gardoni (ed.) *Handbook of Sustainable and Resilient Infrastructure*. Routledge, Abingdon, UK, (in press).

Zhu, B. & Frangopol, D. M. (2012) Reliability, redundancy and risk as performance indicators of structural systems during their life-cycle. *Engineering Structures*, 41, 34–49.

Zhu, B. & Frangopol, D. M. (2013) Incorporation of structural health monitoring data on load effects in the reliability and redundancy assessment of ship cross-sections using Bayesian updating. *Structural Health Monitoring: An International Journal*, 12, 377–392.

Zhu, B. & Frangopol, D. M. (2014a) Effects of postfailure material behavior on system reliability. *ASCE-ASME Journal of Risk and Uncertainty in Engineering Systems, Part A: Civil Engineering*, 1(1), 04014002.

Zhu, B. & Frangopol, D. M. (2014b) Redundancy-based design of nondeterministic systems. In: Frangopol, D. M. & Tsompanakis, Y. (eds.) *Maintenance and Safety of Aging Infrastructure in the Book Series Structures and Infrastructures*. CRC Press/Balkema, Taylor & Francis Group, London. pp. 707–738.

Zhu, B. & Frangopol, D. M. (2015) Effects of post-failure material behaviour on redundancy factor for design of structural components in nondeterministic systems. *Structure and Infrastructure Engineering*, 11(4), 466–485.

Chapter 10

Determination of Remaining Service Life of Reinforced Concrete Bridge Structures in Corrosive Environments after Load Testing

Dimitri V. Val and Mark G. Stewart

Abstract

Reinforced concrete (RC) bridge structures deteriorate with time, and corrosion of reinforcing steel is one of the main causes for that. Load testing alone is unable to provide information about the extent of deterioration and remaining service life of a structure. In this chapter, a framework for the reliability-based assessment of the remaining service life of RC bridge structures in corrosive environments will be described. Existing models for corrosion initiation, corrosion-induced cracking, and effects of corrosion on stiffness and strength of RC members will be considered. Special attention will be paid to the potential effects of a changing climate on corrosion initiation and propagation in these structures. Examples illustrating the framework application will be provided.

10.1   Introduction

Reinforced concrete (RC) bridge structures deteriorate with time, being subject to loads and aggressive, in particular corrosive, environments that affect their strength and serviceability. To ensure adequate performance and safety of RC bridge structures during their service life, the possible effects of deterioration should be considered in the design of new structures and assessment of existing ones. Load testing alone provides very limited information about the state of corrosion in an existing bridge structure, in particular, if corrosion is yet to start or is at an early stage. In principle, a RC bridge needs to be specifically inspected and then constantly monitored if signs of corrosion-induced deterioration have been detected but not addressed through repair and/or protective measures. However, even inspections and monitoring cannot provide complete information for assessment of the state of corrosion in the bridge and prediction of its remaining service life. Thus predictive models describing corrosion initiation and propagation, including its structural effects, are needed. Taking into account numerous uncertainties associated with material properties, loads, environmental conditions, inspection/monitoring data, and the models themselves, it has become increasingly accepted that such models should be probabilistic/stochastic and analysis/assessment of structures be reliability based (e.g. Stewart et al., 2001; Val, 2007).

The initiation and rate of deterioration depend not only on material and structural properties, construction processes and imposed loads but also on the surrounding climatic environment. Climate change modifies this environment due to (1) gradual changes in average temperature, humidity, precipitation, atmospheric CO_2, and so forth; and (2) larger variations of the environmental parameters during seasonal cycles and an increase in frequency

DOI: https://doi.org/10.1201/9780429265969

and intensity of extreme weather. These phenomena may lead to the premature initiation of deterioration and an increase in its rate (Stewart et al., 2011).

The remaining service life can be predicted based on probabilistic methods using serviceability and/or ultimate strength limit states. For instance, predicting the likelihood and extent of corrosion damage to bridges (serviceability limit state) can be used to optimize inspection, maintenance, and repair of bridges, which may then lead to a service life extension (e.g. Val and Stewart, 2003; Mullard and Stewart, 2009, 2012). On the other hand, the effect of proof loading and service-proven performance of bridges can be used to update the probability of collapse for bridges, which in many cases results in an enhancement of structural reliability and, thus, an extension to service life (e.g. Stewart and Val, 1999; Faber et al., 2000). Moreover, Bayesian methods may use the observation of visual corrosion damage to update ultimate strength structural reliabilities and changes to service life predictions (e.g. Suo and Stewart, 2009).

In this chapter a framework for reliability-based assessment of durability and remaining service life of RC bridge structures in corrosive environments will be described. Existing models for corrosion initiation, corrosion-induced cracking, and effects of corrosion on stiffness and strength of RC members will be considered, including their applicability to probabilistic analysis in terms of data availability for quantification of relevant uncertainties and required computational effort. Since the design working life of RC bridges is 100–120 years and the actual service life can be even longer, new and existing RC bridges will definitely experience the long-term effects of climate change. Hence special attention will be paid to the potential effects of a changing climate on corrosion initiation and propagation in these structures. Another important aspect addressed in the chapter is the influence of the concrete composition, namely the partial replacement of portland cement with supplementary cementitious materials (SCMs) (e.g. fly ash [FA], blast furnace slag, and silica fume), on corrosion initiation and propagation. Such concretes have become increasingly popular because their use leads to a reduction of carbon dioxide emissions associated with concrete production. Examples illustrating the framework application will be provided.

10.2   Deterioration of RC structures in corrosive environments

There are a number of causes of deterioration of RC structures over time (e.g. alkali-aggregate reaction, sulfate attack and freeze-thaw cycles). However, the most common cause of deterioration is corrosion of embedded reinforcing steel due to carbonation or chloride contamination of concrete. The development of corrosion with time is usually considered a two-phase process involving corrosion initiation and then propagation; deterioration is associated with the second phase (e.g. Tuutti, 1982).

Corrosion initiation involves:

- Penetration of aggressive agents (CO_2, chloride ions) through the concrete cover
- Direct ingress of aggressive agents through cracks in the concrete cover

Corrosion propagation leads to:

- Loss of the cross-sectional area of reinforcing or prestressing steel
- Cracking and then spalling/delamination of the concrete cover

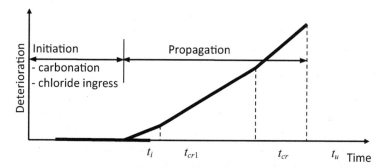

Figure 10.1 Schematic description of corrosion-induced deterioration in a RC member

- Reduction of bond between concrete and reinforcing steel
- Changes in ductility and mechanical properties of reinforcing/prestressing steel.

Typical development of corrosion-induced deterioration in a RC member over time is shown in Figure 10.1, where t_i is the time to corrosion initiation, t_{cr1} is the time of initial cracking of the concrete cover (i.e. hairline cracks of width less than 0.05 mm appearing on the concrete surface), t_{cr} is the time of excessive cracking and spalling/delamination (i.e. width of corrosion-induced cracks exceeds a specified limit value), and t_u is the time when an ultimate limit state is no longer satisfied due to strength reduction. It should be noted that the description of the deterioration process in Figure 10.1 is a schematic one, that is, the deterioration rates within each stage are not necessarily constant and differences between the rates at different stages may be more significant.

The deterioration processes depend on a large number of influencing variables which may be correlated, including concrete properties such as porosity, compressive and tensile strength, and water-cement ratio; thickness of the concrete cover; type of reinforcing steel; reinforcement spacing; quality of concrete constituents; exposure to aggressive agents (CO_2, chlorides); service loading; repair quality, and so forth. In addition, they depend on ambient environmental conditions (e.g. temperature, relative humidity), which may be affected in the future by climate change. Hence in order to correctly predict corrosion-induced deterioration of new and existing RC structures during their intended service life, it is essential to take the effects of a changing climate into account. The deterioration processes are highly variable in time and space and there are numerous uncertainties associated with the influencing variables and predictive models. For this reason, probabilistic/stochastic models are employed to predict temporal and spatial effects of corrosion-induced deterioration on serviceability and strength of RC members.

10.3 Reliability-based approach to structural assessment

Probabilistic/stochastic models are usually used within a reliability-based approach when the condition of a structure or its member(s) is assessed in terms of the probability of failure, P_f, or related to the reliability index, $\beta = -\Phi^{-1}(P_f)$, where Φ denotes the standard normal distribution function. For calculation of P_f, uncertainties associated with material properties, loads, environmental conditions, models, and so forth need to be quantified so that these parameters can be treated as random variables, processes, or fields. The failure of a structure should then be formally defined and expressed as a function of relevant random variables, processes, or fields, which is referred to as the limit state function, G. Conventionally, this function is defined such that $G \leq 0$ denotes failure of the structure, $G > 0$ denotes its survival, and $G = 0$ is the boundary between the failure and safe states,

$$G(t, \mathbf{x}, \boldsymbol{\theta}) = \mathbf{R}(t, \mathbf{x}, \boldsymbol{\theta}) - \mathbf{S}(t, \mathbf{x}, \boldsymbol{\theta}) \qquad (10.1)$$

where $\mathbf{R}$ denotes capacity and $\mathbf{S}$ demand, that is, the case of an ultimate limit state – resistance and load effect, respectively. Generally, $\mathbf{R}$ and $\mathbf{S}$ may include several random variables, processes or fields, which depend on time, t, and spatial coordinates, $\mathbf{x}$. Very often complete statistical information about random quantities is unavailable and certain assumptions (e.g. about their mean, variance, and covariance) need to be made to describe them in terms of probabilistic models. The uncertainties associated with this can be represented by a vector of random parameters, $\boldsymbol{\theta}$. Moreover, the development of deterioration in a structure may depend on loads acting on it, which means that $\mathbf{R}$ may depend on $\mathbf{S}$.

For a given location within/on the surface of a structure, $\mathbf{x}_c$, the probability of failure over time interval $(0, t_c)$ is

$$P_f(t_c, \mathbf{x}_c, \boldsymbol{\theta}) = \Pr(G(t, \mathbf{x}_c, \boldsymbol{\theta}) \leq 0 | \ t \in [0, t_c]) \qquad (10.2)$$

This probability depends on values of the parameters $\boldsymbol{\theta}$ while for decision-making a unique definition of the probability of failure is needed. The latter can be obtained as the expected value of Equation 10.2

$$P_f(t_c, \mathbf{x}_c) = E(P_f(t_c, \mathbf{x}_c, \boldsymbol{\theta})) = \int_{\boldsymbol{\theta}} P_f(t_c, \mathbf{x}_c, \boldsymbol{\theta}) f_{\boldsymbol{\Theta}}(\boldsymbol{\theta}) d\boldsymbol{\theta} \qquad (10.3)$$

where $E(\cdot)$ is the expectation operator and $f_{\boldsymbol{\Theta}}(\boldsymbol{\theta})$ is the joint probability density of the parameters $\boldsymbol{\theta}$. A decision is then made by comparing the obtained P_f (or β) with the corresponding target values (i.e. $P_{f,T}$ [or β_T]). The uncertainty associated with parameters $\boldsymbol{\theta}$ is usually the highest at the design stage, when the structure does not yet exist so that material properties, dimensions or loads cannot be actually measured. For existing structures additional information can be collected by load testing and/or on-site inspections and then used for updating probabilistic models based on a Bayesian approach (e.g. Zhang and Mahadevan, 2000; Ma et al., 2013; Tran et al., 2015). The remaining service life of the structure after testing/inspection can then be determined as the length of time interval over which the probability of failure remains less than its target value.

10.4   Corrosion initiation modeling

10.4.1   Carbonation-induced corrosion

Carbonation is a process involving two mechanisms: diffusion of carbon dioxide from the atmosphere into concrete and its reaction with the alkaline cement hydration products. The reaction leads to reduction of the alkalinity, pH, of the concrete pore solution from 12 to 13 to below 9, which destroys a thin passive oxide layer on the surface of steel reinforcement and makes the steel susceptible to corrosion.

CO_2 penetrates into concrete mainly due to gaseous diffusion through interconnected air-filled pores; its diffusion through the water that fills pores and convection with the water through the pores can be neglected. Relatively detailed numerical models, which account not only for the CO_2 diffusion but also for the transport of moisture and calcium ions as well for the change in the concrete porosity (e.g. Bary and Sellier, 2004; Zhang, 2016), are computationally intensive and by that reason difficult to use in probabilistic analysis. To obtain simpler closed-form solutions the diffusion of CO_2 into concrete is usually described by Fick's first law of diffusion (e.g. CEB, 1997), which for one-dimensional diffusion is expressed as

$$J_{CO_2} = -D_{CO_2} \frac{\partial C_{CO_2}}{\partial x} \tag{10.4}$$

where J_{CO_2} is the flux of CO_2, C_{CO_2} the concentration of CO_2, and D_{CO_2} the diffusion coefficient of CO_2 in concrete. Taking into account that $J_{CO_2} = dQ_{CO_2}/Adt$, Equation 10.4 can be written as

$$dQ_{CO_2} = AD_{CO_2} \frac{\Delta C_{CO_2}}{x} dt \tag{10.5}$$

where Q_{CO_2} is the mass of diffusing CO_2, A the surface area, ΔC_{CO_2} the difference in the CO_2 concentration in the atmosphere and at the carbonation front, and x the distance from the concrete surface to the carbonation front.

CO_2 that diffuses into concrete reacts with the alkaline cement hydration products. This reaction is called carbonation and it may simply be described by the reaction between Ca $(OH)_2$ and CO_2 (e.g. Taylor, 1997):

$$Ca(OH)_2 + CO_2 \overset{H_2O}{\rightarrow} CaCO_3 + H_2O \tag{10.6}$$

It should be noted that Equation 10.6 represents the final result of the reaction between Ca $(OH)_2$ and CO_2, which in reality involves a number of intermediate reactions not shown here for the sake of simplicity. As a result of the reaction, practically insoluble $CaCO_3$ and water are formed (as noted previously, these reaction results influence the carbonation process but are not taken into account in simple carbonation models). The reaction of carbonation removes free CO_2 from the mass balance that can be described by the following equation

$$dQ_{CO_2} = a_{CO_2} Adx \tag{10.7}$$

where a_{CO_2} is the CO_2 binding capacity of concrete, which depends mainly on the composition of the concrete.

If to assume that the reaction of carbonation, Equation 10.6, occurs instantaneously withdrawing CO_2 from the concrete pores and preventing it from diffusing further, the mass balance equation at the carbonation front can then be formulated using Equations 10.5 and 10.7 as

$$a_{CO_2} Adx = AD_{CO_2} \frac{\Delta C_{CO_2}}{x} dt \tag{10.8}$$

Solution of this equation is

$$x_c(t) = \sqrt{\frac{2D_{CO_2} \Delta C_{CO_2}}{a_{CO_2}}} \sqrt{t} \qquad (10.9)$$

where x_c is the depth of the carbonation front at time t.

The solution in Equation 10.9 has provided a basis for many semi-empirical models of concrete carbonation (e.g. Papadakis et al., 1991; Yoon et al., 2007; Yang et al., 2014; Ta et al., 2016). In all the models ΔC_{CO_2} is represented by the ambient CO_2 concentration. However, the estimation of D_{CO_2} and a_{CO_2} varies between different models. For example, the CO_2 binding capacity of concrete, a_{CO_2}, is often assessed as (e.g. CEB, 1997; Pade and Guimaraes, 2007; Yoon et al., 2007)

$$a_{CO_2} = 0.75C[CaO]\alpha_H \frac{M_{CO_2}}{M_{CaO}} \qquad (10.10)$$

where C is the cement content in concrete, CaO the amount of CaO per weight of cement, α_H the degree of hydration and M_{CO_2} (44 g/mol) and M_{CaO} (56 g/mol) the molar masses of CO_2 and CaO, respectively. Equation 10.10 is based on the assumption that the cement paste in concrete has been completely carbonated. To account for different cement types, in particular possible addition of SCMs, Ta et al. (2016) suggested modifying this formula by introducing the cement clinker content, φ_{cl}:

$$a_{CO_2} = 0.75\varphi_{cl}C[CaO]\alpha_H \frac{M_{CO_2}}{M_{CaO}} \qquad (10.11)$$

Typical values of φ_{cl} for different cement types are 0.975 for CEM I 52.5 N, 0.87 for CEM II A/L 52.5 N, 0.56 for CEM III/A 42.5 N, and 0.34 for CEM III/B 42.5 N. The degree of hydration can be estimated as (Yang et al., 2014)

$$a_H(t) = \frac{t}{2+t}\alpha_\infty \qquad (10.12)$$

$$a_\infty = \frac{1.031W/C_{eff}}{0.194 + W/C_{eff}} \qquad (10.13)$$

where t is the time/age of concrete (in days); α_∞ the final degree of hydration (i.e. at $t = \infty$); W/C_{eff} the effective water/cement ratio, $C_{eff} = C + kP$; W, C, and P the water, cement, and SCM content in the concrete (kg/m^3), respectively; and k the efficiency factor for SCM (e.g. for FA, $k = 0.5$). As can be seen from Equation 10.12, for concrete more than one year old, $\alpha_H \approx \alpha_\infty$.

The diffusivity of CO_2 depends on a number of parameters including concrete properties (e.g. porosity, composition), ambient conditions (e.g. relative humidity, temperature), and technological factors (e.g. curing). To take these parameters into account various functions of these parameters have been introduced to modify D_{CO_2}. A brief overview of such functions as well as various formulas proposed for the estimation of a_{CO_2} can be found in (Ta et al., 2016).

To account for the effect of SCM additions on the CO_2 diffusivity an empirical formula proposed by Papadakis and Tsimas (2002) can be used:

$$D_{e,CO_2} = 6.1 \times 10^{-6} \left[\frac{(W - 0.267C_{eff})/1000}{\dfrac{C_{eff}}{\rho_C} + \dfrac{W}{\rho_W}} \right]^3 (1 - RH)^{2.2} \tag{10.14}$$

where D_{e,CO_2} is the effective diffusivity of CO_2 in carbonated concrete (m²/s); ρ_C (= 3120 kg/m³) and ρ_W (= 1000 kg/m³) the cement and water densities, respectively; and RH the ambient relative humidity. The last term in the formula accounts for a decrease in the CO_2 diffusivity as the moisture content of concrete (i.e. RH) increases since the rate of CO_2 diffusion in water is very slow. However, the rate of carbonation also decreases when concrete becomes too dry because in this case the rate of CO_2 dissolution in pore water slows down. To take this into consideration a function whose value decreases with a decrease in RH is introduced (Salvoldi et al., 2015):

$$k_{RH} = \left(\frac{RH}{RH_{ref}} \right)^{2.6} \tag{10.15}$$

where RH_{ref} is the reference value of relative humidity. The optimum values of RH for concrete carbonation are between 0.50 and 0.70; since the empirical formula for D_{e,CO_2}, Equation 10.14, was derived based on experimental data obtained at $RH = 0.65$ in this study RH_{ref} is also set equal to 0.65.

The dependence of the CO_2 diffusivity in concrete on the temperature is usually described by Arrhenius's law (e.g. Stewart et al., 2012; Talukdar et al., 2012; Ta et al., 2016)

$$k_T = \exp\left[\frac{E_a}{R} \left(\frac{1}{T_{ref}} - \frac{1}{T} \right) \right] \tag{10.16}$$

where E_a is the activation energy of the CO_2 diffusion process, R (8.314 × 10^{-3} kJ/mol·K) the gas constant, and T and T_{ref} the ambient temperature (in Kelvin) and its reference value, respectively. It is based on the assumption that the internal temperature within a concrete member is equal to the ambient temperature, which is commonly used for analysis of concrete carbonation. The reference temperature is assumed to be equal to 25°C (i.e. $T_{ref} = 298$ K), which was the temperature of the tests used to derive Equation 10.14. The activation energy of the CO_2 diffusion process (in kJ/mol) can be evaluated by the following empirical formula (Li et al., 2013):

$$E_a = -8.7W/C_{eff} + 24 \tag{10.17}$$

The effect of curing on the CO_2 diffusion in concrete can be described by the following function (fib, 2006):

$$k_c = \left(\frac{t_c}{7} \right)^{b_c} \tag{10.18}$$

where t_c is the period of curing in days and b_c the exponent of regression. Following the model in (fib, 2006), the concrete resistivity to carbonation can be represented by a single parameter, which accounts for both diffusivity and binding capacity of concrete,

$$R_{NAC,0}^{-1} = \frac{D_{e,CO_2}}{a_{CO_2}} \tag{10.19}$$

where $R_{NAC,0}^{-1}$ is the inverse carbonation resistance of concrete determined in natural conditions (in $[(m^2/s)/(kg/m^3)]$ if a_{CO_2} is in kg/m^3).

Thus, the depth of carbonation (in meters) is evaluated as

$$x_c(t) = \sqrt{2k_{RH}k_Tk_cR_{NAC,0}^{-1}[CO_2]t} W_f(t) \tag{10.20}$$

where $[CO_2]$ is the concentration of CO_2 in the surrounding air in (kg/m^3) and $W_f(t)$ the weather function. According to available evidence, CO_2 concentration in rural, suburban, and urban areas is higher than the global atmospheric concentration measured in remote locations, $[CO_2]_a$. To account for that, for concrete structures located in non-remote environments $[CO_2]_a$ will be increased by the factor k_{site} (Stewart et al., 2011; Peng and Stewart, 2016), that is, $[CO_2] = k_{site}[CO_2]_a$, where k_{site} is 1.05, 1.07, and up to 1.15 for rural, suburban, and urban areas, respectively (Peng and Stewart, 2016). Since the average atmospheric concentration is usually given in parts per million by volume (ppm) it can be converted into kg/m^3 using the following formula:

$$[CO_2]_a = \frac{M_{CO_2}[CO_2]_{a,ppm}p}{RT} \times 10^{-9} \; (kg/m^3) \tag{10.21}$$

where p (101,325 Pa) is the atmospheric pressure and $[CO_2]_{a,ppm}$ the average atmospheric concentration of CO_2 measured in remote locations, which is currently about 400 ppm.

The weather function, $W_f(t)$, takes into account the micro-climate conditions of the considered concrete surface and is described by the following formula (fib, 2006):

$$W_f(t) = \left(\frac{t_0}{t}\right)^w \tag{10.22}$$

where t_0 is the reference time, t the time of exposure, and w the weather exponent. The latter is estimated as

$$w = \frac{(p_{sr}ToW)^{b_w}}{2} \tag{10.23}$$

where ToW is the time of wetness, p_{sr} the parameter representing the probability of driving rain, and b_w the exponent of regression. ToW represents the proportion of rainy days per year; a day is defined as rainy if the amount of precipitation water during this day $\geq$ 2.5 mm, that is,

$$ToW = \frac{\text{days per year with rainfall} \geq 2.5 \text{ mm}}{365} \tag{10.24}$$

ToW can be estimated using data from the nearest weather station; p_{sr} represents the average distribution of the wind direction when it rains: for vertical elements it should

Table 10.1 Description of random variables related to corrosion initiation due to carbonation

Variable	Mean	COV	Distribution type
$R_{NAC,0}^{-1}$	Eq. 10.19	0.25	Normal
E_a	Eq. 10.17	0.05	Normal
b_c	−0.567	0.042	Normal
b_w	0.446	0.365	Normal
k_{site}	1.15	0.10	Normal
c	Nominal	SD = 8–10 mm	Lognormal

be evaluated from the nearest weather station data; for horizontal elements $p_{sr} = 1$; for interior and sheltered elements, $p_{sr} = 0$.

There is significant uncertainty associated with the prediction of the carbonation depth, which needs to be taken into consideration. In order to do that a number of the parameters of the model presented above will be treated as random variables. According to fib (2006), $R_{NAC,0}^{-1}$ can be modeled as a normal random variable and its coefficient of variation (COV) varies approximately between 0.22 and 0.28, depending on the concrete composition. Thus, in this study $R_{NAC,0}^{-1}$ is treated as normal random variable with mean given by Equation 10.19 and COV = 0.25. Based on the experimental data presented in Li et al. (2013), E_a is represented by a normal random variable with mean given by Equation 10.17 and COV of 0.05. The exponent of regression in Equation 10.18, b_c, is described by a normal distribution with mean of −0.567 and COV of 0.042; the exponent of regression in Equation 10.23, b_w, can be described by a normal distribution with mean of 0.446 and COV of 0.365 (fib, 2006). Finally, the factor k_{site} representing an increase in the CO_2 atmospheric concentration in urban environments is assumed normally distributed with mean of 1.15 and COV of 0.10 (Stewart et al., 2011). The description of the random variables is summarized in Table 10.1.

The model of carbonation is used to estimate the probability of corrosion initiation, P_{corr}, which occurs when the carbonation front reaches reinforcing steel, that is,

$$P_{corr}(t) = Pr[c - x_c(t) \le 0] \tag{10.25}$$

where c is the thickness of the concrete cover, which can be modeled as a lognormal random variable with mean equal to its nominal value and standard deviation (SD) of 8–10 mm (without particular execution requirements) and 6 mm (with additional execution requirements) (fib, 2006).

10.4.2 Chloride-induced corrosion

Chloride-induced corrosion is generally recognized as more dangerous for RC structures compared to that induced by carbonation. Corrosion of this type starts when the chloride concentration near reinforcing steel exceeds a certain threshold value. Since the initial concentration of chlorides in reinforced concrete must be very low, the time of corrosion initiation is mainly controlled by ingress of chloride ions from external sources (e.g. seawater, deicing salts).

The ingress of chloride ions into concrete is a complex process involving different transport mechanisms such as ionic diffusion and convection (mainly in the form of capillary suction). As such it depends on a large number of parameters, for example, composition of concrete, its porosity and microstructure, degree of pore saturation, temperature, and exposure conditions. A number of these parameters are interdependent and time-dependent, in particular due to chemical reactions which proceed in concrete over time. Since only free chloride ions can penetrate into concrete, the chloride ingress depends also on binding of chlorides with the cement paste hydration products, which in turn depends on the chloride concentration.

Physically based models with various levels of complexity have been proposed to describe this process. The models can be broadly divided into those that take into account the interaction of chlorides with other ions present in the concrete pore solution and those that neglect this phenomenon. Models from the first group are usually based on the so-called multi-species approach, with the flux of different ions described by the Nernst-Planck equation (e.g. Khitab et al., 2005). Due to their complexity and lack of data for estimation of their parameters, in particular uncertainties associated with these parameters, the models are currently not used in practice and not suitable for probabilistic analysis. Models from the second group usually take into account chloride binding, coupled chloride, and moisture diffusion, and some of them also heat transfer (e.g. Meijers et al., 2005). Although these models are simpler than those from the first group, they are still rather complex and available experimental data are still insufficient for evaluation of their parameters, in particular quantification of their uncertainties, that limits their applicability within a reliability-based framework. Val and Trapper (2008) have proposed a relatively simple probabilistic model belonging to the second group, which takes into account chloride binding and coupled chloride and moisture diffusion and can describe both one- (1D) and two-dimensional (2D) ingress of chloride ions. The model has been extended in (Bastidas-Arteaga et al., 2011) to account for heat transfer. Such models are very suitable for considering effects of climate change (Bastidas-Arteaga et al., 2013); however, quantification of uncertainties associated with their parameters is still an unresolved issue that may cast some doubt on the validity of the obtained results.

Models usually used in practice are semi-empirical models based on Fick's second law of diffusion since it has been observed that chloride profiles (at least for internal layers of concrete where the influence of convection is insignificant) may be approximated sufficiently well by solutions of the following differential equation representing this law

$$\frac{\partial C_{Cl}}{\partial t} = D_{Cl}\frac{\partial^2 C_{Cl}}{\partial x^2} \tag{10.26}$$

where C_{Cl} is the total concentration of chloride ions at distance x from the surface after the time t of exposure to chlorides and D_{Cl} the chloride diffusion coefficient. To complete the problem formulation boundary and initial conditions need to be defined:

$$C_{Cl}(\Delta x, t > 0) = C_{Cl,s}(t)$$
$$C_{Cl}(x > \Delta x, t) = C_{Cl,i}(x) \tag{10.27}$$

where Δx is the depth of the convection zone (i.e. the surface concrete layer in which chloride penetration is mainly governed by convection and Fick's second law is inapplicable), $C_{Cl,s}(t)$ the chloride concentration at depth Δx, and $C_{Cl,i}(x)$ the initial chloride concentration.

Initially, in order to obtain a solution of Fick's second law it was typical to assume that D_{Cl} and the surface chloride concentration, $C_{Cl,s}$, were time-invariant constants. However, experimental evidence has been collected indicating that D_{Cl} and $C_{Cl,s}$ are time-dependent, and so a number of empirical models taking into account the time-dependency of these parameters have been proposed (e.g. Maage et al., 1996; Ann et al., 2009). Time-dependence of the chloride diffusion coefficient is usually described by a power function (e.g. DuraCrete, 2000; fib, 2006):

$$D_{Cl,a}(t) = D_{Cl,a,ref}\left(\frac{t_{ref}}{t}\right)^m \tag{10.28}$$

where $D_{Cl,a}$ denotes the apparent diffusion coefficient, $D_{CL,a,ref}$ its value at the reference time t_{ref}, which also depends on temperature, and m the age factor. It is important to stress that according to its definition $D_{Cl,a}$ remains constant within a time period t and changes only when the duration of the time period changes. Therefore, $D_{Cl,a}$ does not represent a physical property of concrete at a particular moment in time, but rather characterizes an average diffusivity over the time period $(0, t)$. The time-dependency of $C_{Cl,s}$ may be neglected when the long-term performance of a RC structure is considered since there are indications that $C_{Cl,s}$'s built-up periods are usually relatively short.

Assuming that the surface chloride concentration remains constant with time and the time-dependence of the diffusion is described by Equation 10.28, the solution of Equation 10.26 can be expressed as (e.g. DuraCrete, 2000; fib, 2006)

$$C_{Cl}(x, t) = C_{Cl,i} + (C_{Cl,s} - C_{Cl,i})\text{erfc}\left(\frac{x - \Delta x}{2\sqrt{tD_{Cl,a}(t)}}\right) \tag{10.29}$$

where erfc(.) is the error function complement. $D_{Cl,a,ref}$ can be estimated as

$$D_{Cl,a,ref} = k_T D_{RCM,0} \tag{10.30}$$

where $D_{RCM,0}$ is the reference value determined as the chloride migration coefficient measured by the rapid chloride migration (RCM) method (NT Build 492) and k_T the temperature parameter described by Arrhenius's law

$$k_T = \exp\left[b_T\left(\frac{1}{T_{ref}} - \frac{1}{T}\right)\right] \tag{10.31}$$

where b_T is a regression parameter, T_{ref} the reference temperature (293 K), and T the ambient temperature. Statistical description of the parameters of this simple model, which are treated as random variables, can be found in (fib, 2006) and summarized in Table 10.2.

It should be noted that this solution is applied under the assumption that the concrete cover is uncracked. However, cracks have a significant influence on the chloride diffusion coefficient and, subsequently, on t_i. According to results presented in (Takewaka et al., 2003) only hairline cracks up to 0.05 mm wide do not affect the chloride diffusivity of concrete, those between 0.05–0.1 mm increase the diffusivity by one order of magnitude, and between 0.1–0.2 mm by three orders.

The probability of corrosion initiation, P_{corr}, is obviously time-dependent and can be expressed as

$$P_{corr}(t) = Pr[C_{Cl,crit} - C_{Cl}(x = c, t) < 0] \tag{10.32}$$

Table 10.2 Description of random variables related to corrosion initiation due to chloride contamination

Variable	Mean	COV	Distribution type
$D_{RMC,0}$ (W/C_{eff} = 0.55*)			
CEM I 42.5 R	19.7×10^{-12} m²/s	0.20	Normal
CEM I 42.5 R + 22% FA	10.9×10^{-12} m²/s	0.20	Normal
CEM III/B 42.5 R	3.0×10^{-12} m²/s	0.20	Normal
m			
CEM I 42.5 R	0.30	0.40	Beta on [0,1]
CEM I 42.5 R + 22% FA	0.60	0.25	Beta on [0,1]
CEM III/B 42.5 R	0.45	0.44	Beta on [0,1]
b_T	4800 K	0.146	Normal
Δx			
splash zone	9 mm	0.60	Beta on [0,50]
atmospheric zone	0 mm		Deterministic
$C_{Cl,crit}$	0.6 wt% cement	0.25	Beta on [0.2,2]

*Values of $D_{RMC,0}$ for other values of W/C_{eff} can be found in (fib, 2006).
FA – fly ash.

where $C_{Cl,crit}$ is the threshold chloride concentration and c the thickness of the concrete cover. The threshold chloride concentration is a highly variable parameter, which according to (fib, 2006) can be described by a beta distribution (see Table 10.2).

10.5   Corrosion propagation modeling

10.5.1   Corrosion rate

To predict the development of corrosion-induced deterioration with time the rate of corrosion needs to be estimated. The latter is usually described in terms of either the corrosion current density, i_{corr}, or the corrosion penetration, V_{corr}. According to DuraCrete (2000) and LIFECON (2003), V_{corr} can be estimated as

$$V_{corr} = V \times ToW \,[\mu m/year] \qquad (10.33)$$

where V is the corrosion rate. The corrosion rate is expressed via the concrete resistivity, $\rho(t)$

$$V = \frac{m_0}{\rho(t)} F_{Cl} \qquad (10.34)$$

where $m_0 = 882$ μm·Ωm/year and $F_{Cl} = 1 + k_{Cl}(C_{Cl} - C_{Cl,crit})$ the chloride corrosion rate factor ($F_{Cl} \geq 1$). The concrete resistivity is time-variant and depends on a number of factors

$$\rho(t) = \rho_0 k_{R,T} k_{R,RH} k_{R,Cl} \left(\frac{t}{t_0}\right)^n \qquad (10.35)$$

where ρ_0 is the specific electrical resistivity of concrete at time t_0, $k_{R,T} = \exp[b_{R,T}(1/T_{real} - 1/T_0)]$ the temperature factor, $b_{R,T}$ the regression variable, $T_0 = 293$ K, $k_{R,RH}$ the relative humidity factor, $k_{R,Cl} = 1 - (1 - a_{Cl})C_{Cl}/2$ the chloride factor, $t_0 = 28$ days, t the age of concrete (≤ 1 year), and n the age factor for resistivity. The following parameters are treated as random variables: k_{Cl}, ρ_0, $b_{R,T}$, $k_{R,RH}$, $k_{R,Cl}$ (or a_{Cl}) and n, whose statistical properties

Table 10.3 Description of random variables related to estimation of corrosion rate

Variable	Mean	COV	Distribution type
ρ_0			
f_{ck} = 30 MPa	116 Ωm	0.16	Normal
f_{ck} = 40 MPa	134 Ωm	0.16	Normal
f_{ck} = 50 MPa	155 Ωm	0.16	Normal
k_{Cl}	2.63	1.335	Shift. lognormal, min 1.09
$b_{R,T}$	3815 K	0.15	Lognormal
$k_{R,RH}$			
splash zone	1.0	–	Deterministic
atmospheric zone	1.07	0.13	Lognormal
α_{Cl} (C_{Cl} < 2%)	0.72	0.153	Normal
$k_{R,Cl}$ ($C_{Cl} \geq$ 2%)	0.72	0.153	Normal
n	0.23	0.174	Normal

are assigned in accordance to DuraCrete (2000) and LIFECON (2003) and shown in Table 10.3.

To determine ρ_0 a value of the characteristic compressive strength of concrete, f_{ck}, is needed. The latter can be estimated using a formula proposed by Papadakis and Tsimas (2002) for the mean compressive strength of concrete (in MPa) with added SCMs

$$f_{cm} = 38.8 \left(\frac{1}{W/C_{eff}} - 0.5 \right)$$
(10.36)

so that (fib, 2013)

$$f_{ck} = f_{cm} - 8$$
(10.37)

The compressive strength of concrete is modeled as a lognormal random variable with mean given by Equation 10.36 and COV of 0.15. It has been found that based on the values presented in Table 10.3 the relationship between the mean value of ρ_0, $\mu_{\rho 0}$, and f_{ck} can be accurately described by the following formula

$$\mu_{\rho 0} = 75.085 \exp(0.0145 f_{ck})$$
(10.38)

Thus, in probabilistic analysis ρ_0 can be described by a normal distribution with mean given by Equation 10.38 and COV of 0.16.

10.5.2   Cracking of concrete cover

Corrosion products occupy a larger volume than the consumed steel. As they form, they exert pressure on the concrete surrounding a corroding reinforcing bar. This eventually leads to cracking of the concrete cover, which poses a serious problem for serviceability of reinforced concrete structures. There are two important parameters characterizing the performance of reinforced concrete structures in relation to corrosion-induced cracking: (1) the time of appearance of a corrosion-induced crack on the concrete surface, t_{cr1}; and (2) the time of excessive cracking, t_{cr}. In the following, probabilistic models for their estimation are described.

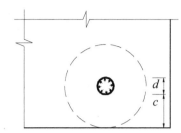

Figure 10.2 Thick-walled cylinder model

10.5.2.1 Time to crack initiation

The time to crack initiation equals

$$t_{cr1} = t_i + \Delta t_{cr1} \tag{10.39}$$

where t_i is the time to corrosion initiation and Δt_{cr1} the time between corrosion initiation and cover cracking. The latter is estimated using the so-called thick-walled cylinder model (e.g. El Maaddawy and Soudki, 2007). In this model the concrete around a corroding reinforcing bar is represented by a hollow thick-walled cylinder with the wall thickness equal to the thickness of the concrete cover, c, and the internal diameter equal to the diameter of the reinforcing bar, d. Expansion of the corrosion products is represented by uniform pressure applied to the inner surface of the cylinder (see Fig. 10.2).

A relationship between the expansion of corrosion products Δd (i.e. increase of the bar diameter) and the pressure, P, caused by it is given by the following formula

$$\Delta d = \frac{d}{E_{c,ef}} \left[1 + v_c + \frac{d^2}{2c(c+d)} \right] P \tag{10.40}$$

where v_c is the Poisson's ratio of concrete, $E_{c,ef} = E_c/(1+\varphi)$ the effective modulus of elasticity of concrete, E_c the modulus of elasticity of the concrete at age of 28 days, and φ the concrete creep coefficient.

The concrete cover is fully cracked when the average tensile stress in it becomes equal to the tensile strength of concrete, f_{ct}. The average tensile stress is estimated as the average tangential stress in the cylinder wall so that the internal pressure causing the concrete cover cracking, P_{cr}, equals

$$P_{cr} = \frac{2cf_{ct}}{d} \tag{10.41}$$

Mean values of f_{ct} and E_c can be found based on f_{cm} or f_{ck} (fib, 2013) as

$$\begin{aligned} f_{ctm} &= 0.3 f_{ck}^{2/3} & f_{ck} \leq 50 \text{ MPa} \\ f_{ctm} &= 2.12 \ln(1 + 0.1 f_{cm}) & f_{ck} > 50 \text{ MPa} \end{aligned} \tag{10.42}$$

$$E_{cm} = E_{c0}(0.1 f_{cm})^{1/3} \tag{10.43}$$

where $E_{c0} = 21.5 \times 10^3$ MPa. To account for uncertainty associated with f_{ct} and E_c and their evaluation they are modeled as lognormal random variables with COVs of 0.15 and 0.05, respectively.

The expansion of corrosion products can be estimated as (e.g. Chernin and Val, 2011)

$$\Delta d = 2(\alpha_v - 1)V_{corr}\Delta t \ [\mu m] \tag{10.44}$$

where α_v is the volumetric expansion ratio of corrosion products, V_{corr} is given by Equation 10.33, and Δt the time since corrosion initiation in years.

However, not all corrosion products contribute to the pressure exerted on the surrounding concrete since a part of them diffuses into concrete pores and microcracks. Denote the relative fraction of the corrosion products diffused into concrete as η, then the actual expansion of the corrosion products around a corroding reinforcing bar is $\Delta d(1 - \eta)$ (Chernin and Val, 2011). Substituting the last result into Equation 10.40 and using Equations 10.41 and 10.44, the time between corrosion initiation and cover cracking is found as

$$\Delta t_{cr1} = \frac{cf_{ct}}{(\alpha_v - 1)(1 - \eta)E_{c,ef}V_{corr}}\left[1 + v_c + \frac{d^2}{2c(c + d)}\right] \ [\text{years}] \tag{10.45}$$

10.5.2.2   Time to excessive cracking

The time of excessive cracking is defined as the time when the width of corrosion-induced cracks on the surface of a RC element reaches its critical value of w_{lim} (e.g. $w_{lim} = 0.3$ mm). The time is estimated using the empirical model proposed by Mullard and Stewart (2009)

$$t_{cr} = t_{cr1} + k_R\frac{w_{lim} - 0.05}{k_{con}ME_{rcr}r_{cr}}\left(\frac{0.0114}{i_{corr}}\right) \ [\text{years}] \tag{10.46}$$

where i_{corr} is the corrosion current density (in $\mu A/cm^2$; note that $i_{corr} = V_{corr}/11.6$), k_R is the rate of loading correction factor (introduced because experimental data for the model were obtained for $i_{corr,exp} = 100 \ \mu A/cm^2$), k_{con} the confinement factor that accounts for faster crack growth near external reinforcing bars due the lack of concrete confinement, r_{cr} the rate of crack growth and ME_{rcr} the model error.

$$r_{cr} = 0.0008\exp(-1.7\psi_{cr,p}) \tag{10.47}$$

$$\psi_{cr,p} = \frac{c}{df_{ct}} \quad 0.1 \leq \psi_{cr,p} \leq 1 \tag{10.48}$$

$$k_R = 0.95\left[\exp\left(-\frac{0.3i_{corr,exp}}{i_{corr}}\right) - \frac{i_{corr,exp}}{2500i_{corr}} + 0.3\right] \quad 0.25 \leq k_R \leq 1 \tag{10.49}$$

where $\psi_{cr,p}$ is the cover cracking parameter. If a reinforcing bar is in an internal location then $k_{con} = 1$; for reinforcing bars located at edges or corners of RC elements k_{con} is between 1.2 and 1.4. The model error, ME_{rcr}, can be treated as a lognormal random variable with mean of 1.04 and COV of 0.09. Statistical description of the random variable related to modeling corrosion-induced cracking is summarized in Table 10.4.

Table 10.4 Description of random variables related to corrosion-induced cracking of the concrete cover

Variable	Mean	COV	Distribution type
f_{ct}	Eq. 10.42	0.15	Lognormal
E_c	Eq. 10.43	0.05	Lognormal
α_v	3.0	0.30	Beta on [2,6.4]
η	0.70	0.30	Beta [0,0.9]
ME_{rcr}	1.04	0.09	Lognormal

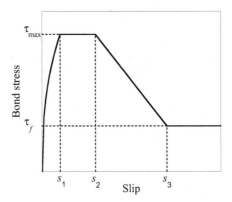

Figure 10.3 Bond stress-slip relationship

The probability of serviceability failure due to excessive cracking, $P_{f,s}$, at time t (in years) can then be calculated as

$$P_{f,s}(t) = \Pr[t_{cr} - t \le 0]$$ (10.50)

10.5.3   *Effect of corrosion on bond between concrete and reinforcing steel*

Corrosion influences a number of parameters which control the interaction between concrete and reinforcing steel, such as the level of confinement, adhesion, and friction between concrete and steel and the height of the ribs in deformed bars that decreases as corrosion propagates. These affect both strength and stiffness of the bond between reinforcing steel and concrete. These effects can be applied to the fib Model Code 2010 model shown in Figure 10.3 (fib, 2013), which relates bond stress, τ, with slip, s:

$$\tau = \begin{cases} \tau_{max} \cdot \left(\dfrac{s}{s_1}\right)^\alpha & 0 \le s \le s_1 \\ \tau_{max} & s_1 \le s \le s_2 \\ \tau_{max} - \dfrac{\tau_{max} - \tau_f}{s_2 - s_3}(s - s_3) & s_2 \le s \le s_3 \\ \tau_f & s \ge s_3 \end{cases}$$ (10.51)

Values of the model parameters (τ_{max}, τ_f, α, s_1, s_2, and s_3) can be found in (fib, 2013) and depend on whether concrete is treated as unconfined or confined. Concrete can be treated as unconfined when the ratio of transverse reinforcement to longitudinal reinforcement, ρ_r, does not exceed 0.25; confined concrete corresponds to $\rho_r > 1$. This ratio is defined as

$$\rho_r = \frac{A_{st}}{n_l A_s} \tag{10.52}$$

where A_{st} is the cross-sectional area of stirrups (two legs) over the anchorage length (usually between $15d$ and $20d$ with d denoting the diameter of longitudinal bars), n_l the number of longitudinal bars enclosed by stirrups, and A_s the cross-sectional area of one longitudinal bar.

Uncertainty associated with the model, that is, Equation 10.51 can be described by a normal random variable with the mean value equal to unity and COV of 0.3 (i.e. values of τ obtained from Equation 10.51 should be multiplied by this random variable). The unloading branch of the bond stress-slip relationship is linear; its slope is independent of s and can be taken equal to 200 N/mm³.

According to available experimental data at low levels of corrosion the bond strength initially increases and then starts to decrease as the corrosion propagates (the decrease usually occurs after the formation of corrosion-induced cracks in the concrete cover). Results of tests also indicate that there is residual bond strength (about 15% of its initial value) even at very high levels of corrosion (up to 80% of cross-sectional loss) (for more detail see Val and Chernin (2009)). Based on these observations the following relationship between the normalized bond strength, $\tau_{max}/\tau_{max,0}$, representing the ratio of the bond strength with corrosion, τ_{max}, to the initial bond strength, $\tau_{max,0}$, and the corrosion penetration (i.e. the reduction in radius of the uncorroded reinforcing steel), p_{corr}, is proposed (see also Fig. 10.4)

$$\frac{\tau_{max}}{\tau_{max,0}} = \begin{cases} 1 + (k_1 - 1)\dfrac{p_{corr}}{p_{corr,cr1}}, & p_{corr} \le p_{corr,cr1} \\[2mm] max[k_1 - k_2(p_{corr} - p_{corr,cr1}); \quad 0.15], & p_{corr} > p_{corr,cr1} \end{cases} \tag{10.53}$$

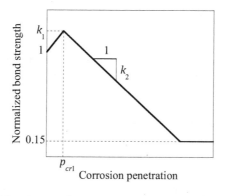

Figure 10.4 Normalized bond strength versus corrosion penetration

where $p_{corr,cr1}$ is the corrosion penetration corresponding to crack initiation in the concrete cover.

If $\tau_{max} < \tau_f$ then Equation 10.51 becomes

$$\tau = \begin{cases} \tau_{max}\left(\dfrac{s}{s_1}\right)^\alpha & 0 \leq s \leq s_1 \\ \\ \tau_{max} & s > s_1 \end{cases} \tag{10.54}$$

The corrosion penetration, p_{corr}, at time t (in years) since corrosion initiation and the corrosion penetration corresponding to crack initiation, $p_{corr,cr1}$, can be estimated as

$$p_{corr} = V_{corr}t \ [\mu m] \tag{10.55}$$

$$p_{corr,cr1} = V_{corr}\Delta t_{cr1} \ [\mu m] \tag{10.56}$$

where V_{corr} is calculated using Equation 10.33 and Δt_{cr1} – Equation 10.45.

The value of k_1 representing the initial increase in the bond strength after corrosion initiation apparently depends on the level of confinement provided by the concrete cover and transverse reinforcement (i.e. stirrups). The former is usually characterized by the ratio of the concrete cover to the reinforcing bar diameter, c/d, while the latter can be represented by the ratio ρ_r (see above). Thus, in order to determine k_1 the maximum values of the bond strength between corroding reinforcement and concrete for different values of c/d and ρ_r need to be known. However, such data are not available; published experimental results provide values of the bond strength at different levels of corrosion, most of which are at high levels of corrosion when corrosion-induced cracks have been already formed. There are just a few experimental results (mostly for specimens without transverse reinforcement, i.e., $\rho_r = 0$), which provide values (not necessarily maximum) of the bond strength before the formation of corrosion-induced cracks. Based on these results the following bilinear relationship between k_1 and c/d is proposed (Val and Chernin, 2009)

$$k_1 = \begin{cases} 1, & c/d \leq 1 \\ 1 + 0.085(c/d - 1), & c/d > 1 \end{cases} \tag{10.57}$$

It is likely that k_2, which represents the rate of decrease in the bond strength after the formation of corrosion-induced cracks, depends on the confinement provided by transverse reinforcement (i.e. on ρ_r) and on the remaining rib height of a reinforcing bar (that is related to the bar diameter, d), while the influence of the c/d ratio should be relatively insignificant. However, with respect to d available experimental results do not allow to make any decisive conclusion. Therefore, it is assumed that for unconfined concrete (i.e. $\rho_r \leq 0.25$) k_2 does not depend on d (or c/d) and its mean value equals $0.005 \ \mu m^{-1}$. There are not sufficient experimental data on the effect of corrosion on the bond strength for confined concrete (i.e. $\rho_r > 1$). It is suggested to adopt for confined concrete $k_2 = 0.0025 \ \mu m^{-1}$, which is the slope obtained by regression analysis of experimental data from Al-Sulaimani et al. (1990). Values of k_2 for $0.25 < \rho_r \leq 1$ can be found by interpolation. Thus, the relationship

Table 10.5 Description of random variables related to bond between concrete and corroded reinforcement

Variable	Mean	COV	Distribution type
Bond model uncertainty	1.0	0.30	Normal
k_1	Eq. 10.57	0.20	Normal, truncated at 1
k_2	Eq. 10.58	0.20	Normal

between k_2 (in μm^{-1}) and ρ_r can be summarized as (Val and Chernin, 2009)

$$k_2 = \begin{cases} 0.005 & \rho_r \leq 0.25 \\ 0.005 - \dfrac{\rho_r - 0.25}{300} & 0.25 < \rho_r \leq 1 \\ 0.0025 & \rho_r > 1 \end{cases} \tag{10.58}$$

Statistical properties of the random variables representing uncertainties associated with the bond stress-slip relationship of corroded reinforcing steel are given in Table 10.5.

10.5.4 Effect of corrosion on reinforcing steel

Two types of corrosion (general and pitting) are possible. General corrosion affects a substantial area of reinforcement with more or less uniform metal loss over the perimeter of reinforcing bars. Pitting (or localized) corrosion, in contrast to general corrosion, concentrates over small areas of reinforcement.

10.5.4.1 Loss of cross-sectional area due to general corrosion

The cross-sectional area of a reinforcing bar with the original diameter d after t years of general corrosion will be

$$A_s(t) = \frac{\pi[d - 2 \times 10^{-3} V_{corr} t]^2}{4} \geq 0 \, [mm^2] \tag{10.59}$$

According to available experimental data, general corrosion does not affect such properties of reinforcing steel as strength and ductility.

10.5.4.2 Loss of cross-sectional area due to pitting corrosion

Pitting corrosion, in contrast to general corrosion, concentrates over small areas of reinforcement. According to results of Gonzales et al. (1995), the maximum penetration of pitting, p_{max}, on the surface of a rebar is about 4 to 8 times the average penetration, p, of general corrosion. The results were obtained for 125 mm specimens of 8 mm diameter reinforcing bars. These results are in broad agreement with those reported by Tuutti (1982), who received the ratio $R = p_{max}/p$ between 4 and 10 for 5 mm and 10 mm reinforcing bars of 150 mm to 300 mm length. Thus, the depth of a pit, p_{max} (which is equivalent to the maximum penetration of pitting) after t years since corrosion initiation is

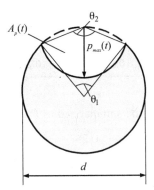

Figure 10.5 Pit configuration

evaluated as

$$p_{max}(t) = V_{corr} \times 10^{-3}tR \,[\text{mm}] \tag{10.60}$$

Based on a hemispherical model of a pit suggested by Val and Melchers (1997) (see Fig. 10.5), the cross-sectional area of a pit, A_p, in a reinforcing bar with a diameter d after t years of corrosion can be calculated as

$$A_p(t) = \begin{cases} A_1 + A_2, & p_{max}(t) \le \dfrac{d}{\sqrt{2}} \\[2ex] \dfrac{\pi d^2}{4} - A_1 + A_2, & \dfrac{d}{\sqrt{2}} < p_{max}(t) \le d \\[2ex] \dfrac{\pi d^2}{4}, & p_{max}(t) > d \end{cases} \tag{10.61}$$

where

$$A_1 = \frac{1}{2}\left[\theta_1 \left(\frac{d}{2}\right)^2 - a\left|\frac{d}{2} - \frac{p_{max}(t)^2}{d}\right|\right], \quad A_2 = \frac{1}{2}\left[\theta_2 p_{max}^2(t) - a\frac{p_{max}^2(t)}{d}\right] \tag{10.62}$$

$$a = 2p_{max}(t)\sqrt{1 - \left[\frac{p_{max}(t)}{d}\right]^2} \tag{10.63}$$

$$\theta_1 = 2\arcsin\left(\frac{a}{d}\right), \quad \theta_2 = 2\arcsin\left(\frac{a}{2p_{max}(t)}\right) \tag{10.64}$$

The cross-sectional area of a reinforcing bar after t years of pitting corrosion is estimated as

$$A_s(t) = \frac{\pi d^2}{4} - A_p(t) \ge 0 \tag{10.65}$$

There is significant uncertainty associated with R, that is, the ratio between the maximum pit depth and the average corrosion penetration. At the same time, a popular approach to modeling pit depth is based on statistical characterization of maximum pit depth using extreme value theory, in particular the Gumbel distribution Thus, R is treated as a random variable modeled by the Gumbel distribution.

$$F(R) = \exp\left\{-\exp\left[-\frac{(R-\mu)}{\alpha}\right]\right\} \tag{10.66}$$

where α and μ are the parameters of the distribution. According to Stewart (2004), for an 8 mm diameter bar of 125 mm length the parameters of the Gumbel distribution are $\mu_0 = 5.08$ and $\alpha_0 = 1.02$. For a reinforcing bar with different dimensions the parameters of the Gumbel distribution are determined as

$$\mu = \mu_0 + \frac{1}{\alpha_0}\ln\left(\frac{A}{A_0}\right), \quad \alpha = \alpha_0 \tag{10.67}$$

where A is the surface area of the given bar and A_0 the surface area of an 8 mm diameter bar of 125 mm length. For pitting statistics for other reinforcing and prestressing steel, see Stewart and Al-Harthy (2008) and Darmawan and Stewart (2007).

According to available experimental data, pitting corrosion may also affect strength and ductility of reinforcing steel. It has been suggested that yield strength, f_y, of a steel reinforcing bar reduces linearly with steel loss caused by pitting corrosion (Stewart, 2009)

$$f_y(t) = [1 - \alpha_y Q_{corr}(t)]f_{y,0} \tag{10.68}$$

where $f_{y,0}$ is the yield strength of uncorroded reinforcing bar, α_y an empirical coefficient and $Q_{corr}(t)$ the percentage corrosion loss, which is measured in terms of reduced cross-sectional area, i.e.,

$$Q_{corr}(t) = \frac{A_p(t)}{\pi d^2/4} \times 100\% \tag{10.69}$$

A review of experimental studies shows that α_y can be in a range from 0.0 to 0.017, with average around 0.005. Thus, α_y can be described by a Beta distribution on [0.0, 0.017] with mean of 0.005 and COV of 0.20.

There is strong evidence that the mechanical behavior of reinforcing bars changes from ductile to non-ductile (brittle) as pitting corrosion loss increases (Stewart, 2009). While there is a gradual transition from ductile to brittle behavior with an increase in corrosion loss, for simplicity it can be assumed that the complete loss of ductility in corroding reinforcing bars occurs after $Q_{corr(t)}$ exceeds a threshold value, Q_{limit}. This leads to two types of mechanical behavior:

Ductile behavior : $Q_{corr} \le Q_{limit}$

Brittle behavior : $Q_{corr} > Q_{limit}$ \tag{10.70}

The literature shows that it is reasonable to set $Q_{limit} = 20\%$, although more research is needed to more accurately quantify this important variable.

10.6 Effect of spatial variability on corrosion initiation and propagation

There is overwhelming evidence from laboratory and field observations of deteriorated structures that the deterioration process is spatially and time-dependent in nature. In other words, corrosion damage is not homogenous along a structure, but is highly spatially variable (heterogeneous) due to the spatial variability of concrete and steel material properties, environment, moisture, concrete cover, surface cracking, pitting corrosion, and so forth. For instance, Fazio et al. (1999) measured corrosion parameters and concrete properties of a 39-year-old bridge deck and found significant spatial variability in chloride content, concrete cover, concrete strength and corrosion damage. For example, corrosion rate was found to differ by up to two orders of magnitude across the bridge deck, and concrete cover varied from less than 25 mm to over 50 mm. While much work has progressed on the time-dependent structural reliability of deteriorating structures, there is relatively little work on spatial modeling of deterioration processes.

Not including spatial variability in a reliability analysis oversimplifies structural characterization and may lead to significant underestimation of failure probabilities in RC structures (e.g. Stewart, 2004). Recent work has shown the advantage of incorporating this spatial variability into stochastic models to predict the likelihood and extent of corrosion damage in RC structures and their effect on structural reliability (e.g. Stewart and Mullard, 2007; Sudret et al., 2008; Suo and Stewart, 2009). Of considerable benefit is that a spatial time-dependent reliability model allows the extent of corrosion damage to be predicted and as such, it is especially useful in estimating the time at which maintenance actions will be required and their cost. By incorporating maintenance strategies into an existing spatial time-dependent reliability model (Stewart and Mullard, 2007), a comparison can be made of differing maintenance strategies in terms of damage and repair criteria as defined by the asset owner.

The probabilistic analysis of structures considering random spatial variability involves the discretization of the corresponding random fields into sets of spatially correlated random variables. Random fields are an effective way to represent spatial variability (e.g. Stewart and Mullard, 2007) and in the case of a RC concrete surface, a 2D random field is used. The surface is discretized into a number of k identical square elements of size Δ and a random variable is used to represent the random field over each element. Each of the random variables within the random field are statistically correlated based on the correlation function of the corresponding random field (Vanmarcke, 1983).

The correlation function $\rho(\Delta)$ determines the correlation coefficient between two elements separated by distance Δ and is representative of the spatial correlation between the elements. As the distance between correlated elements increases the correlation coefficient reduces. Vanmarcke (1983) describes various types of correlation functions but the validation of these requires large amounts of spatial data that is often not available. The Gaussian (or squared exponential) correlation function, however, is commonly used in engineering applications (including spatial modeling of RC structures) and is defined as

$$\rho(\Delta) = \exp\left[-\left(\frac{|\Delta_x|^2}{d_x^2}\right) - \left(\frac{|\Delta_y|^2}{d_y^2}\right)\right] \tag{10.71}$$

where $d_x = \theta_x/\sqrt{\pi}$; $d_y = \theta_y/\sqrt{\pi}$; θ_x and θ_y are the scales of fluctuation for a two dimensional random field in x and y directions, respectively (Vanmarcke, 1983); and $\Delta_x = x_i - x_j$ and

$\Delta_y = y_i - y_j$ are the distances between the centroids of element i and element j in the x and y directions, respectively.

Data with which to characterize the spatial correlation of concrete dimensional, material property and corrosion parameters are limited (Na et al., 2012; Stewart and Mullard, 2007; Sudret, 2008). The scale of fluctuation defines the distance over which correlation persists in a random field. Vanmarcke (1983) proposed various methods for the calculation of the scale of fluctuation for stationary random fields. These methods, however, require large amounts of spatial data and, although this information has been collected in some cases, the scale of fluctuation is typically estimated using engineering judgment. O'Connor and Kenshel (2013) estimated the scale of fluctuation of chloride surface concentration and the diffusion coefficient of chloride ions from field work on an aging RC bridge structure located in a marine environment in Ireland. The scales of fluctuation for the two investigated deterioration variables were found to be 1.5–5.0 times higher than previously reported. However, it is estimated that the scales of fluctuation for diffusion coefficient, binding capacity, concrete cover and concrete strength are in the range of 1–2 m (Peng and Stewart, 2014). Nonetheless, more effort is needed to characterize the scale of fluctuation, as well as the shape of the correlation function, with more confidence.

10.7 Influence of climate change

The 2014 Intergovernmental Panel for Climate Change Fifth Assessment (AR5) Synthesis Report concluded that the "Warming of the climate system is unequivocal, and since the 1950s, many of the observed changes are unprecedented over decades to millennia. The atmosphere and ocean have warmed, the amounts of snow and ice have diminished, sea level has risen, and the concentrations of greenhouse gases have increased" (IPCC, 2014). What is less certain is the impact that rising temperatures will have on rainfall, wind patterns, sea-level rise, and other phenomena.

Future climate is projected by defining carbon emission scenarios in relation to changes in population, economy, technology, energy, land use and agriculture – a total of four scenario families were used in the IPCC's Third and Fourth Assessment Reports in 2001 and 2007, respectively. The A1 scenarios indicate very rapid economic growth, a global population that peaks in mid-century and declines thereafter, and the rapid introduction of new and more efficient technologies, as well as substantial reduction in regional differences in per capita income. Sub-categories of A1 scenarios include A1FI and A1B, which represent the energy in terms of fossil intensive and a balance across all sources, respectively. The B1 scenarios are more integrated and more ecologically friendly than A1 scenarios with rapid changes towards a service and information economies, and reductions in material intensity and the introduction of clean and resource efficient technologies.

The IPCC Fifth Assessment Report (IPCC, 2013) uses Representative Concentration Pathways (RCPs). The four RCPs (RCP 2.6, RCP 4.5, RCP 6, and RCP 8.5) are named after a possible range of radiative forcing values in the year 2100 (2.6, 4.5, 6.0, and 8.5 W/m^2, respectively), where RCP 8.5, RCP 6.0 and RCP 4.5 are roughly equivalent to A1FI, A1B, and A1B to B1 emission scenarios, respectively. The selected RCPs were considered to be representative of the literature, and include a strict mitigation scenario leading to a low forcing level (RCP 2.6) with CO_2 concentrations reaching 421 ppm by

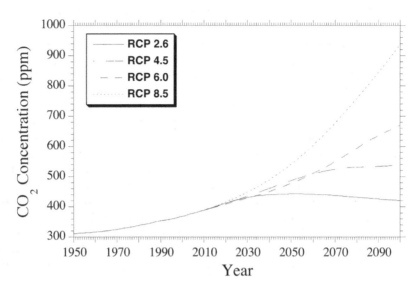

Figure 10.6 Projected CO_2 concentrations

the end of the century, two (medium) stabilization scenarios (RCP 4.5 – CO_2 concentration of 538 ppm, RCP 6.0 – CO_2 concentration of 670 ppm) and one scenario with very high greenhouse gas emissions (RCP 8.5 – CO_2 concentration of 936 ppm). The RCPs can thus represent a range of 21st-century climate policies, and are shown in Figure 10.6. The COV of CO_2 atmospheric concentrations is approximately 0.06 for projections at 2100.

Atmosphere-ocean general circulation models (GCMs) are currently the main tool for climate change studies. AOGCMs are numerical models based on differential equations, which describe physical processes in the atmosphere and ocean (and usually land-surface and sea ice as well) and interactions between them. Uncertainties associated with future emission scenarios are usually not quantified and future climate projections are produced separately for individual scenarios (Stewart et al., 2014). Selecting a GCM to be used in an impact assessment is not a trivial task, given the variety of models. The effect of these uncertainties on climate projections can be considerable. Figure 10.7 shows that RCP 4.5 climate projections for temperature and relative humidity vary considerably when six climate models are compared for Sydney in Australia: CSIRO-Mk3.6.0, ACCESS model (Australia), IPSL CM5A-LR (France), MICRO5 (Japan), bcc-csm1–1 (China) and CNRM-CM5 (European Center).

One of the consequences of an increase in CO_2 concentration and temperature, and changes in relative humidity and rainfall, is an acceleration of deterioration (corrosion) processes that consequently affect the durability, safety, and serviceability of existing concrete, steel, and timber infrastructure. Until recently all corrosion research assumed constant average climatic conditions for the development of models. For example, for atmospheric corrosion models CO_2 levels, time of wetness, temperature, humidity, and so forth typically are modeled as stationary variables. This is still the case but some efforts have been made to consider the effect of time and spatially dependent changes in the parameters involved. For example, short-term corrosion rates for steel reinforcement increase by up to 15% if the

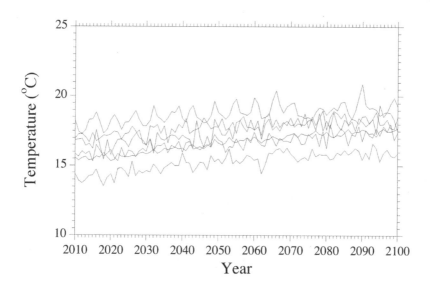

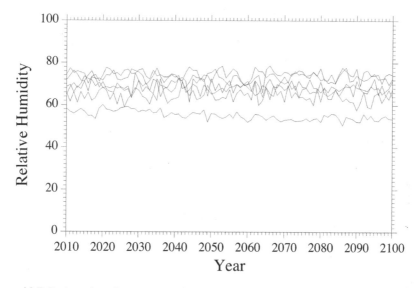

Figure 10.7 Projected median temperatures and relative humidities for six GCM predictions for RCP 4.5 emissions scenario for Sydney (Australia)

atmospheric temperature increases by only 2°C (Stewart et al., 2012). The RCP8.5 emission scenario predicts CO_2 concentrations increasing by 130% to 936 ppm by 2100, and this will increase carbonation depths of concrete by up to 40% in Australia and China (Stewart et al., 2012; Peng and Stewart, 2016). A changing climate will increase the likelihood of chloride-induced corrosion of RC structures in Australia and France by up to 100% (Stewart et al., 2012; Bastidas-Arteaga and Stewart, 2015).

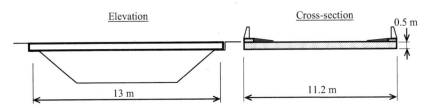

Figure 10.8 Simple-span bridge: basic dimensions

10.8 Illustrative examples

10.8.1 Simple-span RC bridge – case study description

To illustrate the application of the models described above the reliability assessment of a simple-span RC bridge shown in Figure 10.8 will be carried out. The bridge consists of a RC slab which is 11.2 m wide, 0.5 m thick and spans 13 m. The slab is made of concrete with the W/C ratio of 0.55 ($W = 203.5$ kg/m³, $C = 370$ kg/m³), which is produced with blended CEM III/B cement (contains 66% of ground granulated blast furnace slag) and has characteristic compressive strength around 43 MPa (in accordance to Equations 10.36 and 10.37). The bottom longitudinal reinforcement of the slab consists of 140 25 mm diameter reinforcing bars with characteristic yield strength of 500 MPa; the nominal thickness of the concrete cover for the bottom reinforcement is 40 mm.

The main ultimate limit state for the bridge is related to its bending resistance. For simplicity, only permanent load due to self-weight of the bridge deck components and variable load due to traffic are taken into account. The bending failure is considered at the bridge midspan, where the maximum bending moment is expected. The bending resistance, M_R, of the bridge slab is estimated as (e.g. Val, 2007)

$$M_R = \min\left[A_s f_y\left(d_e - \frac{A_s f_y}{1.7 f_c b}\right); \ \frac{1}{3} f_c b d_e^2\right] \tag{10.72}$$

where b is the cross-sectional width of the bridge slab, d_e its effective depth, f_c the compressive strength of concrete, and f_y the yield strength of reinforcing steel. The limit state function, G, is then formulated as

$$G = m_{MR} M_R - M_P - M_T \tag{10.73}$$

where m_{MR} is a random variable representing the uncertainty associated with the bending resistance model, that is, Equation 10.72, M_P the bending moment due to permanent load and M_T the bending moment due to traffic load. According to Lu et al. (1994), m_{MR} can be described by a normal distribution with mean of 1.10 and COV of 0.12. The bending resistance depends on two random variables f_c and f_y, which both follow a lognormal distribution. The mean value of f_c is given by Equation 10.36, while its COV = 0.15. The standard deviation of f_y is 30 MPa, while its mean value equals its characteristic value plus two standard deviations (JCSS, 2001).

The bending moment due to permanent load is treated as a normal random variable with COV of 0.10 and mean equal to 1.05 times its nominal value (Val, 2007); for the bridge

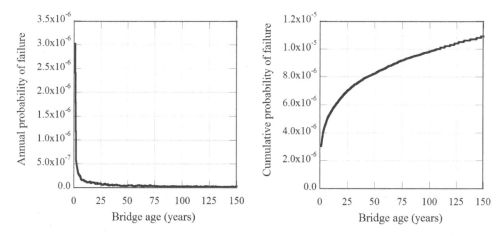

Figure 10.9 Probabilities of failure of non-deteriorating bridge

under consideration the latter has been estimated as 3870 kNm. The maximum annual traffic load on a short-span bridge can be modeled by a Gumbel distribution with the following parameters (e.g. Soriano et al., 2017)

$$a_N = \sqrt{2\ln(N)}; \quad u_N = a_N - \frac{\ln(\ln(N)) + \ln(4\pi)}{2a_N} \tag{10.74}$$

where N is the number of heavy trucks crossing the bridge per year. It is assumed that 100 heavy trucks cross the bridge per working day, so that $N = 100 \times 250 = 25000$, where 250 is the number of working days per year. The mean value of the bending moment caused by one truck is assumed to be equal to 1290 kNm, while its COV = 0.40. The assumed values ensure that the bending moment due to characteristic traffic load on the bridge can be exceeded with probability of 0.10 in 100 years.

The probability of the bridge bending failure, $P_f = \Pr[G \leq 0]$, is calculated by Monte Carlo simulation. The bending moments due to traffic load are independently generated for each year. Results of the analysis, that is, the annual and cumulative probabilities of failure for 150 years of the bridge life, obtained using 10^9 simulations trials, are shown in Figure 10.9. Since possible deterioration of the bridge has not been considered in this analysis the annual probability of failure decreases over time due to a positive effect of satisfactory past performance (e.g. Val and Stewart, 2002); the decrease is especially noticeable in the first couple of years after the bridge opening. The cumulative probability of failure after 100 years, that is, typical design life of a bridge, is 9.8×10^{-6} that corresponds to the reliability index of 4.27.

10.8.2 Reliability-based assessment of remaining service life of the bridge subject to carbonation

It is now considered that the bridge is subject to carbonation that can lead to corrosion of reinforcing steel. It is worth to note that carbonation leads to general corrosion of

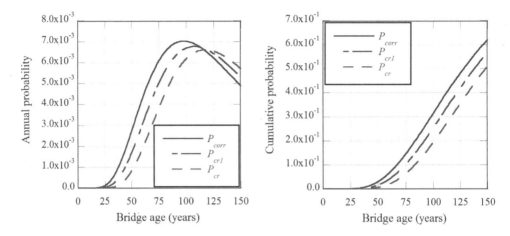

Figure 10.10 Probabilities of corrosion initiation and corrosion-induced cracking due to carbonation

the steel. The following ambient conditions are assumed for the bridge: average annual temperature of 12°C, relative humidity of 78% and CO_2 concentration of 400 ppm. The standard deviation of the thickness of the concrete cover is set equal to 8 mm; all other relevant random variables are described in Tables 10.1, 10.3, and 10.4. The reduction of bond strength between concrete and corroding reinforcing bars is neglected. The analysis is carried out using Monte Carlo simulation (10^9 simulation trials) and the following probabilities, both annual and cumulative, are calculated: probability of corrosion initiation, P_{corr}; probability of crack initiation, P_{cr1}; and probability of excessive cracking, P_{cr}, that is, the probability that the corrosion-induced crack width exceeds $w_{lim} = 0.3$ mm. Results of these calculations are shown in Figure 10.10. In addition, like in the previous section, the probabilities of bending failure are also obtained (see Fig. 10.11). As can be seen from the results, in this case carbonation-induced corrosion affects the bridge serviceability (e.g. the cumulative probability of excessive cracking after 100 years is 0.187) but has a negligible effect on the probability of bending failure; the values of the latter are very similar to those obtained for the non-deteriorating bridge (compare Figs. 10.9 and 10.11).

Since the bridge may show signs of corrosion-induced deterioration it is also assumed that a proof load test is carried out when the bridge is 60 years old. The load applied in the test together with the permanent load creates in the bridge the maximum bending moment of 9500 kNm. The probability of the bridge failure during the test is 1.36×10^{-4} that is much higher than the annual probability of failure at that time ($\approx 4 \times 10^{-8}$) and actually much higher than the cumulative probability of failure of the untested bridge after 150 years (11.1×10^{-6}). If the bridge survives the proof load tests, the probabilities of the bridge failure after the test are also shown in Figure 10.11; as can be seen, the annual probabilities of bending failure after the test become less than 1×10^{-9}. It may also be worth to note that the proof load test has no effect on the probabilities of corrosion-induced cracking of the concrete cover, which remain the same as shown in Figure 10.10.

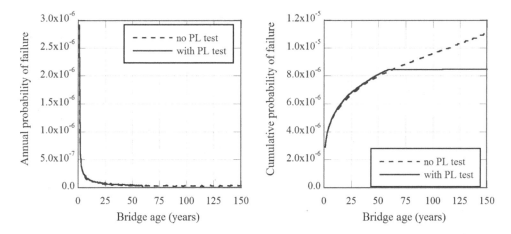

Figure 10.11 Probabilities of failure of the bridge subjected to carbonation without and with proof load (PL) test

10.8.3 Reliability-based assessment of remaining service life of the bridge subject to chloride contamination

To illustrate the application of the models related to chloride-induced corrosion it is assumed that the bridge is located right on the coast but not in direct contact with seawater. The surface chloride concentration is then presented by a lognormal random variable with mean of 2.95 kg/m^3 and COV of 0.70 (Val and Stewart, 2003). The average annual temperature and relative humidity for the bridge location are the same as in the previous section, i.e. 12°C and 78%, respectively. It is known that chloride contamination usually leads to pitting corrosion. However, it is often assumed that this corrosion affects more or less uniformly all surface of reinforcing bars and, therefore, can be treated similar to general corrosion. Thus, the analysis in this section is performed for two types of corrosion (general and pitting), for which the probabilities of corrosion initiation and bending failure are calculated. In the case of general corrosion the probabilities of corrosion-induced crack initiation and excessive cracking are also obtained; it should be noted that these probabilities are not relevant for pitting corrosion since in this case the models of corrosion-induced cracking described in Section 10.5.2 are not applicable.

In the case of pitting corrosion in order to model the random variable R representing the ratio between the maximum pit depth and the average corrosion penetration the surface area of the given reinforcing bar is calculated for 1 m length, that is, it is assumed that only pits occurring within 1 m of the bridge midspan affect its bending failure. It is also assumed that the maximum depths of pits in different reinforcing bars across the bridge width are independent random variables, that is, values of R are generated separately for each of the 140 bars of the bottom reinforcement. The probabilistic analysis is carried out using Monte Carlo simulation with 10^9 simulation trials. The effect of a proof load test on the bridge reliability is investigated for both general and pitting corrosion; similar to the previous section it is assumed that the proof load test is performed when the bridge is 60 years old and the maximum bending moment created in the test is 9500 kNm. The probabilities of corrosion initiation, crack initiation and excessive cracking are shown in Figure 10.12. The

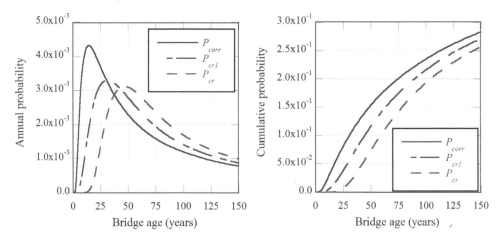

Figure 10.12 Probabilities of corrosion initiation and corrosion-induced cracking due to chloride ingress

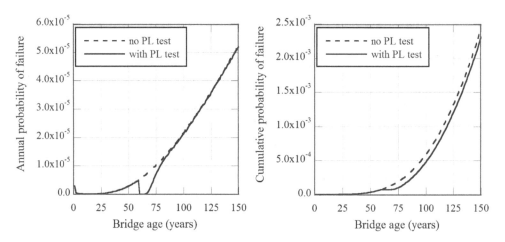

Figure 10.13 Probabilities of failure of the bridge subjected to chloride-induced uniform corrosion without and with proof load (PL) test

probabilities of the bridge bending failure without and with the proof load test for general and pitting corrosion are presented in Figures 10.13 and 10.14, respectively.

As can be seen, the chloride-induced corrosion, especially pitting corrosion, has much larger influence on the bridge reliability in the context of the ultimate limit state than the carbonation-induced corrosion. This occurs because the chloride-induced corrosion starts earlier than the carbonation-induced one and its rate is higher due to the presence of chloride ions; since pitting corrosion concentrates on smaller areas of reinforcing bars it leads to larger losses of their cross-sectional area. The proof load tests have a short-term influence on the assessed probabilities of the bridge failure: in the case of uniform corrosion for about 15 years (see Fig. 10.13), while in the case of pitting corrosion for less than 10 years (Fig. 10.14). At the same time, the probabilities of bending failure during the

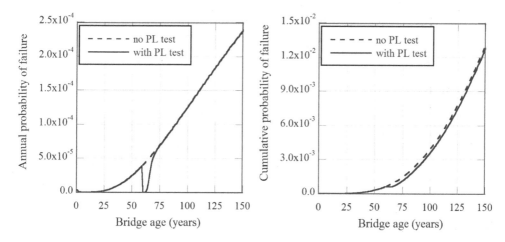

Figure 10.14 Probabilities of failure of the bridge subjected to chloride-induced pitting corrosion without and with proof load (PL) test

test are significantly higher than the corresponding annual probabilities in the same year: in the case of uniform corrosion 2.53×10^{-4} vs. 5.27×10^{-6}, while in the case of pitting corrosion 4.58×10^{-4} vs. 3.86×10^{-5}. The latter results illustrate the fact that the smaller the difference between the probability of failure a structure during a proof load test and that of the structure without the test the lesser the influence of the test on the assessed reliability of the structure.

10.8.4 Concluding remarks

The main aim of the examples is to illustrate the application of the previously presented models in combination with proof load testing. However, based on the limited results which have been obtained a few tentative recommendations can be drawn. First, it should be clear that proof load testing by itself does not provide sufficient information to assess the level of corrosion-induced damage to a RC bridge and the bridge reliability. A time-dependent probabilistic analysis of the bridge, which covers its intended life from the moment of the bridge completion and employs relevant models of corrosion initiation and propagation, is essential for such an assessment. The analysis is also needed to determine the magnitude of load used in the proof load test to give an acceptable probability of failure during the test. Second, a successful proof load test (i.e. bridge is undamaged by the test) has a limited effect on the assessment of the bridge reliability and only in the context of ultimate limit states. The effect depends on the magnitude of load applied in the test and the corresponding probability of bridge failure. The larger the load and the higher the probability of failure during the test, the more significant and longer-term increase in the predicted bridge reliability after the test can be achieved. Thus, it is important to find the right balance between the risk taken in the test and expected benefits, that is, an increase in the predicted bridge reliability. This can be undertaken by carrying out a risk-benefit analysis. Third, since carbonation-induced corrosion mainly affects bridge serviceability, it is not expected that a proof load test will be of much benefit in this case. Finally, although this has not been considered in the examples, an on-site inspection of the bridge, even just a visual one, may be very useful for assessing the level of corrosion and associated with it damage, especially in combination with corrosion-predictive models.

10.9 Summary

Probabilistic models for initiation and propagation of both carbonation- and chloride-induced corrosion have been presented. It has been explained and demonstrated by examples how these models can be used to assess the time-dependent reliability of RC bridges subject to corrosion, in the context of both serviceability and ultimate limit states. It has also been shown how these models and based on them the reliability assessment of a bridge can be combined with results of a proof load test. Such important issues as the influence of spatial variability and the impact of a changing climate on the reliability of RC bridges subject to carbonation or chloride contamination have also been considered.

References

Al-Sulaimani, G. J., Kaleemullah, M., Basumbul, I. A. & Rasheeduzzafar (1990) Influence of corrosion and cracking on bond behavior and strength of reinforced concrete members. *ACI Structural Journal*, 87(2), 220–231.

Ann, K. Y., Ahn, J. H. & Ryou, J. S. (2009) The importance of chloride content at the concrete surface in assessing the time to corrosion of steel in concrete structures. *Construction and Building Materials*, 23, 239–245.

Bary, B. & Sellier, A. (2004) Coupled moisture–carbon dioxide–calcium transfer model for carbonation of concrete. *Cement and Concrete Research*, 34, 1859–1872.

Bastidas-Arteaga, E., Chateauneuf, A., Sánchez-Silva, M., Bressolette, P. & Schoefs, F. (2011) A comprehensive probabilistic model of chloride ingress in unsaturated concrete. *Engineering Structures*, 33, 720–730.

Bastidas-Arteaga, E., Schoefs, F., Stewart, M. G. & Wang, X. (2013) Influence of global warming on durability of corroding RC structures: A probabilistic approach. *Engineering Structures*, 51, 259–266.

Bastidas-Arteaga, E. & Stewart, M. G. (2015) Damage risks and economic assessment of climate adaptation strategies for design of new concrete structures subject to chloride-induced corrosion. *Structural Safety*, 52, 40–53.

CEB (1997) New approach to durability design – An example for carbonation induced corrosion. *CEB Bulletin*, 238, CEB, Paris.

Chernin, L. & Val, D. V. (2011) Prediction of corrosion-induced cover cracking in reinforced concrete structures. *Construction and Building Materials*, 25, 1854–1869.

Darmawan, M. S. & Stewart, M. G. (2007) Effect of pitting corrosion on capacity of prestressing wires. *Magazine of Concrete Research*, 59(2), 131–139.

DuraCrete (2000) *DuraCrete–Final Technical Report*. General Guidelines for Durability Design and Redesign, Document BE95–1347/R17, The European Union–Brite EuRam III.

El Maaddawy, T. & Soudki, K. (2007) A model for prediction of time from corrosion initiation to corrosion cracking. *Cement and Concrete Composites*, 29, 168–175.

Faber, M. H., Val, D. & Stewart, M. G. (2000) Proof load testing for assessment and upgrading of bridges. *Engineering Structures*, 22(12), 1677–1689.

Fazio, R., Mirza, M. S., McCafferty, E., Andrews, R. J., Masheer, P. A. & Long, A. E. (1999) In-situ assessment of corrosion induced damage of the Dickson Bridge deck. *Proceedings of the 8th*

International Conference on Durability of Building Materials and Components: 8 DBMC, NRC Research Press, Ottawa, Vol. 1, pp. 269–279.

fib (2006) *Model Code for Service Life Design*. fib Bulletin 34, fib, Lausanne.

fib (2013) *The fib Model Code for Concrete Structures 2010*. fib, Lausanne.

Gonzales, J. A., Andrade, C., Alonso, C. & Feliu, S. (1995) Comparison of rates of general corrosion and maximum pitting penetration on concrete embedded steel reinforcement. *Cement and Concrete Research*, 25(2), 257–264.

IPCC (2013) *Climate Change 2013: The Physical Science Basis*. Contribution of Working Group I to the Fifth Assessment Report of the Intergovernmental Panel on Climate Change [Stocker, T. F., D. Qin, G.-K. Plattner, M. Tignor, S. K. Allen, J. Boschung, A. Nauels, Y. Xia, V. Bex & P. M. Midgley (eds.)]. Cambridge University Press, Cambridge, UK and New York, NY, USA, 1535 pp.

IPCC (2014) *Climate Change 2014: Impacts, Adaptation, and Vulnerability. Part A: Global and Sectoral Aspects*. Contribution of Working Group II to the Fifth Assessment Report of the Intergovernmental Panel on Climate Change [Field, C. B., V. R. Barros, D. J. Dokken, K. J. Mach, M. D. Mastrandrea, T. E. Bilir, M. Chatterjee, K. L. Ebi, Y. O. Estrada, R. C. Genova, B. Girma, E. S. Kissel, A. N. Levy, S. MacCracken, P. R. Mastrandrea & L. L. White (eds.)]. Cambridge University Press, Cambridge, UK and New York, NY, USA, 1132 pp.

JCSS (2001) *Probabilistic Model Code: Part III–Resistance Models*. [Online] Joint Committee on Structural Safety. Available from: www.jcss.byg.dtu.dk/Publications/Probabilistic_Model_Code.aspx [accessed 2nd April 2018].

Khitab, A., Lorente, S. & Ollivier, J. P. (2005) Predictive model for chloride penetration through concrete. *Magazine of Concrete Research*, 57, 511–520.

Li, G., Yuan, Y., Du, J. & Ji, Y. (2013) Determination of the apparent activation energy of concrete carbonation. *Journal of Wuhan University of Technology – Materials Science Edition*, 28(5), 944–949.

LIFECON (2003) *Deliverable D 3.2: Service Life Models*. Project G1RD-CT-2000-00378. European Community: Fifth Framework Program.

Lu, R. H., Luo, Y. H. & Conte, J. P. (1994) Reliability evaluation of reinforced concrete beams. *Structural Safety*, 14(4), 277–298.

Ma, Y., Zhang, J., Wang, L. & Liu, Y. (2013) Probabilistic prediction with Bayesian updating for strength degradation of RC bridge beams. *Structural Safety*, 44, 102–109.

Maage, M., Helland, S., Poulsen, E., Vennesland, Ø. & Carlsen, J. E. (1996) Service life prediction of existing concrete structures exposed to marine environment. *ACI Materials Journal*, 93, 602–608.

Meijers, S. J. H., Bijen, J. M. J. M., de Borst, R. & Fraaij, A. L. A. (2005) Computational results of model for chloride ingress in concrete including convection, drying-wetting cycles and carbonation. *Materials and Structures*, 38, 145–154.

Mullard, J. A. & Stewart, M. G. (2009) Stochastic assessment of the timing and efficiency of maintenance for corroding RC structures. *Journal of Structural Engineering, ASCE*, 135(8), 887–895.

Mullard, J. A. & Stewart, M. G. (2012) Life-cycle cost assessment of maintenance strategies for RC structures in chloride environments. *Journal of Bridge Engineering, ASCE*, 17(2), 353–362.

Na, U. J., Kwon, S., Chaudhuri, S. R. & Shinozuka, M. (2012) Stochastic model for service life prediction of RC structures exposed to carbonation using random field simulation. *KSCE Journal of Civil Engineering*, 16(1), 133–143.

O'Connor, A. & Kenshel, O. (2013) Experimental evaluation of the scale of fluctuation for spatial variability modeling of chloride-induced reinforced concrete corrosion. *Journal of Bridge Engineering, ASCE*, 18(1), 3–14.

Pade, C. & Guimaraes, M. (2007) The CO_2 uptake of concrete in a 100 year perspective. *Cement & Concrete Research*, 37, 1348–1356.

Papadakis, V. G., Vayenas, C. G. & Fardis, M. N. (1991) Fundamental modeling and experimental investigation of concrete carbonation. *ACI Materials Journal*, 88, 363–373.

Papadakis, V. G. & Tsimas, S. (2002) Supplementary cementing materials in concrete part 1: Efficiency and design. *Cement & Concrete Research*, 32(10), 1525–1532.

Peng, L. & Stewart, M. G. (2014) Spatial time-dependent reliability analysis of corrosion damage to concrete structures under a changing climate. *Magazine of Concrete Research*, 66(22), 1154–1169.

Peng, L. & Stewart, M. G. (2016) Climate change and corrosion damage risks for reinforced concrete infrastructure in China. *Structure and Infrastructure Engineering*, 12(4), 499–516.

Salvoldi, B. G., Beushausen, H. & Alexander, M. G. (2015) Oxygen permeability of concrete and its relation to carbonation. *Construction and Building Materials*, 85, 30–37.

Soriano, M., Casas, J. R. & Ghosn, M. (2017) Simplified probabilistic model for maximum traffic load from weigh-in-motion data. *Structure and Infrastructure Engineering*, 13(4), 454–467.

Stewart, M. G. (2004) Spatial variability of pitting corrosion and its influence on structural fragility and reliability of RC beams in flexure. *Structural Safety*, 26(4), 453–470.

Stewart, M. G. (2009) Mechanical behaviour of pitting corrosion of flexural and shear reinforcement and its effect on structural reliability of corroding RC beams. *Structural Safety*, 31(1), 19–30.

Stewart, M. G. & Al-Harthy, A. (2008) Pitting corrosion and structural reliability of corroding RC structures: Experimental data and probabilistic analysis. *Reliability Engineering and System Safety*, 93(3), 273–382.

Stewart, M. G. & Mullard, J. A. (2007) Spatial time-dependent reliability analysis of corrosion damage and the timing of first repair for RC structures. *Engineering Structures*, 29(7), 1457–1464.

Stewart, M. G. & Val, D. V. (1999) Role of load history in reliability-based decision analysis of ageing bridges. *Journal of Structural Engineering, ASCE*, 125(7), 776–783.

Stewart, M. G., Rosowsky, D. V. & Val, D. V. (2001) Reliability-based bridge assessment using risk-ranking decision analysis. *Structural Safety*, 23, 397–405.

Stewart, M. G., Wang, X. & Nguyen, M. N. (2011) Climate change impact and risks of concrete infrastructure deterioration. *Engineering Structures*, 33(4), 1326–1337.

Stewart, M. G., Wang, X. & Nguyen, M. (2012) Climate change adaptation for corrosion control of concrete infrastructure. *Structural Safety*, 35, 29–39.

Stewart, M. G., Val, D. V., Bastidas-Arteaga, E., O'Connor, A. & Wang, X. (2014) Climate adaptation engineering and risk-based design and management of infrastructure. In: Frangopol, D. M. & Tsompanakis, Y. (eds.) Maintenance and Safety of Aging Infrastructure. CRC Press, The Netherlands. pp. 641–684.

Sudret, B. (2008) Probabilistic models for the extent of damage in degrading reinforced concrete structures. *Reliability Engineering and System Safety*, 93(3), 410–422.

Suo, Q. & Stewart, M. G. (2009) Corrosion cracking prediction updating of deteriorating RC structures using inspection information. *Reliability Engineering and System Safety*, 94(8), 1340–1348.

Ta, V.-L., Bonnet, S., Kiesse, T. S. & Ventura, A. (2016) A new meta-model to calculate carbonation front depth within concrete structures. *Construction and Building Materials*, 129, 172–181.

Talukdar, S., Banthia, N. & Grace, J. R. (2012) Carbonation in concrete infrastructure in the context of global climate change – Part 1: Experimental results and model development. *Cement and Concrete Composites*, 34(8), 924–930.

Takewaka, K., Yamaguchi, T. & Maeda, S. (2003) Simulation model for deterioration of concrete structures due to chloride attack. *Journal of Advanced Concrete Technology*, 1, 139–146.

Taylor, H. F. W. (1997) *Cement Chemistry*. 2nd edition. Thomas Telford, London.

Tran, T.-B., Bastidas-Arteaga, E. & Schoefs, F. (2015) Improved Bayesian network configurations for probabilistic identification of degradation mechanisms: Application to chloride ingress. *Structure and Infrastructure Engineering*, 12(9), 1162–1176.

Tuutti, K. (1982) *Corrosion of Steel in Concrete*. Report CBI Fo 4.82. Swedish Cement and Concrete Research Institute, Stockholm.

Val, D. V. (2007) Deterioration of strength of RC beams due to corrosion and its influence on the beam reliability. *Journal of Structural Engineering, ASCE*, 133(9), 1297–1306.

Val, D. V. & Chernin, L. (2009) Serviceability reliability of reinforced concrete beams with corroded reinforcement. *Journal of Structural Engineering, ASCE*, 135(8), 896–905.

Val, D. V. & Melchers, R. E. (1997) Reliability of deteriorating RC slab bridges. *Journal of Structural Engineering, ASCE*, 123(12), 1638–1644.

Val, D. V. & Stewart, M. G. (2002) Safety factors for assessment of existing structures. *Journal of Structural Engineering, ASCE*, 128(2), 258–265.

Val, D. V. & Stewart, M. G. (2003) Life cycle cost analysis of reinforced concrete structures in marine environments. *Structural Safety*, 25(4), 343–362.

Val, D. V. & Trapper, P. A. (2008) Probabilistic evaluation of initiation time of chloride-induced corrosion. *Reliability Engineering and System Safety*, 93, 364–372.

Vanmarcke, E. (1983) *Random Fields: Analysis and Synthesis*. The MIT Press, Cambridge, MA.

Yang, K.-H., Seo, E.-A. & Tae, S.-H. (2014) Carbonation and CO_2 uptake of concrete. *Environmental Impact Assessment Review*, 46, 43–52.

Yoon, I.-S., Çopuroğlu, O. & Park, K.-B. (2007) Effect of global climatic change on carbonation progress of concrete. *Atmospheric Environment*, 41(34), 7274–7285.

Zhang, Q. (2016) Mathematical modeling and numerical study of carbonation in porous concrete materials. *Applied Mathematics and Computation*, 281, 16–27.

Zhang, R. & Mahadevan, S. (2000) Model uncertainty and Bayesian updating in reliability-based inspection. *Structural Safety*, 22, 145–160.

Chapter 11

Load Testing as Part of Bridge Management in Sweden

Lennart Elfgren, Bjorn Täljsten, and Thomas Blanksvärd

Abstract

Load testing of new and existing bridges was performed regularly in Sweden up to the 1960s. It was then abandoned due to high costs versus little extra information obtained. Most bridges behaved well in the serviceability limit range and no knowledge of the ultimate limit stage could be obtained without destroying the bridge. At the same time the methods for calculating the capacity developed and new numerical methods were introduced. Detailed rules were given on how these methods should be used. Some decommissioned bridges were tested to their maximum capacity to be able to study their failure mechanisms and to calibrate the numerical methods. In this chapter, some examples are given on how allowable loads have increased over the years and of tests being performed. Nowadays, load testing may be on its way back, especially to test existing rural prestressed concrete bridges, where no design calculations have been retained.

11.1 Introduction

Load testing is one of the oldest ways to check the quality of a bridge (Bolle et al., 2010). The deformations of a bridge during loading summarize its general condition and stiffness. Thus, the deformation is identified as a key performance indicator (see e.g. COST 1406 WG1, 2016 and COST 1406 WG2, 2017). Here some examples and experiences are given from load testing in Sweden; how quality control and management of bridges can be improved and how numerical models may be calibrated.

Load testing can be performed at (a) service-load levels and (b) loads to check the ultimate capacity (failure) of a structure. Testing for service-load levels is often divided into two groups (Lantsoght et al., 2017a, 2017b, 2017c):

- Diagnostic tests to update the analytical model of a bridge so that the allowable load can be better defined. Often the stiffness of a bridge is determined in the linear elastic stage.
- Proof loading tests to demonstrate that a bridge can carry the loads it is intended for (Casas and Gómez, 2013). Higher loads are usually used than in diagnostic tests. Recommendations for proof loading are given in some codes and stop criteria are given to prevent damage. The criteria often prescribe maximum values for concrete and steel strains, crack widths and residual deflections. Brittle failures are feared so bridges with a risk for shear failure are usually not allowed to be proof loaded.

DOI: https://doi.org/10.1201/9780429265969

Testing to failure can be used to increase the knowledge of the real function of a type of structure and how well codes can predict the load-carrying capacity (Bagge et al. 2018). Load testing to failure is often more expensive than diagnostic and proof tests where standardized load rigs or trucks with known weight can be used.

11.2   History

11.2.1   Overview of development of recommendations

An example of an early proof loading in 1905 of a railway bridge in southern Sweden is given in Figure 11.1. When a new suspension bridge was built 60 years later in the harbor of Gothenburg, the deflections of the bridge were tested by proof loading with trucks filled with ballast.

The management of the roads in Sweden was taken over by the government in 1944. The development of the design load is summarized in Table 11.1 based on von Olnhausen (1991) and Coric et al. (2018).

Load testing of new and existing bridges was performed in Sweden up to the 1960s. It was then abandoned due to the high costs versus the little extra information it gave. Most bridges behaved well in the serviceability limit range and no knowledge of the ultimate limit stage could be obtained without destroying the bridge. At the same time, the methods for calculating the capacity developed and new numerical methods were introduced. Detailed rules were given on how these methods should be used and these have been updated successively (TRV Capacity Rules, 2017; TRV Capacity Advice, 2017; TRV Design Advice, 2016; TRV Design Rules, 2016).

11.2.2   Which aim of load test is provided

Earlier the aim was to check the capacity and function of a bridge. Today the focus is more on research, to learn about the function of a special bridge type or to follow the

Figure 11.1 Proof loading of the Harlösa Bridge with three locomotives and extra dead load of rail. Kävlinge brook in southern Sweden, 1905

Source: Photo Sven Stridberg.

Table 11.1 Traffic loads on bridges (von Olnhausen, 1991; Coric et al., 2018). Conversion: 1 kN = 0.225 kip

Building year	Axle load, kN	Bogie load, kN	Gross load, kN	Iron ore railway line, axle load, kN
Before 1931	20–60	–	40–120	140
1931–37	28–75	–	36–100	140
1937–44	45–80	100	70–150	140
1944–77	196*	–	330–650	250
1977–88	210*	420*[1]	Ca 720*	250
1988–2009	210*	420*[2]	Ca 720*	300
2009–	270*	540*[2]	Ca 760*	325 (2018)

Notes:
*Includes dynamic amplification. Distance between bogies
[1] 2.5 m (8.2 ft), [2] 1.5 m (4.9 ft).

behavior of special bridges by structural health monitoring (Sundquist, 1998; Karoumi et al., 2007).

An example is the New Svinesund Bridge between Sweden and Norway with a main arch span of 247 m (810 ft). The bridge was constructed 2005–2007 and a monitoring program and load tests were carried out to check that the bridge was built as designed, to better understand the bridge's response to static and dynamic loadings, and to produce an initial database (a footprint) of the undamaged structure that could be used for future condition assessment of the bridge in relation to the initial condition (Karoumi and Andersson, 2006). Temperature and solar radiation were factors of great importance. The data obtained were also used to calibrate numerical models (Jonsson and Johnson, 2007). The original finite element model in some cases gave up to 40% higher deflections than what was measured (14 mm instead of 10 mm, 0.55 in. instead of 0.39 in.).

11.2.3   *Development of recommendations*

The development of the rules that are given is primarily a responsibility of the government. The Swedish government has initiated projects and guidelines for how the design should be carried out using numerical models (Plos et al., 2004; Puurula, 2004; Pacoste et al., 2012). It has also established a web site with growing information of the bridges in Sweden (BaTMan, 2018).

11.3   Present practice

11.3.1   *Inspection regime of structures*

All bridges maintained by the government are inspected regularly at time intervals depending on the condition of a bridge but at least every sixth year (main inspection). If needed, special inspections and damage investigations are also carried out. Guidelines are given and are successively updated (TRV Inspection Rules, 2018; TRV Inspection Advice, 2018). Nowadays the inspections are mostly carried out by consultants.

The main inspection shall comprise all parts and elements that have an influence on the function or the safety of a bridge. The inspection shall be carried out at arm length distance and in daylight conditions. Visual inspection methods that give the same accuracy may be used. The inspection shall be carried out at a time of the year when the bridge is free from snow and ice. Necessary measurements shall be carried out in order to be able to ascertain the amount and function of any defects.

Special inspections are carried out when there is a need to make a more thorough investigation of the function of a bridge due to damages, for example. Special inspections are also carried out when there is a need of special competence or of an inspection between the main inspections.

The rules for inspection of rail and road bridges are nowadays unified. For rail bridge inspections and assessment the recommendations from the International Union of Railways (UIC) are also consulted (UIC 778–3, 2014).

11.3.2   Levels of assessment of structures

Levels of assessment are given in TRV Capacity Rules (2017) and TRV Capacity Advice (2017), The allowable load on an axle A and on a bogie B is to be determined for a load traveling on a road bridge

- In its own lane
- Alone in the middle of the bridge
- In the middle of the bridge and with traffic in opposite direction.

For railway bridges, the axle load Q is to be determined as well as the line load.

Only linear analysis is accepted in ordinary design and assessment of bridges. Nonlinear methods may be used in special projects. In such cases a global safety factor shall be determined using mean values of material properties. Rotational capacity and crack widths must be controlled.

It is specified that all load assessments shall be checked by an independent consultant engineer. When the assessment is finished, a report of it shall be uploaded in the bridge register in BaTMan.

11.3.3   Configuration of the vehicles

A typical load figuration 1994–2003 is given in Figure 11.2 (Sundquist, 1998). After that the Eurocodes started to be applied.

11.3.4   Development of the traffic

The Traffic Work is in Sweden defined as the length of all trips that vehicles are making on a specific part of a road during a specific time, (TRV Traffic, 2016). If the total Traffic Work on the roads in Sweden in 1974 is set to 100, the work had increased to 117 in 1980, to 152 in 1990, to 170 in 2000, to 190 in 2010, and to 200 in 2015. During 2017 the transport work was about 42 giga (10^9) tonkilometer (29 giga short ton miles) per year. For rail, the transport work has been almost constant, about 20 giga tonkilometer (14 giga short ton miles) per year during the period 1990–2017. However, the transported amount of goods on rail

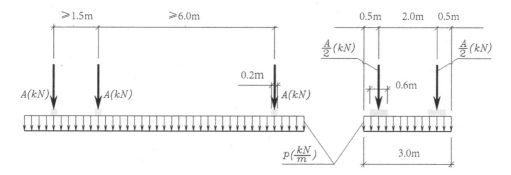

Figure 11.2 Typical load configuration in Sweden 1994–2010. Left: Longitudinal direction, Right: Transverse direction, in the first lane *A* = 250 kN, *p* = 12 kN/m, in the second lane *A* = 170 kN, *p* = 9 kN/m and in the third lane *A* = 0 kN, *p* = 6 kN/m (Sundquist, 1998). Conversion: 1 kN = 0.225 kip, 1 m = 3.3 ft, 1 kN/m = 0.07 kip/ft

has increased from 55 to 70 mega (10^6) tons (61 to 77 mega short tons) per year during the same period (Rail Traffic, 2017).

11.3.5 *Examples of load testing*

In Sweden, quite a few bridge tests have been carried out. Some examples of tests at service-load levels and tests to failure are given in Table 11.2 and Table 11.3, respectively. The overall aim has been to investigate and evaluate the safe function of the bridges for increased loads at the serviceability and ultimate limit states and to improve modeling (Elfgren et al., 2018).

The main result of the tests at service-load levels is that many bridges have a "hidden" capacity and could carry higher loads than what is obtained applying ordinary design rules. In these tests probabilistic analysis was also identified as a viable tool for the assessment.

11.4 Future

11.4.1 *Bridge management*

As traffic increases and our bridge stocks grow older, there will be an increasing demand for more effective and efficient bridge management. Bridge testing can be a tool that can be developed to help achieve this. Monitoring and digital methods are developing and routine procedures may be used to test existing bridges.

In recent years there is an increased interest from Trafikverket to start to make proof loading of rural prestressed concrete bridges where only scare documentation is presented. Such a procedure has been started in Finland (Raunio, 2016).

An idea is also to create digital twin bridges which start their life (*in silico*) during the planning phase of a bridge (Bagge, 2017). The models may be integrated with monitoring of the in-situ bridge to enable model updating and for later assessment of the quality and load-bearing capacity of the bridge during its lifetime. Improved monitoring and numerical

Table 11.2 Examples of load testing at service-load levels in Sweden

Location and type	Photo	Test and results	References
Vindelälven Reinforced concrete (RC) arch railway bridge with a main span of 110 m (361 ft). Built in 1954.		The bridge was to be used for an increased axle load of 250 kN (56 kip). Tests and FEM analyses were carried out with trains having the new load configurations. Deformations, strains, and accelerations were recorded, and it was shown that the bridge could carry the upgraded loads.	(He et al. 2008)
Kalix River RC arch railway bridge with a main span of 87.9 m (288 ft). Built in 1960.		The bridge was to be used for an increased axle load of 250 kN (56 kip). Tests and FEM analyses were carried out with the new load configurations. Deformations, strains, and accelerations were recorded, and it was shown that the bridge could carry the upgraded loads.	(Wang et al., 2016) (Grip et al., 2017)
Haparanda RC double-trough railway bridge with a span of 12.5 m (41 ft). Built in 1959.		The bridge deck was upgraded from 250 to 300 kN (56 to 67 kip) axle load using internal unbonded transversally prestressed steel bars. Tests before and after the strengthening showed acceptable strains and deflections.	(Mainline, 2014) (Nilimaa, 2015)
Lautajokk RC single-trough railway bridge with a span of 7.0 m (23 ft). Built in 1967.		The axle loads on an iron ore railway line were to be increased from 250 to 350 kN (56 to 79 kip). In order to investigate the fatigue capacity, this bridge was tested with an axle load of 360 kN (81 kip) during six million load cycles using hydraulic jacks. The shear capacity was studied of the connection of the slab to the longitudinal beams, where no shear reinforcement was present. After the cyclic loading, not resulting in any detectable damages, the load was increased to yielding of the reinforcement.	(Thun et al., 2000) (Elfgren, 2015)

Location and type	Photo	Test and results	References
Luossajokk Continuous RC single-trough railway bridge with spans of 10.2 m (33 ft) and 6.3 m (21 ft). Built in 1965.		Load testing was carried out and strains and deflections were measured to be a basis in the evaluation of the possibility to increase the axle load from 250 to 300 kN (56 to 67 kip). A study using probabilistic analysis showed that the new load was acceptable.	(Enochsson et al., 2002)
Pite River Continuous steel beam four span road bridge of a length of 257 m (843 ft). Built in 1969.		The capacity of the edge beams of the concrete deck was evaluated by using probabilistic analysis and monitoring. In order to increase the load capacity, the edge beams were strengthened.	(Stenlund, 2008)
Kuivajärvi RC portal frame road bridge with a span of 6.6 m (22 ft). Built in 1934.		A transversal movement of the supports towards each other was detected, which caused cracking of the deck. Based on measurements of strains and deflections, the capacity was identified as sufficient for continued use. Thus, strengthening of the structure was avoided.	(Stenlund, 2008)

methods may in the future be used to determine hidden deterioration (Grip et al., 2017; Huang et al., 2016).

11.4.2 *Numerical tools*

Numerical tools as linear and nonlinear finite element methods have been shown to be useful for assessment, especially combined with material testing and stepwise refinements from linear to nonlinear modeling of bending, shear, and anchorage (Bagge, 2017; Hendriks et al., 2017).

In numerous studies (see e.g. Bagge et al., 2018), large differences have been demonstrated between standard structural assessment methods and more detailed analyses by using nonlinear FEM. The differences have probably partly arisen from redistribution of loads during testing in statically indeterminate structures, from conservative load-carrying models, increased values of material properties and built-in properties of the supports.

The structural analysis and the verification of the required level of structural safety can be carried out at several levels of increasingly complex approximation. In addition to the choice of safety concept, the level of safety is an important issue for bridge assessment and should take into account what is already known about the structure and economical, societal and environmental risks associated with it (Paulsson et al., 2016; Bagge, 2017).

Table 11.3 Examples of load testing to failure in Sweden

Location, Type	Photo	Tests and Results	References
Stora Höga Reinforced concrete (RC) portal frame road bridge with a span 21.0 m (69 ft). Built in 1980.		The bridge was strengthened with externally bonded steel plates to avoid a bending failure. Loads were produced by using four hydraulic jacks anchored in the bedrock. Brittle shear failure at the supporting wall at a load of 4.6 MN (1034 kip). The theoretically assessed capacity with the Swedish code was 48% of the test value.	(Plos et al., 1990) (Täljsten, 1994) (Plos, 1995) (Bagge et al., 2018)
Stora Höga Prestressed concrete (PC) portal frame road bridge with a span 31.0 m (102 ft). Built in 1980.	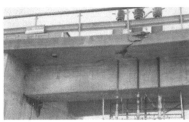	The bridge was cut longitudinally to reduce the load needed to fail the structure. Loads were produced by using four hydraulic jacks anchored in the bedrock. At 8.5 MN (1911 kip) the girder suddenly punched into the support wall. With new boundary conditions a shear-bending failure occurred at 6.3 MN (1416 kip). The theoretically assessed capacity with the Swedish code was 77% of the test value.	(Plos et al., 1990) (Plos, 1995) (Bagge et al., 2018)
Örnsköldsvik Continuous RC single-through railway bridge with spans of 11.9 m (39 ft) and 12.2 m (40 ft). Built in 1955.		The bridge was strengthened with near surface mounted carbon fiber reinforced polymer (CFRP) bars to avoid a bending failure. The loading was produced by using two hydraulic jacks anchored in the bedrock. Brittle bond failure of CFRP was followed by shear-bending-torsion failure at 11.7 MN (2630 kip). The theoretically assessed capacity with codes was 65% to 78% of the test value.	(Sustainable Bridges, 2007) (Puurula et al., 2012, 2014, 2015) (Bagge et al., 2018)

Location, Type	Photo	Tests and Results	References
Åby River Steel truss railway bridge with a span of 33 m (108 ft). Built in 1955.		The bridge was placed beside the original site at Åby river and loaded with two hydraulic jacks anchored in the bedrock. A failure initiated by fatigue was expected but instead buckling occurred in the two longitudinal top girders for 11 MN (2473 kip). The bridge was designed for about 35% of failure load.	(Mainline, 2014) (Häggström, 2016) (Häggström et al., 2017)
Kiruna Continuous PC girder road bridge with five spans of a total length of 121.5 m (399 ft). Built in 1959.		The bridge was monitored for eight years to check settlements due to mining. The bridge was thereafter loaded in the middle of the second span by hydraulic jacks anchored in the bedrock. Longitudinal non-prestressed reinforcement and vertical shear reinforcement yielded. In the final stage, stirrups ruptured and the loading plate punched through the slab. The maximum load was 13.4 MN (3012 kip) and the girder was designed for about 22% of the failure load.	(Bagge, 2017) (Bagge et al., 2018) (Huang et al., 2016)

11.4.3 Fatigue

Fatigue is an important factor when the load on a structure is increased. The rate of damage when the stresses are increased rise with a logarithmic factor that can be three to five times the stress increase. This means than an increase of stress range may cause a proportionally much larger reduction of the number of allowable load cycles. Methods to determine the remaining fatigue capacity would be very valuable both for steel and concrete bridges. The concrete fatigue capacity in shear is not as critical as many codes envision (Elfgren, 2015). Shear stresses are in the design phase often converted to tensile and compressive stresses and the tensile stresses are mostly carried with reinforcement or eliminated by pre-stressing. Furthermore, concrete in compression seldom gives any fatigue problems.

11.4.4 Strengthening

Strengthening with carbon fiber reinforced polymers (CFRP) has been applied successfully in different cases. For instance, the load-carrying capacity was substantially increased in the Örnsköldsvik Bridge (see Table 11.3). Here, nonlinear finite element models of the bridge were calibrated and used to simulate the structural behavior in a good way. It was important to accurately model tension stiffening and support conditions (Puurula et al., 2014, 2015). The concrete tensile strength and fracture energy were also identified as crucial parameters in numerical modeling. Often they are determined from empirical formulae from the concrete compression strength; however, more efforts should be taken to determine these properties directly from assessed existing structures. Guidelines for strengthening were developed in Sustainable Bridges (2007) and Mainline (2014).

11.4.5 *Full-scale failure tests*

Some lessons learned from load testing to failure of concrete bridges are presented in (Bagge et al., 2018). About 28% of full-scale tests on 30 bridges ended with a failure mode different to that predicted. In some cases, this was related to inaccuracies in the methods for determining the load-carrying capacity but, in the majority of cases, it was caused by a lack of insight into aspects shown to be critical, particularly associated with the shear and punching capacities and the boundary conditions. Consequently, there is a need of further studies in order to provide reliable codes and guidelines on how to accurately assess the capacity of bridges.

Many tested structures had a considerable "hidden" capacity which can be disregarded during ordinary assessment processes and which is accounted for neither in standards nor in design guidelines. One reason is the high safety factors that are used both for loads and materials in the construction phase and which may not be necessary in an assessment process where geometry, materials, and load may be better known (Paulsson et al., 2016). Probabilistic methods can be applied successfully to improve the study of reliability and safety of existing structures. More experience and acceptance of reduced reliability factors for different existing structures are needed.

Society may learn and save money from the experiences from "full-scale" failure tests. They can act as a complement to the experiences from unwanted and unexpected failures due to increased loads, scour, corrosion, and other forms of deterioration. It is therefore recommended that additional tests are to be carried out in order to further improve the understanding of existing bridges. The tests should as far as possible be based on realistic load cases, in order to optimize the outcome. Different bridge types can be tested to check their real capacity and give a background for establishing numerical models of them. As the tests are costly it is important that planning, preparations, and analysis are done in a careful way – preferably in international cooperation.

11.5 Conclusions

Load tests are a relatively easy way to get precise information about the behavior of a bridge and also to provide useful information about different bridge types and their typical behavior. Tests need to be designed carefully to achieve useful results and the results need be analyzed and published in order to get a full insight of its implications.

This chapter presents the experiences from a range of tested bridges in Sweden. Most of the bridges had more capacity than the original design calculations indicated, and in several cases, a diagnostic load test was a cost-effective way to avoid strengthening or renewal of the bridge. Tests to failure may show differences, for example, in load distribution, superstructures composite behavior or support conditions, which often result in extra "hidden" capacity to the bridge. In some cases, test results can reveal damages in the bridge superstructure which make the distribution even worse than calculated and the capacity of the bridge weaker.

Additional work is needed regarding recommendations for load testing, proof load levels, test setup and calibration of numerical models. Above all, more tests to failure of different bridge types are suggested to give a better base for reliable assessment of existing bridges in order to improve quality control, a cost efficient bridge management and a sustainable usage of the existing bridge stock.

Acknowledgments

The support from Trafikverket, Sweden, EU FP 6 (Sustainable Bridges, 2007), EU FP 7 (Mainline, 2014), and many companies, institutions, and colleagues is acknowledged with thanks. Discussions with Robert Ronnebrant, Ibrahin Coric, and Håkan Thun at Trafikverket are appreciated.

References

Bagge, N. (2017) *Structural Assessment Procedures for Existing Concrete Bridges: Experiences from failure tests of the Kiruna Bridge*. Doctoral Thesis, Luleå University of Technology. [Online] ISBN 978-91-7583-879-3, 310 pp. Available from: http://ltu.diva-portal.org/ [accessed 9th September 2018].

Bagge, N., Popescu, C. & Elfgren, L. (2018) Failure tests on concrete bridges: Have we learnt the lesson? *Structure and Infrastructure Engineering*, 14(3), 292–319, doi: 10.1080/15732479.2017.1350985.

BaTMan (2018) *Bridge and Tunnel Management in Sweden (BaTMan – Bro och Tunnel MANagement. In Swedish)*. [Online]. Webpage with data on Swedish bridges and tunnels and recommendations for their management, Version 8, Trafikverket. Available from: https://batman.trafikverket.se/externportal# [accessed 9th September 2018].

Bolle, G., Schacht, G. & Marx, S. (2010) Geschichtliche Entwicklung und aktuelle Praxis der Probebelastung. (Loading tests of existing structures–history and present practise. In German). *Bautechnik*, 87(11&12), 700–707 & 784–789.

Casas, J. R. & Gómez, J. D. (2013) Load rating of highway bridges by proof-loading. *KSCE Journal of Civil Engineering*, 17(3), 556–567.

Coric, I., Täljsten, B., Blanksvärd, T., Ohlsson, U. & Elfgren, L. (2018) Railway bridges on the Iron Ore line in Sweden: From Axle loads of 14 to 32,5 ton. *IABSE Conference 2018*, Engineering the Past, to Meet the Needs of the Future, June 25–27, 2018, Copenhagen, Denmark, 8 pp.

COST 1406 WG1 (2016) Performance Indicators for Roadway Bridges of Cost Action TU1406, ISBN: 978-3-900932-41-1, 40 pp. [Online]. Available from: www.tu1406.eu/wp-content/uploads/2016/10/COST_TU1406_WG1_TECH_REPORT.pdf [accessed 9th September 2018].

COST 1406 WG2 (2017) Performance Goals for Roadway Bridges of Cost Action TU1406, ISBN: 978-3-90093x-4x-x, 82 pp. [Online]. Available from: www.tu1406.eu/wp-content/uploads/2017/11/tu1406_wg2.pdf [accessed 9th September 2018].

Elfgren, L. (2015) *Fatigue Capacity of Concrete Structures: Assessment of Railway Bridges*. Research Report, Structural Engineering, Luleå University of Technology, 103 pp. [Online]. Available from: http://ltu.diva-portal.org/ [accessed 9th September 2018].

Elfgren, L., Täljsten, B., Blanksvärd, T., Sas, G., Nilimaa, J., Bagge, N., Tu, Y., Puurula, A., Häggström, J. & Paulsson, B. (2018) Load testing used for quality control of bridges. *COST TU 1406*, Wroclaw, 1–2 March, 6 pp. [Online]. Available from: www.tu1406.eu/ [accessed 9th September 2018].

Enochsson, O., Hejll, A., Nilsson, M., Thun, H., Olofsson, T. & Elfgren, L. (2002) *Bro över Luos-sajokk. Beräkning med säkerhetsindexmetod (In Swedish)*. Luleå University of Technology, Report 2002:06, 93 pp. [Online]. Available from: http://ltu.diva-portal.org/ [accessed 9th September 2018].

Grip, N., Sabourova, N., Tu, Y. & Elfgren, L. (2017) *Vibrationsanalys av byggnader* (Vibration analysis of buildings. In Swedish with a summary and appendices in English), SBUF 13010. [Online]. Available from: http://ltu.diva-portal.org/ [accessed 9th September 2018].

Häggström, J. (2016) *Evaluation of the Load Carrying Capacity of a Steel Truss Railway Bridge: Testing, Theory and Evaluation*. Lic. Thesis, Luleå University of Technology, 142 pp. [Online]. Available from: http://ltu.diva-portal.org/ [accessed 9th September 2018].

Häggström, J., Blanksvärd, T. & Täljsten, B. (2017) *Bridge over Åby River: Evaluation of Full Scale Testing*. Research Report, Div. of Structural Engineering, Luleå University of Technology, 180 pp. [Online]. Available from: http://ltu.diva-portal.org/ [accessed 9th September 2018].

He, G., Zou, Z., Enochsson, O., Bennitz, A., Elfgren, L., Kronborg, A., Töyrä, B. & Paulsson, B. (2008) Assessment of a railway concrete arch bridge by numerical modelling and measurements. *Bridge Maintenance, Safety . . .*, CRC Press/Balkema, ISBN 978-0-415-46844-2, 722, pp 3733–3742. [Online]. Available from: http://ltu.diva-portal.org/ [accessed 9th September 2018].

Hendriks, M.A.N., de Boer, A. & Belletti, B. (2017) Guidelines for Nonlinear Finite Element Analysis of Concrete Structures, Rijkswaterstaat Centre for Infrastructure, Report RTD:1016–1:2017, The Netherlands, 69 pp.

Huang, Z., Grip, N., Sabourova, N., Bagge, N., Tu, Y. & Elfgren, L. (2016) Modelling of damage and its use in assessment of a prestressed bridge. *19th Congress of IABSE in Stockholm*, pp. 2093–2108. [Online]. Available from: http://ltu.diva-portal.org/ [accessed 9th September 2018].

Jonsson, F. & Johnson, D. (2007) *Finite Element Model Updating of the New Svinesund Bridge: Manual Model Refinement with Non-Linear Optimization*. Master's Thesis:130, Concrete Structures, Chalmers University of Technology, Göteborg, 244 pp. [Online]. Available from: http://documents.vsect.chalmers.se/CPL/exjobb2007/ex2007-130.pdf [accessed 9th September 2018].

Karoumi, R. & Andersson, A. (2006) *Load Testing of the New Svinesund Bridge*. Presentation of results and theoretical verification of bridge behaviour. TRITA-BKN Report 96, Structural Engineering. Stockholm: Royal Institute of Engineering, 187 pp. [Online]. Available from: www.diva-portal.org/smash/get/diva2:431866/FULLTEXT01.pdf&sa=U&ei=xMlNU5CuHZCKtQaW6Y-DADw&ved=0CEAQFjAH&usg=AFQjCNHV85mgba2OQIrrBeDs-EZmo6aH1Q [accessed 9th September 2018].

Karoumi, R., Sundquist, H., Andersson, A., Enckell, M., Malm, R., Wiberg, J., Ülker, M., Carlsson, F., Plos, M., Enochsson, O. & Täljsten, B. (2007) *Modern Structural Health Monitoring for Assessment of Bridges*. (Modern mät- och övervakningsteknik för bedömning av befintliga broar. In Swedish). KTH, Chalmers, LTH and LTU. Royal Institute of Technology, Div. of Bridges, TRITA-BKN Report 111, 119 pp. [Online]. Available from: http://kth.diva-portal.org/smash/get/diva2:431868/FULLTEXT01.pdf [accessed 9th September 2018].

Lantsoght, E. O. L., van der Veen, C., de Boer, A. & Hordijk, D. A. (2017a) State-of-the-art on load testing of concrete bridges. *Engineering Structures*, 150, November, 231–241.

Lantsoght, E. O. L., van der Veen, C., Hordijk, D. A. & de Boer, A. (2017b) Development of recommendations for proof load testing of reinforced concrete slab bridges. *Engineering Structures*, 152, December, 202–210.

Lantsoght, E. O. L., van der Veen, C., de Boer, A. & Hordijk, D. A. (2017c) Collapse test and moment capacity of the Ruytenschildt reinforced concrete slab bridge. *Structure and Infrastructure Engineering*, 13(9), 1130–1145.

Mainline (2014) *Maintenance, Renewal and Improvement of Rail Transport Infrastructure to Reduce Economic and Environmental Impacts*. [Online]. A European FP7 Research Project during 2011–2014. Some 20 reports are available from: www.mainline-project.eu/ [accessed 9th September 2018].

Nilimaa, J. (2015) *Concrete Bridges: Improved Load Capacity.* Ph.D. Thesis, Luleå University of Technology, Luleå, Sweden, ISBN 978-91-7583-345-3, 176 pp. [Online]. Available from: http://ltu.diva-portal.org/ [accessed 9th September 2018].

Pacoste, C., Plos, M. & Johansson, M. (2012) *Recommendations for Finite Element Analysis for the Design of Reinforced Concrete Slabs.* KTH, ELU & Chalmers, Stockholm: Royal Institute of Technology, TRITA-BKN Rapport 114, 55 pp. [Online]. Available from http://publications.lib.chalmers.se/records/fulltext/176734/local_176734.pdf [accessed 9th September 2018].

Paulsson, B., Bell, B., Schewe, B., Jensen, J. S., Carolin, A. & Elfgren, L. (2016) Results and Experiences from European Research Projects on Railway Bridges. *19th IABSE Congress Stockholm,* pp. 2570–2578. ISBN 978-2-85748-144-4. [Online]. Available from: http://ltu.diva-portal.org/ [accessed 9th September 2018].

Plos, M. (1995) *Application of Fracture Mechanics to Concrete Bridges: Finite Element Analyses and Experiments.* Ph.D. Thesis, Chalmers University of Technology, Gothenburg.

Plos, M., Gylltoft, K. & Cederwall, K. (1990) Full scale shear tests on modern highway concrete bridges. *Nordic Concrete Research,* 9, 134–144.

Plos, M., Gylltoft, K., Jeppsson, J., Carlsson, F., Thelandersson, S., Enochsson, O. & Elfgren, L. (2004) *Evaluering av bärförmåga hos broar med hjälp av förfinade analysmetoder.* (Assessment of bridges with refined methods. In Swedish). A cooperation between LTH, LTU and Chalmers. Concrete Structures, Report 2004:3. Chalmers University of Technology, Göteborg, 56 pp.

Puurula, A. (2004) *Assessment of Prestressed Concrete Bridges Loaded in Combined Shear, Torsion and Bending.* Licentiate thesis 2004:43, Division of Structural Engineering, Luleå University of Technology, Luleå, 261 pp. [Online]. Available from: www.diva-portal.org/smash/get/diva2:990922/FULLTEXT01.pdf [accessed 9th September 2018].

Puurula, A. (2012) *Load-Carrying Capacity of a Strengthened Reinforced Concrete Bridge: Nonlinear finite element modeling of a test to failure. Assessment of train load capacity of a two span railway trough bridge in Örnsköldsvik strengthened with bars of Carbon Fibre Reinforced Polymers (CFRP),* Doctoral Thesis, Luleå University of Technology, Luleå. ISBN 978-91-7439-433-7, 332 p. [Online]. Available from: http://ltu.diva-portal.org/ [accessed 9th September 2018].

Puurula, A., Enochsson, O., Sas, G., Blanksvärd, T., Ohlsson, U., Bernspång, L. & Elfgren, L. (2014) Loading to failure and 3D nonlinear FE modelling of a strengthened RC bridge. *Structure & Infrastructure Engineering,* 10(12), 1606–1619.

Puurula, A., Enochsson, O., Sas, G., Blanksvärd, T., Ohlsson, U., Bernspång, L., Täljsten, B., Carolin, A., Paulsson, B. & Elfgren, L. (2015) Assessment of the strengthening of an RC railway bridge with CFRP utilizing a full-scale failure test and finite-element analysis. *Journal of Structural Enginnering, American Society of Civil Engineering (ASCE),* 141(1), D4014008, 11 p. Available from: https://ascelibrary.org/doi/10.1061/%28ASCE%29ST.1943-541X.0001116 [open access].

Rail Traffic (2017) *Bantrafik 2017.* Trafikverket, Sweden. [Online]. Available from: www.trafa.se/globalassets/statistik/bantrafik/bantrafik/2017/bantrafik-2017_juni.pdf [accessed 9th September 2018].

Raunio, H. (2016) Bearing Capacity of Existing Bridges: Calculations and Load Tests. *19th IABSE Congress Stockholm 21–23 September 2016,* Zürich, pp. 206–213. ISBN 978-2-85748-144-4.

Stenlund, A. (2008) *Load Carrying Capacity of Bridges: Three Case Studies of Bridges in Northern Sweden where Probabilistic Methods Have been Used.* Lic. Thesis:118, Luleå University of Technology, 306 pp. [Online]. Available from: http://ltu.diva-portal.org/ [accessed 9th September 2018].

Sundquist, H. (1998) *Loads and Load Effects on Bridges: A Literature Study* (Laster och lasteffekter av trafik på broar. Litteraturstudie, In Swedish). Royal Institute of Technology, KTH, Brobyggnad rapport 98:16, Stockholm, 40 pp. [Online]. Available from: www.trafikverket.se/contentassets/8425704897db43059cb747ece4421b95/forstudie_till_ramprojektet_utvardering_av_tillaten_trafiklast_bilaga_laster_och_lasteffekter_av_trafik_pa_broar.pdf [accessed 9th September 2018].

Sustainable Bridges (2007) *Assessment for Future Traffic Demands and Longer Lives*. A European FP 6 Integrated Research Project during 2003–2007. [Online]. Four guidelines and 35 background documents are available from: www.sustainablebridges.net [accessed 9th September 2018].

Täljsten, B. (1994) *Plate Bonding Strengthening of Existing Concrete Structures with Epoxy Bonded Plates of Steel or Fibre Reinforced Plastics*. Ph.D. Thesis, Lulea University of Technology, Lulea. 205 pp. [Online]. Available from: http://ltu.diva-portal.org/ [accessed 9th September 2018].

Thun, H., Ohlsson, U. & Elfgren, L. (2000) *Fatigue Capacity of Small Railway Concrete Bridges: Prevision of the Results of Swedish Full-Scale Tests*. ERRI D216, Luleå University of Technology, 99 pp. [Online]. Available from: http://ltu.diva-portal.org/ [accessed 9th September 2018].

TRV Capacity Advice (2017) Recommendations for analysis of the load-carrying capacity of bridges (Råd. Bärighetsberäkning av broar. In Swedish). Swedish Transport Organisation (Trafikverket) TDOK 2013:0273, Version 4.0, 2017–02–20, 34 pp. [Online]. Available from: http://trvdokument.trafikverket.se/ [accessed 9th September 2018].

TRV Capacity Rules (2017) Rules for analysis of the load-carrying capacity of bridges (Krav. Bärighetsberäkning av broar. In Swedish). Swedish Transport Organisation (Trafikverket) TDOK 2013:0267, Version 4.0, 2017–02–20, 142 pp. [Online]. Available from: http://trvdokument.trafikverket.se/ [accessed 9th September 2018].

TRV Design Advice (2016) Recommendations for design and building of bridges. (Råd Brobyggande. In Swedish). Swedish Transport Organisation (Trafikverket) TDOK 2016:0204, Version 1.0, 2016–10–03, 353 pp. [Online]. Available from: http://trvdokument.trafikverket.se/ [accessed 9th September 2018].

TRV Design Rules (2016) Rules for design and building of bridges. (Krav Brobyggande. In Swedish). Swedish Transport Organisation (Trafikverket) TDOK 2016:0203, Version 1.0, 2016–10–03, 197 pp. [Online]. Available from: http://trvdokument.trafikverket.se/ [accessed 9th September 2018].

TRV Inspection Advice (2018) Recommendations for inspection of structures. (Råd inspektion av byggnadsverk. In Swedish). Swedish Transport Organisation (Trafikverket) TDOK 2018:0180, 2018–06–227. [Online]. Available from: http://trvdokument.trafikverket.se/ [accessed 9th September 2018].

TRV Inspection Rules (2018) Rules for inspection of structures. (Krav inspektion av byggnadsverk. In Swedish). Swedish Transport Organisation (Trafikverket) TDOK 2018:0179, 2018–06–227. [Online]. Available from: http://trvdokument.trafikverket.se/ [accessed 9th September 2018].

TRV Traffic (2016) Trafikarbetets förändring 2015–2016 (The change of Traffic Work 2015–2016. In Swedish). Publication 2017:178 by Thomas Vestman, Stockholm: Trafikverket, 52 pp. [Online]. Available from: https://trafikverket.ineko.se/Files/sv-SE/31751/Ineko.Product.Related-Files/2017_178_trafikarbetets_forandring_2015_2016.pdf [accessed 9th September 2018].

UIC 778–3 (2014) *Recommendations for the Inspection, Assessment and Maintenance of Masonry Arch Bridges*. UIC Code 778–3 R, 3rd Ed., International Union of Railways, Paris, to 167 pp.

Von Olnhausen, W. (1991) *Swedish Road Bridges: Past–Present–Future*. (Sveriges vägbroar. Återblick-Nuläge -Utblick. In Swedish). Tekniska museéts årsbok, Daedalus, pp. 33–53.

Wang, C., Wang, Z., Zhang, J., Tu, Y., Grip, N., Ohlsson, U. & Elfgren, L. (2016) FEM-based research on the dynamic response of a concrete railway arch bridge. *19th IABSE Congress Stockholm*, Zürich, pp. 2472–2479. [Online]. Available from: http://ltu.diva-portal.org/ [accessed 9th September 2018].

Chapter 12

Load Testing as Part of Bridge Management in the Netherlands

Ane de Boer

Abstract

In the Netherlands, field tests on existing bridges have been carried out over the past 15 years. The tests consists of SLS load level tests (before the onset of nonlinear behavior, also known as proof load tests) on existing bridges, ULS tests (collapse tests) on existing bridges, and laboratory tests on beams sawn from the existing bridges. The goal of these experiments is to have a better assessment of the existing bridges. The assessment is combined with material sampling (concrete cores, reinforcement steel samples) for a better quantification of the material parameters, and with nonlinear finite element models. In the future, a framework will be developed in which first the concrete compressive strength, determined from a large number of similar bridges from which core samples are taken, is used as input for an assessment with a nonlinear finite element model according to the Dutch NLFEA guidelines. Then, it can be determined if sampling of the actual viaduct is needed, following recommendations on how to handle concrete cores and how to scan reinforcement, and if a load test should be used to evaluate the bridge under consideration. For this purpose, guidelines for proof load testing of concrete bridges are under development, but further research on the topic of load testing for shear is necessary for the development of such guidelines.

12.1   Introduction

Load testing started in the Netherlands when there were questions about the capacity and structural behavior of Alkali-Silica Reaction (ASR) affected concrete structures (Borsje et al., 2002; Nijland and Siemes, 2002). In 1996, inspections showed that concrete structures in the highway A59 in the province of North Brabant were damaged by ASR. This damage was at that moment new in the Netherlands, so several research projects were started to get a better understanding of the behavior of these structures. This research focused on the following topics:

1   Recommendations for the structural inspection of ASR-affected concrete structures
2   Recommendations regarding the assessment of ASR-affected concrete structures
3   Development of a monitoring system to control the progress of ASR
4   Development of a list of relevant parameters which should be controlled by monitoring.

DOI: https://doi.org/10.1201/9780429265969

A number of viaducts have been monitored since 2000, which resulted in an evaluation and monitoring report in 2013 (Bouwdienst Rijkswaterstaat, 2013). Load testing in situ was at that moment not possible; the equipment and the knowledge around load testing was too poor, so the decision was made to saw some beams out of an affected structure and test them to the Ultimate Limit State (ULS) load level (collapse test) in the laboratory of Delft University of Technology (den Uijl and Kaptijn, 2002).

The first load test has been done in 2008 in the North of the Netherlands in Aduard by the province of Groningen. The tested structure was a masonry bridge where the Serviceability Limit State (SLS) load level was tested (proof load test). The so-called BELFA vehicle from Germany (Bretschneider et al., 2012; Hochschule Bremen – City University of Applied Sciences, 2018a) was used as the equipment vehicle, and this vehicle was later on also used in the third and fourth load test. The total load that can be applied by the BELFA was restricted to 1100 kN (247 kip), which is not sufficient for a lot of structures in the Netherlands in the highway network. Also the maximum span that can be tested with the BELFA is restricted. Figure 12.1 shows a load test with the BELFA vehicle.

The second load test, see Figure 12.2, shows a framework of steel as equipment tool for the load application. This setup took a lot of time to install and re-install. After the fourth load test in the Netherlands, only the Mammoet equipment, a combination of steel girders coupled together as an alternative steel single-span bridge (Fig. 12.3), is used for the load application. Later on, a sliding system was added to this setup to increase the number of tests over the width of the bridge deck, see Figure 12.4.

Over the past 15 years, load tests were done at SLS (proof load test) and ULS load level (collapse test) in the Netherlands. The results of these tests can contribute to improvements of material, load, and geometrical models of finite element (FE) analysis of these structures. The available numerical and analytical models are not only used to simulate the behavior of existing concrete structures in operation, but can also be used for simulations of experiments. So the parallel with FE analysis, especially the nonlinear behavior of the material (concrete) and the bonding of concrete and its (prestressed) reinforcement, is an essential part of the research on load testing and the evaluation of structures through load tests.

Figure 12.1 BELFA vehicle on viaduct Medemblik

Source: Kapphahn (2009); Hochschule Bremen – City University of Applied Sciences (2018c).

Figure 12.2 Load testing with fixed steel frame on viaduct Heidijk
Source: Dieteren and den Uijl (2009).

Figure 12.3 Setup with steel spreader beam, showing hydraulic jacks for application of four wheel prints

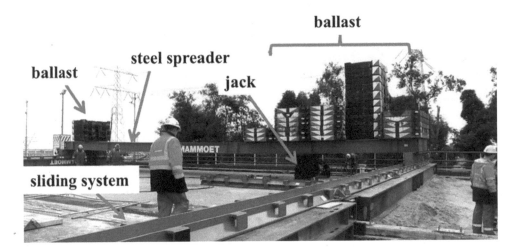

Figure 12.4 Setup with steel spreader beam and sliding system
Source: Reproduced from Ensink et al. (2018).

The current situation is that load testing is not always possible. Sometimes, there are other issues like costs and hindrance of traffic flows that influence the decision to carry out a load test in situ.

The goal of all these research actions together is the development of a recommendation of load testing in a controlled way by setting up a recommendation of load testing to the SLS load levels (proof load testing) for bending moment and shear force behavior. At this moment, load testing for bending moment is possible and leads to reliable results (the scatter is within 10%), but the shear force recommendations need further improvements. Nevertheless, the aim is to have a recommendation for both failure modes and for the equipment to apply the load during a load test as soon as possible. Otherwise, the road authority should close the structure immediately after getting the results of an inspection of a structure with a very negative or critical result.

12.2 Overview of load tests on existing structures

Table 12.1 shows an overview with some details of the load tests on existing structures in the Netherlands that have been carried out over the past 15 years. Figure 12.5 shows the action of sawing beams from a slab bridge for testing in the laboratory.

Table 12.1 shows a mix of SLS (proof load tests) and ULS load tests (collapse tests), but also a lot of different authorities. Most authorities subcontracted parts of the work and analysis to TNO Research and Delft University of Technology. Besides that, there were also other consultancy offices involved, which made their efforts in preparation work, safety aspects on the site, and inspections.

The reason to carry out a load test was often related to the problem at that moment, which is mentioned in the column of remarks in Table 12.1. The Quick and Dirty load test was carried out when there was less preparation time for an assessment of the structure.

Table 12.1 Load tests. ULS load tests are tests to collapse of the structure, SLS load tests are proof load tests

Year	Location	Authority	Load Level	Remarks	Reference
2005	A59	Rijkswaterstaat	ULS	Shear force, Sawn beams	(Den Uijl, 2004)
2006–12	Aduard	Province Groningen	SLS	Masonry	(Hochschule Bremen – City University of Applied Sciences, 2018b)
2007	Heidijk	Rijkswaterstaat	SLS	ASR-affected	(Dieteren and den Uijl, 2009)
2008	Eindhoven	Rijkswaterstaat	ULS	Sawn beams	(Yang et al., 2010)
2008	Maarssenbroek	Rijkswaterstaat	SLS	Quick and Dirty	
2009	Medemblik	Province North Holland		Chloride affected	(Kapphahn, 2009)
2013–11	Vlijmen-Oost	Rijkswaterstaat	SLS	Slab, ASR-affected	(Fennis et al., 2014)
2014–05	Alkmaar	Province North Holland	SLS	Slab	(Fennis and Hordijk, 2014)
2014–11	Ruytenschildt	Province Friesland & Rijkswaterstaat	ULS	Slab & sawn beams	(Lantsoght et al., 2016; Lantsoght et al., 2017d)
2015–06	Zijlweg	Rijkswaterstaat	SLS	Slab, ASR-affected	(Lantsoght et al., 2017b)
2015–11	De Beek	Rijkswaterstaat	SLS	Slab with traffic restriction and further used as reference slab for proof load testing	(Lantsoght et al., 2017c)
2016–10	Vechtbrug	Rijkswaterstaat	ULS	System of prestressed in-situ T-beams and transversely prestressed deck	(Ensink et al., 2018)

The Maarssenbroek viaduct was a load test with a lot of sandbags to simulate a uniformly distributed load, and steel plates as counterweight of a crane simulating a tandem axle load, see Figure 12.6. Only visual inspections afterwards were done to look at possible cracks at the bottom side of the bridge deck.

Beside that there is and was a strong incentive to saw beams out of a slab bridge deck, so that those beams can be tested later on in the laboratory. Such tests in a controlled environment are always helpful afterwards when the evaluation of the load test is done. It is very useful to have some material to test afterwards! The parallel action to develop or improve the Guidelines for the Nonlinear Finite Element Analysis (NLFEA) of Concrete Structures (De Boer et al., 2014; Rijkswaterstaat, 2017a) with the results of the in-situ bridge tests will ask almost always some validation, which can be done by using afterwards those beams for the comparison of the laboratory experiments and the NLFEA models of these experiments.

After the initial concerns regarding the safety of ASR-affected concrete bridges in the late 1990s, the topic was studied again in the 2010s when PhD students at Delft University of Technology continued the first experiments related to ASR (Esposito and Hendriks, 2013; Esposito, 2016). Another goal defined in the 1990s was the continuous monitoring

Figure 12.5 Sawing of beams from slab bridge in Eindhoven for additional laboratory testing

of ASR-affected viaducts. However, monitoring temperature and humidity over 10 years couldn't give a direct sufficient answer for the ASR phenomena.

Even though a large number of in-situ tests have been carried out, the Dutch recommendation isn't updated yet and is still in progress. More attention is paid to the development of limitations (such as stop criteria) of the load test. The bending moment behavior can be satisfactorily monitored during a load test, and the correspondence between the behavior measured in-situ and the representative NLFEM is good. The shear force behavior is still under progress, but the scatter is rather narrowing. The expectations are still high to get a good result at the end of this research program.

12.3 Inspections and re-examination

In the Netherlands all infrastructures in the highway network are inspected every six years. If there is a need to inspect a structure more accurately, then a so-called focused technical inspection is set up. The inspections are carried out according to CUR 117:2015 (CUR, 2015), and additional handbooks for the inspection of concrete (Waltje and Gorkum, 2017), steel (Lodema, 2015), masonry (CUR, 2018), bearings (Booij, 2017a), and joints (Booij, 2017b) are available, with a handbook for timber forthcoming. When there are doubts about the load-carrying capacity, an analytical re-examination of the structure is done. For that reason a decision has to be made to drill concrete cores from the bridge deck (Rijkswaterstaat, 2014) and/or take some samples of the reinforcement out of the bridge deck to get an actual compression strength of the concrete and a stress-strain relation

Figure 12.6 "Quick and Dirty" load test on viaduct Maarssenbroek

of the reinforcement. By drilling cores at a large number of structures built before 1976, default values of the compressive strength are available. With this material input, a realistic re-examination can be set up. There are different levels of assessment, with increasing requirements for computational time and effort from a linear static 1D model to a nonlinear 3D model with solid elements. A lot of refinements on the models as compared to a new structure can be done. It is not only refinements on the geometry of the structure but also on the material model, the load model, and the constraints.

Nevertheless, a load restriction on the bridge can be the result of an inspection or a re-examination. Strengthening of a structure is the simplest way to keep a structure in operation on the network. The SLS load test (proof load test) can validate this decision to keep the structure in operation.

The development of the traffic loads on a structure over time needs to be discussed as well. Since 1963, the Netherlands have already a vehicle load of 600 kN (135 kip), but the vehicle configuration was different from the actual Eurocode 1991 (CEN, 2003) vehicle configuration. As a result, there are always some doubts about the safety of the structure. The introduction of code provisions for the shear capacity in the Netherlands in 1974 results in doubts regarding the structures built before 1976. When the Unity Check (ratio

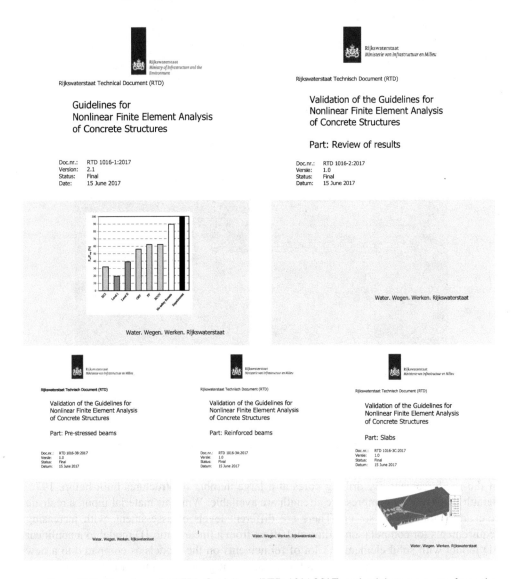

Figure 12.7 Overview of NLFEA Guidelines (RTD 1016:2017) and validation reports from the Netherlands (Rijkswaterstaat, 2017a, 2017d, 2017c, 2017b, 2017e)

of load effect to capacity) has a value far below 1.0, there is no need to reevaluate the structure. Otherwise a deeper re-examination is needed (Lantsoght et al., 2017a).

The introduction of Safety Formats to nonlinear analysis of structures in the Eurocode (Schlune, 2011; Schlune et al., 2011; Blomfors et al., 2016) as well as in the Model Code 2010 (fib, 2012) gives the engineering offices tools for determining the extra load-carrying capacity of the structure. Setting up a validated NLFEA Guideline (Rijkswaterstaat, 2017a) will help the engineer and the checking engineer to fulfill a nonlinear analysis in a robust and trustful way. Collapse tests on real existing structures give measurement data which helps to get extra bearing capacity (Lantsoght et al., 2017d). Results of only laboratory samples would be too narrow. However, there is time needed to evaluate these load tests measurements in a structured way. Some results are already really promising, but still improvements are needed (Lantsoght et al., 2018). Nevertheless, nonlinear analysis of the field tests in the way proposed in the NLFEA Guideline (Rijkswaterstaat, 2017a) shows stable and uniform results. Figure 12.7 shows the NLFEA guidelines and validation reports.

When these models are stable enough, other damage processes can be added, like ASR and chloride ingress into the concrete surface on a structural level. Fatigue of the concrete itself is checked in the re-examination, as well as fatigue of the (prestressed) reinforcement bars. Until now there are not so many problems with the fatigue failure in concrete existing structures in the Netherlands. Load tests in the laboratory on beams and beam systems have shown that there is sufficient capacity of fatigue in transversely prestressed reinforced slab panels (Lantsoght et al., in press). The lower limit is of course a non-prestressed panel.

Another element that needs to be considered is the type of reinforcement. Most experiments nowadays are carried out on specimens with ribbed reinforcement bars, whereas many existing bridges in the Netherlands have plain bars. The change from plain bars to ribbed bars however shows some discrepancies on beams with a larger height (Yang et al., 2016). This research is in progress and will soon lead to a quantification of the size effect for elements with plain bars failing in shear, which is closely related to the sectional capacity at intermediate support lines of slab bridges that have a larger height over the support.

12.4   Conclusions and outlook

The mix of load tests in situ and in the laboratory, together with the development of numerical tools for the re-examination of existing concrete structures shows a lot of extra load-carrying capacity of existing concrete structures.

The mix of tests in the laboratory and the in-situ proof load tests can be used for the improvement of the existing recommendations of the existing concrete structures.

The controlled execution of the proof load test until the SLS load state of the structure during proof load testing is already documented in an acceptable way. However, for shear-critical structures, more research is needed and more proof load tests should be done. The cooperation between Delft University of Technology, Consultancy Companies, TNO Research, and the Ministry of Infrastructure is needed to improve the recommendations for proof load testing of shear-critical structures.

The experience from the pilot proof load tests shows that replacing the fixed steel frames with the steel spreader sliding system reduces the overall proof load test time from 12 working days to 5 working days on site.

Meanwhile, the number of drilled cores from existing concrete structures in the highway network is large enough to predict the compressive strength of concrete structures built before 1976 of which no samples are taken. Therefore, default values can be taken for the compression strength of concrete for the re-examination. The default values are divided by structure types. Today, reinforced and prestressed reinforced concrete structures have their own default values. In the future, there will be more dedicated sets related to construction type with their own default values for the compressive strength of concrete. If the re-examination shows that more strength is needed, samples can be taken from the structure to find the actual value of the concrete compressive strength.

Re-examination of concrete structures is based on the NLFEA Guideline, which is validated by laboratory experiments. This number of experiments will be expanded in the future to tackle all kinds of failure mechanisms. Also, closely related experiments can be categorized in a group of experiments to get a better understanding of the variation of results.

The results of the proof load in-situ tests underline the results of the laboratory experiments, and as such they improve the knowledge of the behavior of concrete structures.

References

Blomfors, M., Engen, M. & Plos, M. (2016) Evaluation of safety formats for non-linear finite element analyses of statically indeterminate concrete structures subjected to different load paths. *Structural Concrete*, 17, 44–51.

Booij, N. (2017a) *Handbook Inspection Bearings: Conform CUR-Recommendation 117*. SBRCUR, Gouda, the Netherlands.

Booij, N. (2017b) *Handbook Inspection Joints: Conform CUR-Recommendation 117*. SBRCUR, Gouda, the Netherlands.

Borsje, H., Peelen, W. H. A., Postema, F. J. & Bakker, J. D. (2002) Monitoring Alkali-Silica Reaction in structures. *Heron*, 47, 96–109.

Bouwdienst Rijkswaterstaat (2013) *Monitoren kunstwerken A59*. Analyserapport 4e kwartaal 2012; Periode van 1 april 2013 (referentiedatum) tot 31 december 2012.

Bretschneider, N., Fiedler, L., Kapphahn, G. & Slowik, V. (2012). Technical possibilities for load tests of concrete and masonry bridges. *Bautechnik*, 89, 102–110 (in German).

CEN (2003) *Eurocode 1: Actions on Structures – Part 2: Traffic Loads on Bridges, NEN-EN 1991–2:2003*. Comité Européen de Normalisation, Brussels, Belgium.

CUR (2015) *CUR-Recommendation 117:2015 Inspection and Advice for Bridges*. Civieltechnisch Centrum Uitvoering Research en Regelgeving, Gouda, the Netherlands.

CUR (2018) *Handbook Inspection Masonry: Conform CUR-Recommendation 117*. SBRCUR, Gouda, the Netherlands.

De Boer, A., Hendriks, M. A. N., Den Uijl, J. A., Belletti, B. & Damoni, C. (2014) *Nonlinear FEA Guideline for Modelling of Concrete Infrastructure Objects*. EuroC. St. Anton am Arlberg, Austria.

Den Uijl, J. A. (2004) *Dwarskrachtdraagvermogen van bestaand plaatviaduct. Stevin Report 25.5 04–07*. Delft University of Technology, Delft, the Netherlands.

Den Uijl, J. A. & Kaptijn, N. (2002) Structural consequences of ASR: An example on shear capacity. *Heron*, 47, 1–13.

Dieteren, G. G. A. & Den Uijl, J. A. (2009) *Evaluation Proof Load Test Heidijk* (in Dutch). TNO Bouw en Ondergrond/TU Delft.

Ensink, S. W. H., Van Der Veen, C., Hordijk, D. A., Lantsoght, E. O. L., Van Der Ham, H. & De Boer, A. (2018) Full-size field test of prestressed concrete T-beam bridge. *European Bridge Conference*. Edinburgh, Scotland.

Esposito, R. (2016) *The Deteriorating Impact of Alkali-Silica Reaction on Concrete: Expansion and Mechanical Properties*. Ph.D. Thesis, Delft University of Technology, Delft, The Netherlands.

Esposito, R. & Hendriks, M. A. N. (2013) *ASR-Affected Viaduct Vlijmen-Oost: Overview of the Mechanical Tests Performed on Drilled Cores*. Delft University of Technology, Delft, The Netherlands.

Fennis, S., Hemert, P. V., Hordijk, D. & Boer, A. D. (2014) Proof loading viaduct Vlijmen Oost. Cement, 5, 40–45 (in Dutch).

Fennis, S. A. A. M. & Hordijk, D. A. (2014) *Proof Loading Halvemaans Bridge Alkmaar* (in Dutch). Stevin Report 25.5–14–05. Delft University of Technology, Delft, The Netherlands.

FIB (2012) *Model Code 2010: Final Draft*. International Federation for Structural Concrete, Lausanne.

Hochschule Bremen – City University of Applied Sciences (2018a) *Belastungsfahrzeug BELFA*. [Online]. Available from: www.hs-bremen.de/internet/de/forschung/transfer/einrichtungen/belfa/.

Hochschule Bremen – City University of Applied Sciences (2018b) *Gewölbebrücke Aduard (Groningen, Niederlande)*. [Online]. Available from: www.hs-bremen.de/internet/de/forschung/transfer/einrichtungen/belfa/referenzen/aduard/.

Hochschule Bremen – City University of Applied Sciences (2018c) *Straßenbrücke Medemblik (Niederlande)*. [Online]. Available from: www.hs-bremen.de/internet/de/forschung/transfer/einrichtungen/belfa/referenzen/medemblik/.

Kapphahn, G. (2009) *Experimentelle Tragsicherheitsbewertung des Überbaues der Brücke Nr. 12 (14H04) im Zuge der Straße N240 bei Medemblik*. ifem, Markkleeberg, Germany.

Lantsoght, E. O. L., Yang, Y., Van Der Veen, C., De Boer, A. & Hordijk, D. (2016) Ruytenschildt Bridge: Field and laboratory testing. *Engineering Structures*, 128, 111–123.

Lantsoght, E. O. L., De Boer, A. & Van Der Veen, C. (2017a) Levels of approximation for the shear assessment of reinforced concrete slab bridges. *Structural Concrete*, 18, 143–152.

Lantsoght, E. O. L., Koekkoek, R. T., Hordijk, D. A. & De Boer, A. (2017b) Towards standardization of proof load testing: Pilot test on viaduct Zijlweg. *Structure and Infrastructure Engineering*, 16.

Lantsoght, E. O. L., Koekkoek, R. T., Veen, C. V. D., Hordijk, D. A. & Boer, A. D. (2017c) Pilot proof-load test on Viaduct De Beek: Case study. *Journal of Bridge Engineering*, 22, 05017014.

Lantsoght, E. O. L., Van Der Veen, C., De Boer, A. & Hordijk, D. A. (2017d) Collapse test and moment capacity of the Ruytenschildt reinforced concrete slab bridge. *Structure and Infrastructure Engineering*, 13, 1130–1145.

Lantsoght, E. O. L., De Boer, A., Van Der Veen, C. & Hordijk, D. A. (2018) *Modelling of the Proof Load Test on Viaduct De Beek*. Euro-C, Austria.

Lantsoght, E. O. L., Van Der Veen, C., Koekkoek, R. T. & Sliedrecht, H. (in press) Fatigue testing of transversely prestressed concrete decks. *ACI Structural Journal*.

Lodema, M. (2015) *Handbook Inspection Steel: Conform CUR-Recommendation 117*. SBRCUR, Gouda, the Netherlands.

Nijland, T. G. & Siemes, A. J. M. (2002) Alkali-Silica reaction in the Netherlands: Experiences and current research. *Heron*, 47, 81–84.

Rijkswaterstaat (2014) *Guidelines for Coring, Transporting, and Testing of Concrete Cores and the Determination of the Compressive and Splitting Tensile Strength* (in Dutch), RTD 1021:2014. Rijkswaterstaat, Utrecht, The Netherlands.

Rijkswaterstaat (2017a) *Guidelines for Nonlinear Finite Element Analysis of Concrete Structures*, RTD 1016-1:2017. Rijkswaterstaat, Utrecht, The Netherlands.

Rijkswaterstaat (2017b) *Validation of the Guidelines for Nonlinear Finite Element Analysis of Concrete Structures – Part 2: Review of Results*, RTD 1016-2:2017. Rijkswaterstaat, Utrecht, The Netherlands.

Rijkswaterstaat (2017c) *Validation of the Guidelines for Nonlinear Finite Element Analysis of Concrete Structures – Part 3A: Reinforced Beams*, RTD 1016-3A:2017. Rijkswaterstaat, Utrecht, The Netherlands.

Rijkswaterstaat (2017d) *Validation of the Guidelines for Nonlinear Finite Element Analysis of Concrete Structures – Part 3B: Pre-stressed Beams*, RTD 1016-3B:2017. Rijkswaterstaat, Utrecht, The Netherlands.

Rijkswaterstaat (2017e) *Validation of the Guidelines for Nonlinear Finite Element Analysis of Concrete Structures – Part 3C: Slabs*, RTD 1016-3C:2017. Rijkswaterstaat, Utrecht, The Netherlands.

Schlune, H. (2011) *Safety Evaluation of Concrete Structures with Nonlinear Analysis*. Ph.D. Thesis, Chalmers University, Gothenburg, Sweden.

Schlune, H., Plos, M. & Gylltoft, K. (2011) Safety formats for nonlinear analysis tested on concrete beams subjected to shear forces and bending moments. *Engineering Structures*, 33, 2350–2356.

Waltje, E. & Gorkum, J. V. (2017) *Handbook Inspection Concrete: Conform CUR-Recommendation 117*. SBRCUR, Gouda, the Netherlands.

Yang, Y., Den Uijl, J. A., Dieteren, G. & De Boer, A. (2010) Shear capacity of 50 years old reinforced concrete bridge deck without shear reinforcement. *3rd fib International Congress*. fib, Washington, DC, USA.

Yang, Y., Van Der Veen, C., Hordijk, D. & De Boer, A. (2016) The shear capacity of reinforced concrete members with plain bars. In: Forde, M. (ed.) *Structural Faults and Repair 2016*, Edinburgh, UK.

Part V

Conclusions and Outlook

Chapter 13

Conclusions and Outlook

Eva O. L. Lantsoght

Abstract

This chapter brings all information from the previous chapters together and restates the main topics that have been covered in this book. The topics to which more research energy has been devoted and the remaining open questions are discussed. Based on the current knowledge and state of the art, as well as the practical experiences presented by the authors in this book, practical recommendations are presented.

13.1 Current body of knowledge on load testing

The current body of knowledge on load testing is reflected by the available codes and guidelines, and the experience gathered with these codes and guidelines reported in project reports and articles. The current codes and guidelines reflect past practice, while the available project reports and articles show new insights that have been successfully applied.

In this book, the current body of knowledge and best practices are gathered in the first five parts, Parts I–III in Volume 12 and Parts I and II in this volume. The first part shows the historical development of load testing, which arose from the need to assure the traveling public that a new bridge is safe for use. The first part also summarizes the main features of the available codes and guidelines from Germany, the United States, the United Kingdom, Ireland, Poland, Hungary, Spain, the Czech Republic and Slovakia, Italy, Switzerland, and France. The second part gives best practices for load testing, regardless of the type of load test and the structure type and construction material. These best practices are subdivided in the general considerations that are important prior to each load testing project, and the preparation, execution, and post-processing of a load test. The third and fourth part (Part I of this volume) then address diagnostic load tests and proof load tests respectively, two types of load tests that serve different purposes. For both parts, a number of case studies are included that can give the reader guidance for the preparation of a load test. The fifth part (Part II of this volume) addresses the current body of knowledge regarding load testing of buildings, based on the experience developed from using the German guideline for load testing over the last two decades.

Currently, structures are designed by designing the structural members, and structures are assessed by evaluating the strength (and sometimes stability) of the individual structural member. While load testing aims at confirming the adequate design of a new structure by measuring responses of the critical structural member, or at assessing the critical structural member through measuring its response, load testing gives us information about the global

DOI: https://doi.org/10.1201/9780429265969

behavior of a structure. By measuring the structural responses of several members, information about the load distribution can be gathered. By measuring the deflections, the overall stiffness of the structure can be evaluated. By measuring deformation profiles in the longitudinal direction, the effect of (unintended) continuity (for example, caused by frozen bearings or caused by continuous decks over simply supported girders) can be evaluated.

At the moment, a load test is typically carried out as an isolated project to evaluate a given structure. After a load test, a report of the test may be developed, but these reports are typically not publicly available. As such, it becomes difficult to learn from the experience of past projects and improve the practice of load tests. With the increased attention paid to improving assessment practices, including load testing, the output from conference sessions and volumes with articles that have been gathered has become available. This book aimed at bringing information, practical recommendations, open questions, and different national/local practices together and becoming a reference work for engineers planning to carry out load tests. This book also shows, in particular in Part II of Volume 13, that unsafe practices for load testing are still used, and warns the reader against the use of unsafe load application and measurement methods.

13.2   Current research and open research questions

This book identifies two main research topics: applying new measurement techniques in load tests (in Part III of this volume), and using concepts of structural reliability and life-cycle assessment and cost optimization together with proof load tests (in Part IV of this volume).

At the moment, instrumenting a structure that will be load tested, can be a time-consuming activity. Contact sensors need to be applied to measure the structural responses, and each sensors has to be calibrated, applied, wired, and its output interpreted. At the moment, the only non-contact equipment that is typically used during a load test is a total station, which measures the deflections. To speed up load testing, there is a need to develop sensor plans that rely on non-contact measurements, and to combine long-term monitoring measurements with field testing based on embedded sensors. This book has shown the application of non-contact measurements such as methods based on photography and video as well as radar techniques. An interesting measurement technique, especially for load tests that involve large load levels when microcracking needs to be recorded, is the use of acoustic emission measurements. Such measurements could be combined and/or expanded in the future with embedded smart aggregates. Finally, optic fiber sensors show to be a promising method to capture strains over a larger distance or surface than what can be covered with traditional strain measurements, which can only capture the structural response at a given position.

In the past, when load tests were used to show the traveling public that a new bridge is safe for use, safety was demonstrated by showing that a bridge can carry a given number of heavily loaded vehicles. Nowadays, our codes and guidelines express safety in terms of a probability of failure. The practice of load testing, and in particular proof load testing, still needs to make the step to move from showing that a bridge can carry a heavy load to quantifying the safety in terms of a probability of failure. For that purpose, concepts of structural reliability should be combined with the practice of load testing. Whenever an assessment of "safety" is required, this assessment should be quantified in terms of a probability of failure. By the same token, a transition from member safety to systems safety (considering the entire structure and the overall structure behavior) is needed and requires research.

In particular for proof load testing, research is needed to determine the target proof load so that it can be concluded that a certain bridge fulfills the code requirements in terms of safety, expressed as a probability of failure. This research should also address which uncertainties are removed with a proof load test and which uncertainties still need to be considered. Moreover, the effect of testing only one span but possibly extending these results to the entire bridge, needs to be addressed.

Since load tests at the moment are usually isolated projects that arise from the need for an improved assessment of a given bridge; they do not form an integral part of a bridge management plan yet. Research to identify the optimal time during the life cycle of a structure to load test is necessary. The optimal time could, for example, be based on cost optimization, take forecasted traffic loadings into account, and consider deterioration mechanisms. Moreover, each bridge should be considered to be part of a road network. The optimal time could be analyzed based on the cost of temporary bridge closure on the road network and the resulting driver delays. This topic requires further research.

At the moment, the current codes and guidelines only allow load testing (especially proof load testing) of structures that are expected to fail in a ductile manner, so that the structure will show signs of distress prior to collapse. However, often structures that may fail in a brittle mode are more difficult to assess, so that proof load testing becomes an interesting option. More research on concrete structures that are critical in shear, punching shear, tension, or torsion is needed, as well as more research on fatigue- and fracture-critical steel and composite structures. Safe stop criteria for these structures need to be developed. More research on the application of load testing to timber, masonry, and plastic composite structures is needed as well to identify the critical structural responses and stop criteria.

13.3 Conclusions and practical recommendations

An engineer preparing a load test in a territory where no codes or guidelines for load testing are available could ask: "Which code or guideline should I follow?" As shown in Part I of Volume 12, different codes and guidelines address load testing from a different perspective. Some codes, such as the *Manual for Bridge Evaluation* cover diagnostic and proof load tests for all bridge types, except for bridges that may fail in a brittle manner and long-span bridges, in a general and qualitative manner, whereas other guidelines, such as the German guideline, focus on proof load tests of flexure-critical building structures only, providing detailed methods to determine the target proof load and detailed stop criteria. However, when preparing a proof load test, it is important to realize that each project is different. Depending on the goals for the load test, a different type of load test may be selected, a different load application system, a different sensor plan, and a different overall planning of the project will result. As such, it is often interesting for the engineer preparing such a load test to gather information from the different codes and guidelines that are available, and to learn from previously published load testing projects, such as the case studies presented in this book.

A general recommendation that holds true for all load tests, regardless of the type of test and structure type, is that a good preparation is important. First and foremost, a feasibility study is necessary to gather the required information to prepare the test. Where possible, missing information and material properties should be gathered and/or measured to have a better understanding of the structure and the expected structural behavior. Then, the goals of the load test should be clearly stated, and it should be studied if load testing is the recommended method to meet these goals. From a practical point of view, a good preparation is important to avoid confusion on the bridge site when time is scarce. A technical inspection prior to the load tests should identify possible site restrictions and limitations. A detailed planning of the on-site activities is necessary to help with finishing the load test within the available time. Prior to the load test, the test engineers should think through possible failures of equipment and personnel. A plan B for each scenario should be developed, and the necessary backup equipment and sensors should be available on site.

A second practical recommendation is that communication and safety on site are of the utmost importance. Good communication helps all parties involved to know what needs to happen when, and when possibly dangerous situations arise. Safety on site is important to protect the structure, personnel involved with the load test, and the traveling public. For the structural safety, an adequate sensor plan is important, as well as thorough preparatory calculations exploring the expected structural behavior. For the safety of the personnel involved with the load test, the local safety regulations should be closely followed and a safety engineer should be dedicated to the safety of personnel and the execution. Finally, the safety of the traveling public should be considered and when necessary, a full closure of the bridge during testing may be necessary. When road closures are necessary, good communication to the affected communities and traveling public is important. During the preparation stage of the load test, the way in which such hindrances for the traveling public will be communicated, and who will be in charge of the communication, should be determined.

Author Index

DOI: https://doi.org/10.1201/9780429265969

Subject Index

Note: numbers in italic indicate figures and numbers in bold indicate tables on the corresponding pages.

DOI: https://doi.org/10.1201/9780429265969

Structures and Infrastructures Series

Book Series Editor: Dan M. Frangopol

ISSN: 1747–7735

Publisher: CRC Press/Balkema, Taylor & Francis Group

DOI: https://doi.org/10.1201/9780429265969